Introduction to Naval Engineering

**Titles in the
Fundamentals of Naval Science Series**

Seamanship

Marine Navigation 1: Piloting

Marine Navigation 2: Celestial and Electronic

Introduction to Shipboard Weapons

Introduction to Naval Engineering

Edited by Edward F. Gritzen
Lieutenant Commander
U.S. Navy

With assistance from
Russell A. McCurdy, Lieutenant, U.S. Navy
James L. Wall, Lieutenant, U.S. Navy
Ellis W. Merschoff, Lieutenant, U.S. Navy

Naval Institute Press
Annapolis, Maryland

Fundamentals of Naval Science Series

Copyright © 1980
by the United States Naval Institute
Annapolis, Maryland.

Fourth printing with corrections, 1982

All rights reserved for copyrighted material.
No copyright is claimed for material in the public domain.

Library of Congress Cataloging in Publication Data
Gritzen, Frederick, 1947–
   Introduction to naval engineering.
    Bibliography: p.
    Includes index.
    1. Marine engineering.   I. Title.
VM600.G74   623.8'7   80–81089
ISBN 0–87021–319–9

Printed in the United States of America

To Frederick Vincent Bollin

Preface

This textbook was designed both to reduce the number of textbooks formerly used in the Naval Reserve Officers Training Corps Naval Ships Systems I (Engineering) course, and to provide more current material than that presently available in them. Although the text covers a broad range of material from basic steam propulsion to damage control, there is sufficient detail for the student to develop a basic understanding of all major shipboard marine engineering systems currently in use on board U. S. naval ships.

The text was edited using an engineering systems approach, both in the presentation of the relevant thermodynamic theory and its practical application in the form of various equipments. It is divided into five parts: Steam Propulsion, Gas Turbine Propulsion, Nuclear Propulsion, Auxiliary and Electrical Systems, and Stability and Damage Control.

The editor is indebted to Lieutenant E. W. Merschoff, USN, Lieutenant R. A. McCurdy, USN, and Lieutenant J. L. Wall, USN, for their assistance in organizing material in their areas of expertise, and to Lieutenant D. E. Fleming, USN, Course Coordinator, EN100, U.S. Naval Academy, for his contributions to the chapter on gas turbine engines. Thanks are also due Constance MacDonald for her fine editorial assistance in preparing the manuscript for printing.

Special thanks for their help in making needed corrections and updates in the fourth printing of this volume are due Lieutenant Commander Donald F. Fortik, USN, and Professor Bruce Johnson of the U.S. Naval Academy Naval Systems Engineering Department, and Lieutenant Steve Merrick, USN, of the University of Wisconsin NROTC unit.

Contents

Preface	vii

Chapter
1.	Thermodynamics	1
2.	The Main Steam Cycle	27
3.	Overview of Oil-Fired Marine Boilers	34
4.	1200 psi D-type Boiler Water and Steam Side Systems	43
5.	1200 psi D-type Boiler Firesides and Support Systems	54
6.	Main Propulsion Turbines, Reduction Gear, Shafting, and Auxiliary Turbines	73
7.	Main Condensate System	102
8.	Main Feed System	113
9.	Machinery Arrangement and Plant Layout	134
10.	Nuclear Power Plants	166
11.	Diesel and Gasoline Engine Theory	189
12.	Gas Turbine Engines	210
13.	Comparison of Marine Propulsion Plants	242
14.	Distilling Plants	251
15.	Air Compressors	258
16.	Refrigeration and Air Conditioning Plants	265
17.	Basic Hydraulics and Applications	279
18.	Steering Systems	286
19.	Fundamental Electrical Theory	291
20.	Shipboard Electrical Distribution	301
21.	Ship Construction	314
22.	Stability and Buoyancy	331
23.	Cross Curves of Stability	343
24.	Causes of Impaired Stability	363
25.	Damage Control and the Damage Control Organization	372
26.	Principles of Practical Damage Control	381
27.	Defense Against NBC Attack	407

Reference Books	420
Index	421

Introduction to Naval Engineering

1
Thermodynamics

The shipboard engineering plant may be thought of as a series of devices and arrangements for the exchange and transformation of energy. The energy transformation of greatest importance in the shipboard plant is the production of mechanical work from thermal energy, since we depend largely upon this transformation to make the ship move through the water. On board steam-driven ships, steam serves the vital purpose of carrying energy to the engines. The source of this energy may be the combustion of a conventional fuel oil or the fission of a radioactive material. In either case, the steam that is generated is the medium by which thermal energy is carried to the ship's engines, where it is converted into mechanical energy which propels the ship. In addition, energy transformations related directly or indirectly to the basic propulsion plant provide power for many vital services such as steering, lighting, ventilation, heating, refrigeration and air conditioning, the operation of various electrical and electronic devices, and the loading, aiming, and firing of the ship's weapons.

In order to acquire a basic understanding of the design of shipboard engineering plants, it is necessary to have some understanding of certain concepts in the field of thermodynamics. In the broadest sense of the term, thermodynamics is the physical science that deals with energy and energy transformations.

This chapter deals with certain thermodynamic concepts that are particularly necessary as a basis for understanding the shipboard engineering plant. The information given here is introductory in nature; obviously, it is not in any sense a complete or thorough exploration of the subject. Insofar as possible, we will depend upon verbal description rather than mathematical analysis to develop our understanding of the laws and principles of energy exchanges and transformations.

It should perhaps be noted that many of the terms used in this chapter, including such basic terms as energy and heat, have more specialized and more precise meanings in the study of thermodynamics than they do in everyday life or even in the study of general physics. This is only to be expected; thermodynamics is a highly specialized branch of physics, and, like any other specialty, it requires a certain refinement of terminology. If any difficulty arises from the fact that familiar terms are used in a somewhat unfamiliar sense, the difficulty can be largely minimized by paying particular attention to the exact meaning of each term as defined here, rather than depending upon a general knowledge for an understanding of the terms.

Although energy has a general meaning to almost everyone, it is not easy to define the word in a completely satisfactory way. Energy is intangible and is largely known through its effects. Because energy is so often manifested by the production of work, energy is commonly defined as "the capacity for doing work." However, this is not entirely adequate as a definition, since work is not the only effect that is produced by energy. For example, heat can flow from one body to another without doing any work at all, but the heat must still be considered as energy, and the process of heat transfer must be recognized as a process that has produced an effect. A broader definition, then, and one that satisfies more of the conditions under which we know energy to exist, is "the capacity for producing an effect."

Energy exists in many forms. For convenience, we usually classify energy according to the size and nature of the bodies or particles with which the energy is associated. Thus we say that mechanical energy is the energy associated with large bodies or objects, usually things that are big enough to see. Thermal energy is energy associated with molecules. Chemical energy is energy that arises from the forces that bind the atoms together in a molecule. Chemical energy is demonstrated whenever combustion or any other chemical reaction takes place. Electrical energy, light, X-rays, and radio waves are examples of energy associated with particles that are even smaller than atoms.

Each of these types of energy must be further classified as (1) stored energy or (2) energy in transition. Stored energy can be thought of as energy that is actually "contained in" or "stored in" a substance or system. There are two kinds of stored energy: (1) potential energy and (2) kinetic energy. When energy is stored in a system because of the relative velocities of two or more objects or particles, we call it kinetic energy. It should be emphasized that all stored energy is either potential or kinetic energy.

Energy in transition is, as the name implies, energy that is in the process of being transferred from one object or system to another. All energy in transition begins and ends as stored energy.

In order to understand any form of energy, then, we need to know the relative size of the bodies or particles in the energy system, and to know whether the energy is stored or in transition. Bearing in mind these two modes of classification, let us now examine mechanical and thermal energy, the two forms of energy that are of particular interest in practically all aspects of shipboard engineering.

Mechanical and Thermal Energy

Consider first the two stored forms of mechanical energy. Mechanical *potential* energy exists because of the relative positions of two or more objects. For example, a rock resting on the edge of a cliff in such a position that it will fall freely if pushed has mechanical potential energy. A sled that is being held at the top of an icy hill has mechanical potential energy.

Mechanical *kinetic* energy exists because of the relative velocities of two or more objects. If you push that rock, open the gate of the dam, or let go of the sled, something will move. The rock will fall; the water will flow; the sled will slide down the hill. In each case mechanical potential energy will be changed to mechanical kinetic energy. Another way of saying this is to say that energy of position will be changed to energy of motion.

In these examples, an external source of energy is used to get things started. Energy from some outside source is required to push the rock, open the gate of the dam, or let go of the sled. All real machines and processes require this kind of "boost" from an energy source outside the system. For example, there is a tremendous amount of chemical energy stored in fuel oil; but this energy will not raise steam in the boiler until some energy is expended to start the oil to burn.

Similarly, the energy in any one system affects other energy systems that might be involved in or affected by each energy process. In the examples given in this chapter, therefore, we will consider only one energy process or energy system at a time, disregarding both the energy "boosts" that may be received from outside systems and the energy transfers that may take place between the system we are considering and other systems.

Notice that both mechanical potential energy and mechanical kinetic energy are stored forms of energy. It is easy to see why we regard mechanical potential energy as being stored, but it is not so easy to see the same thing about mechanical kinetic energy. Part of the trouble comes about because mechanical kinetic energy is often referred to as "the energy of motion," thus leading to the false conclusion that "energy in transition" is somehow involved. This is not the case, however. Work is the only form of mechanical energy that can properly be considered energy in transition.

If you have trouble with the idea that mechanical kinetic energy is stored rather than in transition, think of it this way: A bullet that has been fired from a gun has mechanical kinetic energy because it is in motion. The faster the bullet is moving, the more kinetic energy it has. There is no doubt in anybody's mind that the bullet has the capacity to produce an effect; so we may safely say that it has energy. But although the bullet is in transition, the energy of the bullet is not transferred to any other object or system (except for energy absorption due to air and vapor friction) until the bullet strikes some object that resists its passage. When the bullet strikes against a resisting object, then and only then can we say that energy in transition exists in the form of heat and work.

In this example, we are ignoring the fact that some work is done against the resistance of the air and that some heat results from the passage of the bullet through the air. But this does not change the basic idea that kinetic energy is stored energy rather than energy in transition. The air must merely be regarded as a "resisting object" that causes some of the stored kinetic energy of the bullet to be converted into energy in transition (heat and work) while the bullet is passing through the air. However, the major part of the stored kinetic energy does not become energy in transition until the bullet strikes some object that resists its passage more strongly than the air resists it.

Mechanical energy in transition is called work. When an object is moved through a distance against a resisting force, we say that work has been done.

Mechanical potential energy, mechanical kinetic energy, and work are all measured in the same unit, foot-pounds. One foot-pound of work is done when a force of 1 pound acts through a distance of 1 foot. One foot-pound of mechanical potential energy or mechanical kinetic energy is the amount of energy that would be required to accomplish 1 foot-pound of work.

The amount of work done is independent of the time it takes to do it. When you lift a weight of 1 pound through a distance of 1 foot, you have done 1 foot-pound of work, whether you do it in half a second or half an hour. The rate at which work is done is called *power*. The common unit of measurement for power is horsepower (hp).

Energy associated primarily with systems of molecules is called thermal energy. Like other kinds of energy, thermal energy may exist in stored form (in which case it is called internal energy) or as energy in transition (called heat).

In common usage, the term heat is often used to include all forms of thermal energy. However, this lack of distinction between heat and the stored forms of thermal energy can lead to serious confusion. In this text, therefore, the term internal energy is used to describe the stored forms of thermal energy, and the term heat is used only to describe thermal energy in transition.

Internal energy, like all stored forms of energy, exists either as potential energy or as kinetic energy.

Internal potential energy is the energy associated with the forces of attraction that exist between molecules. The magnitude of internal potential energy is dependent upon the mass of the molecules and the average distance by which they are separated, in much the same way that mechanical potential energy depends upon the mass of the bodies in the system and the distance by which they are separated. The force of attraction between molecules is greatest in solids, less in liquids and yielding substances, and least of all in gases and vapors. Whenever something happens to change the average distance between the molecules of a substance, there is a corresponding change in the internal potential energy of the substance.

Internal kinetic energy is the energy associated primarily with the activity of molecules, just as mechanical kinetic energy is the energy associated with the velocities of relatively large bodies. It is important to note that the temperature of a substance arises from and is proportional to the molecular activity with which internal kinetic energy is associated.

For most purposes, we will not need to distinguish between the two stored forms of internal energy. Instead of referring to internal potential energy and internal kinetic energy, therefore, we may often simply use the term internal energy. When used in this way, without qualification, the term should be understood to mean the sum total of all internal energy stored in the substance or system by virtue of the motion of molecules or by virtue of the forces of attraction between molecules.

Although the term heat is more familiar than the term internal energy, it may be more difficult to arrive at an accurate definition of heat. Heat is thermal energy in transition. Like work, heat is a transitory energy form existing between two or more forms of energy.

Since the flow of thermal energy can occur only when there is a temperature difference between two objects or regions, it is apparent that heat is not a property or an attribute of any one object or substance. If a person accidentally touches a hot stove, he may understandably feel that heat is a property of the stove. More accurately, however, he might reflect that his hand and the stove constitute an energy system, and that thermal energy flows from the stove to his hand because the stove has a higher temperature than his hand.

As another example of the difference between heat and internal energy, consider two equal lengths of piping, made of identical materials and containing steam at the same pressure and temperature. One pipe is well insulated; one is not. From everyday experience, we expect more heat to flow from the uninsulated section of pipe than from the insulated section. When the two pipes are first filled with steam, the steam in one pipe contains exactly as much internal energy as the steam in the other. We know this is true because the two pipes contain equal volumes of steam at equal pressures and temperatures.

After a few minutes, the steam in the uninsulated pipe will contain much less internal energy than the steam in the insulated pipe, as we can tell by reading the pressure and temperature gages on each pipe. What has happened? Stored thermal energy, internal energy, has moved from one place to another, first from the steam to the pipe, then from the uninsulated pipe to the air. It is this movement, or this flow, of energy that should properly be called heat. Temperature is a reflection of the amount of internal kinetic energy possessed by an object or a substance, and is therefore an attribute or property of the substance. The movement or flow of thermal energy (i.e., heat) is an attribute of the energy system rather than of any one component of it.

Modes of Heat Transfer

Heat flow, or the transfer of thermal energy from one body, substance, or region to another, always takes place from a region of higher temperature to a region of lower temperature. In thermodynamics, the high-temperature region may be called the source or the emitting region; the low-temperature region may be called the sink, the receiver, or the receiving region.

Although three modes of heat transfer—conduction, radiation, and convection—are commonly recognized, we will find it easier to understand heat transfer if we make a distinction between conduction and radiation, on the one hand, and convection, on the other. Conduction and radiation may be regarded as the primary modes of heat flow. Convection may best be thought of as a related but basically different and special kind of process that involves the movement of a mass of fluid from one place to another.

Conduction. Conduction is the mode by which heat flows from a hotter to a colder region when there is physical contact between the two regions. For example, consider a metal bar that is held so that one end of it is in boiling water. In a very short time the end of the bar that is not in the boiling water will have become too hot to hold. We say that heat has been conducted from molecule to molecule along the entire length of the bar. The molecules in the layer nearest the source of heat become increasingly active as they receive thermal energy. Since each layer of molecules is bound to the adjacent layers by cohesive forces, the motion is passed on to the next layer which, in turn, sets up increased activity in the next layer. The process of conduction continues as long as there is a temperature difference between the two ends of the bar.

Radiation. Thermal radiation is a mode of heat transfer that does not involve any physical contact between the emitting region and the receiving region. A person sitting near a hot stove is warmed by thermal radiation from the stove, even though the air in between remains relatively cold. Thermal radiation from the sun warms the earth without warming the space through which it passes. Thermal radiation passes through any transparent substance—air, glass, ice—without warming it to any extent because transparent materials are very poor absorbers of radiant energy.

All substances—solids, liquids, and gases—emit radiant energy at all times. We tend to think of radiant energy as something that is emitted only by extremely hot objects such as the sun, a stove, or a furnace, but this is a limited view of the nature of radiant energy. Although it absorbs radiant energy emitted by the sun, the earth in turn radiates energy to the stars. A stove radiates energy to everything surrounding it, but at the same time all the surrounding objects are radiating energy to the stove. A child standing near a snowman may well believe that the snowman is "radiating cold" rather than emitting radiant energy; actually, however, both the child and the snowman are emitting radiant energy. The child, of course, is radiating far more energy than the snowman; so the net effect of this energy exchange is that the snowman grows warmer and the child grows colder. We are literally surrounded by, and a part of, such energy exchanges at all times. As we consider these energy exchanges, we may arrive at a new view of thermal equilibrium: when objects are radiating precisely as much thermal energy as they are receiving, in any given period of time, they are in thermal equilibrium.

Thermal radiation is an electromagnetic wave phenomenon, differing from light, radio waves, and other electromagnetic phenomena merely in the wavelengths involved. When the wavelengths are in the infrared part of the electromagnetic spectrum—that is, when they are just below the range of visible light waves—we refer to the radiated energy as thermal radiation.

Convection. At the molecular or submolecular level, heat transfer takes place through the processes of conduction and radiation. If we use the term heat transfer in a somewhat different way, we may also include convection as a mode of heat transfer; but it is important to understand the difference between convection and the basic heat transfer processes of conduction and radiation.

If we put a hot brick into a wheelbarrow and wheel it across the street, we have in one sense "transferred" heat. However, any heat transfer that takes place between the brick and its surroundings while we are wheeling it across the street will be by conduction and by radiation. Therefore, it would really be more accurate to say that we have "transferred" the brick and all its contained thermal energy from one side of the street to the other.

Convection does not occur in solids, such as the brick we have just transported in the wheelbarrow, but only in fluids, liquids, gases, and vapors. Convection is the transportation or movement of some portions within a mass of fluid. As this movement occurs, the moving portions of the fluid transfer their contained thermal energy from one part of the fluid to another. The effect of convection is thus to mix the various parts of the fluid; the part that was at the bottom of the container may move to the top, or the part that was at one side may move to the other side. As this mixing takes place, heat transfer occurs from one part of the fluid to another and between the fluid and its surroundings. But this heat transfer, like any other heat transfer, takes place by conduction and by radiation. In other words, convection transports portions of the fluid; conduction and radiation transfer the thermal energy.

What causes this transportation of a mass of fluid? In the case of natural convection, the movement is caused by differences in the density of different parts of the fluid. The differences in density are usually caused by unequal temperatures within the mass of fluid. For example, as the air over a hot radiator is heated, it becomes less dense and thus begins to rise. Cooler, heavier air is drawn in to replace the heated air, so that convection currents are set up. Another example of natural convection is the circulation of water in a natural circulation boiler. As the water in the generating tubes is heated, it expands and becomes much lighter (less dense) than the cooler water in the downcomers. The hotter and lighter water rises, while the cooler and heavier water flows downward. The resulting circulation of water in the boiler is a clear illustration of convection currents established by differences in temperature (and therefore differences in density) in various parts of the fluid.

In the case of forced convection, some mechanical device such as a pump or a fan produces the movement of the fluid. When the main feed pump moves feedwater toward the boiler, the water is transported by forced convection. The flow of combustion gases through a boiler is partly by natural convection and partly by forced convection. Natural convection occurs because the gases of combustion are hotter and lighter than air, so they tend to rise and go up the stack. Forced convection is also involved in this process, however, because the forced draft blowers supply an air pressure that increases the rate at which the combustion gases travel across the tubes and up the stack. When you stir a cup of hot coffee, you are forcing convection and thus increasing the rate of heat transfer. Natural convection currents would be set up in the coffee if you did not stir it, because differences in density would occur as some portion of the coffee cooled before others. If you want to cool the coffee rapidly, use forced convection (stirring). If you do not want it to cool so rapidly, wait for natural convection to do the job.

In summary, then, we use the term convection to describe the transportation, or, loosely, the "heat transfer," of a mass of fluid and its contained thermal energy. But the processes by which any substance gains or loses thermal energy are most accurately described in terms of conduction and radiation.

Effects of Heat Transfer

The terms sensible heat and latent heat are often used to indicate the effect that the transfer of heat has upon a substance. The flow of heat from one substance to another is normally reflected in a temperature change in each substance—that is, the hotter substance becomes cooler, and the cooler substance becomes hotter. However, the flow of heat is not reflected in a temperature change in a substance that is in the process of changing from one physical phase to another. When the flow of heat is reflected in a temperature change, we say that sensible heat has been added to or removed from a substance. When the flow of heat is reflected in the changing physical phase of a substance, we say that latent heat has been added or removed.

Since heat is defined as thermal energy in transition, we must not infer that sensible heat and latent heat are really two different kinds of heat. Instead, the terms serve to distinguish between two different kinds of effects produced by the transfer of heat; and, at a more fundamental level, they indicate something about the manner in which the thermal energy was or will be stored. Sensible heat involves internal kinetic energy, and latent heat involves internal potential energy.

The three fundamental physical phases of all matter are solid, liquid, and gas (or vapor). The physical phase of a substance is closely related to the distance between molecules. The molecules are closest together in solids, farther apart in liquids, and farthest apart in gases. When the flow of heat to a substance is not reflected in a temperature change, we know that the energy is being used to increase the distance between the molecules of the substance and thus change it from a solid to a liquid or from a liquid to a gas. In other words, the addition of heat to a substance that is in process of changing from solid to liquid or from liquid to gas results in an increase in the amount of internal potential energy stored in the substance, but it does not result in an increase in the amount of internal kinetic energy. Only after the change of phase has been fully accomplished does the addition of heat result in a change in the amount of internal kinetic energy stored in the substances; hence, there is no temperature change until after the change of phase is complete.

In a sense, we may think of latent heat as the energy price that must be paid for a change of phase from solid to liquid or from liquid to gas. But the energy is not lost; rather, it is stored in the substance as internal potential energy. The energy price is "repaid," so to speak, when the substance changes back from gas to liquid or from liquid to solid; during these changes of phase, the substance gives off heat without any change in temperature.

Figure 1–1 shows the relationship between sensible heat and latent heat for one substance, water, at atmospheric pressure. The same kind of chart could be drawn up for other substances; however, different amounts of thermal energy would of course be required for each change of temperature and for each change of physical phase. (Note: Atmospheric pressure is the pressure exerted by the atmosphere in all directions as indicated by a barometer. Standard atmospheric pressure is considered to be 14.7 pounds per square inch, which is equivalent to 29.92 inches of mercury.)

If we start with one pound of ice at 0°F, we must add 16 Btu in order to raise the temperature of the ice to 32°F; to convert it to a pound of water at 32°F, we must add 144 Btu (the latent heat of fusion). [One Btu (British Thermal Unit) is equal to the quantity of heat required to raise one pound of water one degree F.] There will be no change in temperature while the ice is melting. After all the ice has melted, however, the temperature of the water will be raised as additional heat is supplied. If we add 180 Btu (that is, 1 Btu for each degree of temperature between 32°F and 212°F), the water will reach its boiling point; we must add 970 Btu (the latent heat of vaporization) to convert the water to steam. After all the water has been converted to steam, the addition of 42 Btu will cause an increase

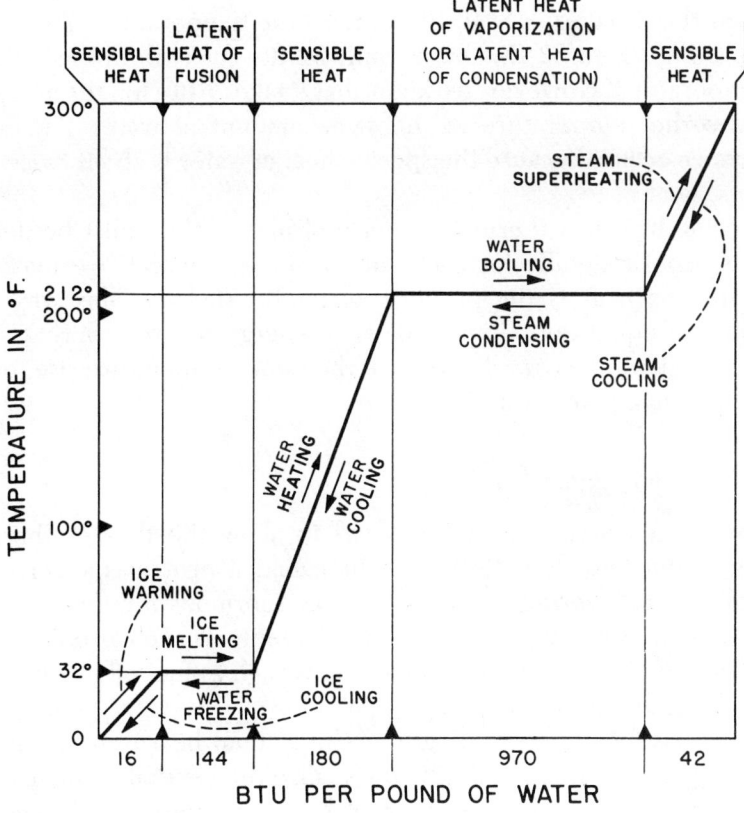

Figure 1-1. Relationship between sensible heat and latent heat for water at atmospheric pressure.

in the temperature of the steam which will superheat the steam of 300°F.

The same relationships apply when heat is being removed. The removal of 42 Btu from the pound of steam that is at 300°F will cause its temperature to drop to 212°F. As the pound of steam at 212°F changes to a pound of water at 212°F, 970 Btu are given off. When a substance is changing from a gas or vapor to a liquid, we usually use the term latent heat of condensation for the heat that is given off. Notice, however, that the latent heat of condensation is exactly the same as the latent heat of vaporization. The removal of another 180 Btu of sensible heat will lower the temperature of the pound of water from 212°F to 32°F. As the pound of water at 32°F changes to a pound of ice at 32°F, 144 Btu are given off without any accompanying change of temperature. Further removal of heat causes the temperature of the ice to decrease.

Thermodynamics 11

In the discussion of sensible heat and latent heat, you may have noticed that it takes only 16 Btu to raise the temperature of 1 pound of ice from 0°F to 32°F—that is, only ½ Btu for each degree of rise in temperature. However, we know that it takes 1 Btu (on the average) to raise the temperature of the same amount of water 1°F. This difference occurs because the specific heat of water is about twice the specific heat of ice.

Specific heat is a thermal property of matter that must be determined experimentally for each substance. In general, we may say that specific heat is the property of matter that explains why the addition of equal quantities of thermal energy to two different substances will not necessarily produce the same temperature rise, even when no change of phase is involved.

Heat-Transfer Apparatus

Any device or apparatus designed to allow the flow of thermal energy from one fluid to another is called a heat exchanger. The shipboard engineering plant contains an enormous number and variety of heat exchangers, ranging from large items such as boilers and main condensers to relatively small items such as fuel oil heaters and lubricating oil coolers.

As a basis for understanding something about heat transfer in real heat exchangers, it is necessary to visualize the general configuration of the most commonly used type of heat exchanger. With few exceptions, heat exchangers used on board ship are of the indirect or surface type—that is, heat flows from one fluid to another through some kind of tube, plate, or other "surface" that separates the two fluids and keeps them from mixing. Most surface heat exchangers are of the shell-and-tube type, consisting of a bundle of metal tubes that fit inside a shell. One fluid flows through the inside of the tubes and the other flows through the shell, around the outside of the tubes.

The transfer of heat in a heat exchanger involves the flow of heat from the hot fluid to the tube metal and from the tube metal to the cold fluid. In addition, heat must also be transferred through two layers of fluid (one on the inside and one on the outside of the tube) that are not flowing with the remainder of the fluid but are almost motionless. These relatively stagnant layers, known as boundary layers or fluid films, are extremely small in size but have an extremely important effect on heat transfer.

Most fluids are very poor transferrers of heat. As a fluid is flowing, however, convection and mechanical mixing of the fluid bring the molecules into such intimate contact that heat transfer can and does

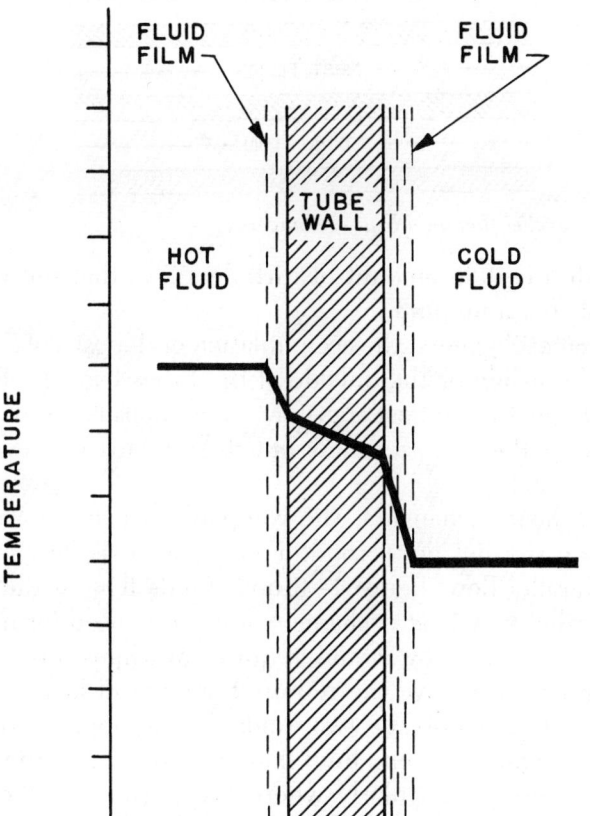

Figure 1–2. Effect of fluid film on heat transfer.

occur. Other things being equal, increasing the velocity of fluid flow increases heat transfer.

Because the fluid film is almost motionless, heat transfer through the film is very poor. The effect of fluid films on heat transfer is shown in figure 1–2. The temperature line indicates the changes in temperature that occur as heat is transferred from the hot fluid to the fluid film, from this fluid film to the tube metal, from the tube metal to the other fluid film, and from this fluid film to the cold fluid. As may be seen, the major part of the temperature drop occurs in the fluid films rather than in the tube metal. Note, also, that the thicker fluid film is more resistant to heat transfer than the thinner fluid film.

The velocity of flow and the amount of turbulence in the flow affect heat transfer by altering the thickness of the fluid film. Increasing the velocity of flow diminishes the thickness of the fluid film and thus increases heat transfer. Although there are some obvious disadvantages to excessive turbulence, many heat exchangers are designed to

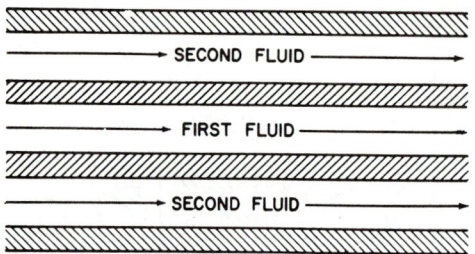

Figure 1–3. Parallel flow in a heat exchanger.

operate with a certain amount of turbulence so that the fluid films will be kept to a minimum.

In real heat exchangers, the accumulation or deposits of scale, soot, or dirt on the inside or the outside of the tubes has a profound and detrimental effect upon heat transfer. Such deposits not only reduce the efficiency of the heat exchanger but also tend to cause overheating of the tube metal.

In surface heat exchangers, the components may be arranged so as to provide parallel flow, counter flow, or cross flow of the two fluids. In parallel flow (figure 1–3) both fluids flow in the same direction. Parallel-flow heat exchangers are rarely used for naval service, largely because they would require an impossibly long heat-transfer surface to achieve the required amount of heat transfer. In counter flow (figure 1–4) the two fluids flow in opposite directions. In cross flow (figure 1–5) one fluid flows at right angles to the other. Cross flow is used particularly where the purpose of the heat exchanger is to remove latent heat and thus change the physical state of a substance. Main and auxiliary condensers are typically of the cross-flow type, as are several other smaller shipboard condensers.

Surface heat exchangers are referred to as single-pass units, if each fluid passes the other only once, or as multipass units, if one fluid passes the other more than once. Multipass flow may be obtained by the arrangement of the tubes and of the fluid inlets and outlets, or it may be obtained by using baffles to guide a fluid so that it passes the other fluid more than once before it leaves the heat exchanger.

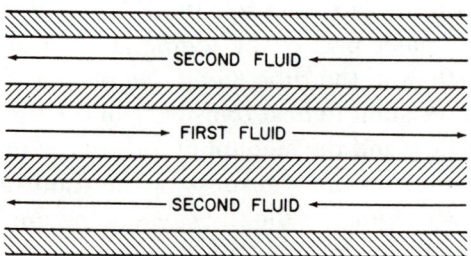

Figure 1–4. Counter flow in a heat exchanger.

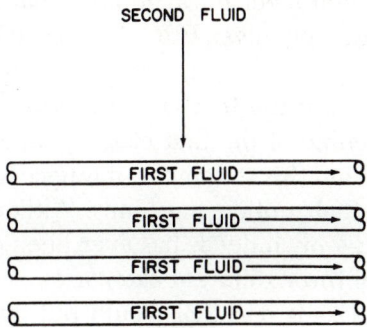

Figure 1-5. Cross flow in a heat exchanger.

The First Law of Thermodynamics

In the previous discussion of energy, we occasionally assumed a general principle that must now be stated. That principle, which is called the principle of the conservation of energy, may be stated in several ways. Most commonly, perhaps, it is stated as the fact that energy can be neither destroyed nor created, but only transformed. Another statement is that energy may be transformed from one form to another, but the total energy of any body or system of bodies is a quantity that can neither be increased nor diminished by the action of the body or bodies. Still another way of stating this principle is by saying that the total quantity of energy in the universe is always the same. Regardless of the mode of expression, the principle of the conservation of energy applies to all kinds of energy.

Energy equations for many thermodynamic processes are based directly upon the principle of the conservation of energy. The first law of thermodynamics is a special statement of the principle of the conservation of energy for a given system. For a closed (fixed mass) system, the first law states that the net heat flow into the system minus the net work out of the system equals the change in stored energy within the system.

If mass can cross the boundaries of the system it is called an open system (control volume). For an open system, the first law states that the net heat flow into the system minus the net work out of the system equals the change in stored energy within the system plus the net stored energy associated with the mass entering and leaving the system. If the heat flow, mechanical energy, and work terms in the equation are all converted to common energy units, a special case of the first law known as the general energy equation for steady flow systems results:

$$\text{total energy in} = \text{total energy out,}$$

valid provided there is no change in stored energy within the system.

The equivalence between mechanical and thermal energy in English units is 778 foot-pounds equals 1 Btu. (In metric units, 1 Newton-meter equals 1 Joule.)

Another consequence of the first law of thermodynamics is that a perpetual motion machine of the first class is impossible. To understand the rationale behind this statement, it is necessary to understand the concept of perpetual motion machines. Although no perpetual motion machine exists—or, indeed, has ever been constructed—it is possible to conceive of three different categories. A perpetual motion machine of the first class is one that would put out more energy in the form of work than it absorbed in the form of heat. Since such a machine would actually create energy, it would violate the first law of thermodynamics and the principle of the conservation of energy. A perpetual motion machine of the second class would permit the reversal of irreversible processes and would thus violate the second law of thermodynamics, as discussed presently. A machine of the third class would be one in which absolutely no friction existed. Interestingly, there are no theoretical grounds for declaring that a machine of the third class is completely impossible; however, such a machine would be entirely contrary to our experience and would violate some of the most profound convictions about the nature of energy and matter.

The Concept of Reversibility

When we put a pan of water on the stove and turn on the heat, we expect the water to boil rather than freeze. After we have mixed hot and cold water, we do not expect the resulting mixture to resolve itself into two separate batches of water at two different temperatures. When we open the valve on a cylinder of compressed air, we expect compressed air to rush out; we would be quite surprised if atmospheric air rushed into the cylinder and compressed itself. When a shaft is rotating, we expect a temperature rise in the bearings; when the shaft has been stopped, we would be truly amazed to observe internal energy from the bearings flowing to the shaft and causing it to start rotating again. When we drag a block of wood across a rough surface, we expect some of the mechanical energy expended in this act to be converted into thermal energy—that is, we expect a storage of internal energy in the wooden block and in the rough surface, as evidenced by temperature rises in these materials. But if this stored internal energy should suddenly turn to mechanical energy and move the wooden block back to its original position, our incredulity would know no bounds.

All these illustrations merely show that we have certain expectations, based on experience, as to the direction in which processes will move. The reasonableness of our expectations is attested to by the fact that in all recorded history there is no report of water freezing instead of boiling when heat is applied; there is no report of a lukewarm fluid unmixing itself and separating into hot and cold fluids; there is no report of a gas compressing itself without the agency of some external force; there is no report of the heat of friction being spontaneously utilized to perform mechanical work.

Are these actions really impossible? The first law of thermodynamics says that mechanical energy and thermal energy are mutually convertible, but it says nothing about the direction of such conversions. If we consider only the first law, all the improbable actions just mentioned are perfectly possible, and all processes could be thought of as being reversible. In an absolute sense, perhaps, we cannot guarantee that water will never freeze instead of boil when it is placed on a hot stove; but we are certainly safe in saying that this or any other completely reversible thermodynamic process is at the outer limits of probability. For all practical purposes, then, we will say that there is no such thing as a completely reversible process.

The Second Law of Thermodynamics

Since the first law of thermodynamics does not deal with the direction of thermodynamic processes, and since experience indicates that actual processes are not reversible, it is apparent that the first law must be supplemented by some statement of principle that will limit the direction of thermodynamic processes. The second law of thermodynamics is such a statement. Although the second law is perhaps more empirical than the first, and perhaps something less of a "law" in an absolute sense, it is of enormous practical value in the study of thermodynamics.

The second law of thermodynamics may be stated in various ways. One statement is that no process is possible where the sole result is the removal of heat from a low-temperature reservoir and the absorption of an equal amount of heat by a high-temperature reservoir. Among other things, this statement indicates that water will not freeze when heat is applied. Note that the statement includes and goes somewhat beyond the common observation that heat flows only from a hotter to a colder substance.

That no process is possible where the sole result is the removal of heat from a single reservoir and the performance of an equivalent amount of work is another statement of the second law. Among other things, this statement says that we cannot expect the heat of friction to reverse itself and perform mechanical work. More broadly, this statement indicates a certain one-sidedness that is inherent in thermodynamic processes. Energy in the form of work can be converted entirely to energy in the form of heat; but energy in the form of heat can never be entirely converted to energy in the form of work.

A very important inference to be drawn from the second law is that no engine, actual or ideal, can convert all the heat supplied to it into work, since some heat must always be rejected to a receiver that is at a lower temperature than the source. In other words, there can be no heat flow without a temperature difference, and there can be no conversion of heat to work without a flow of heat. A further inference from this inference is sometimes given as a statement of the second law: no thermodynamic cycle can have a thermal efficiency of 100 percent.

The question now arises as to how much heat must be rejected to a receiver that is at a lower temperature than the source. Or, looking at it another way, what is the maximum thermal efficiency that could theoretically be achieved by a heat engine operating without friction and without any other of the irreversible processes that must occur in all real machines?

To answer this question, Carnot, a French engineer, developed an imaginary and completely reversible cycle. In the Carnot cycle, all heat is supplied at a single high temperature, and all heat that must be rejected is rejected at a single low temperature. The cycle is fully reversible. When proceeding in one direction, the Carnot cycle takes in a certain amount of heat, rejects a certain amount of heat, and puts out a certain amount of work. When the cycle is reversed, the quantity of work that was originally the output of the cycle is now put into the cycle; the amount of heat that was originally taken in is now the amount rejected. When thus reversed, the cycle is called a Carnot refrigeration cycle.

Obviously, no real machine is capable of such complete reversibility, but the concept of the Carnot cycle is nonetheless an extremely useful one. By analysis of the Carnot cycle, it can be proved that no engine, actual or ideal, can be more efficient than an ideal, reversible Carnot cycle. The thermal efficiency of the Carnot cycle is given by the equation:

$$\text{thermal efficiency} = \frac{\text{work output}}{\text{heat input}} = \frac{T_s - T_r}{T_s}$$

where T_s equals the absolute temperature at which heat flows from the source to the working fluid, and T_r equals the absolute temperature at which heat is rejected to the receiver.

The implications of this statement are of profound importance because it establishes the fact that thermal efficiency depends only upon the temperature difference between the heat source and the heat receiver. Thermal efficiency does not depend upon the properties of the working fluid, the type of engine used in the cycle, or the nature of the process (combustion, nuclear fission, etc.) that produces the heat at the heat source. The basic principle thus established by analysis of the Carnot cycle is called the Carnot principle, and may be stated as follows: The motive power of heat is independent of the agents employed to realize it, its quantity being fixed solely by the temperatures of the bodies between which the transfer of heat occurs.

Thermodynamic Cycles

A thermodynamic cycle is a recurring series of thermodynamic processes through which an effect is produced by the transformation or redistribution of energy. In other words, a cycle is a series of processes repeated over and over again in the same order.

All thermodynamic cycles may be classified as being open cycles or closed cycles. An open cycle is one in which the working fluid is taken in, used, and then discarded. A closed cycle is one in which the working fluid never leaves the cycle, except through accidental leakage; instead, the working fluid undergoes a series of processes that are of such a nature that the fluid is returned periodically to its initial state and then reused.

The open cycle is exemplified by the internal combustion engine, in which atmospheric air supplies the oxygen for combustion and the exhaust products are returned to the atmosphere. In fact, another way to describe an open cycle is to say that it is one that includes the atmosphere at some point.

The closed cycle is exemplified by the condensing steam power plant used for ship propulsion on board many naval ships. In such a cycle, the working substance (water) is changed to steam in the boilers. The steam performs work as it expands through the turbines, and is then condensed to water again in the condenser. The water is returned to the boilers as boiler feed, and is thus used over and over again.

Thermodynamic cycles are also classified as heated-engine cycles or as unheated-engine cycles, depending upon the point in the cycle at which heat is added to the working substance. In a heated-engine cycle, heat is added to the working substance in the engine itself. An internal combustion engine has a heated-engine cycle. In an unheated-engine cycle, the working substance receives heat in some device that is separate from the engine. The condensing steam power plant has an unheated-engine cycle, since the working substance is heated separately in the boilers and then piped to the engine (steam turbines).

There are five basic elements in any thermodynamic cycle: (1) the working substance; (2) the engine; (3) a heat source, or high-temperature region; (4) a heat receiver, or low-temperature region; and (5) a pump.

The working substance is the medium by which energy is carried through the cycle. The engine is the device that converts the thermal energy of the working substance into useful mechanical energy in the form of work. The heat source supplies heat to the working substance, whereas the heat receiver absorbs heat from the working substance. The pump moves the working substance from the low-pressure side of the cycle to the high-pressure side.

The essential elements of a closed, unheated-engine cycle are shown in figure 1–6. This is the basic plan of the typical condensing steam power plant.

In an open, heated-engine cycle such as that of an internal combustion engine, the essential elements are all present but are arranged in a somewhat different order. In this type of cycle, atmospheric air and fuel are both drawn into the cylinder of the engine. Combustion takes place in the cylinder, either by compression or by spark, and the resulting internal energy of the working substance is transformed into work by which the piston is moved. Because the space above the piston is a high-pressure area when the piston is near the top of its stroke and a low-pressure area when the piston is near the bottom, the piston may be thought of as a pump in the sense that it "pumps" the working fluid from the low-pressure to the high-pressure side of

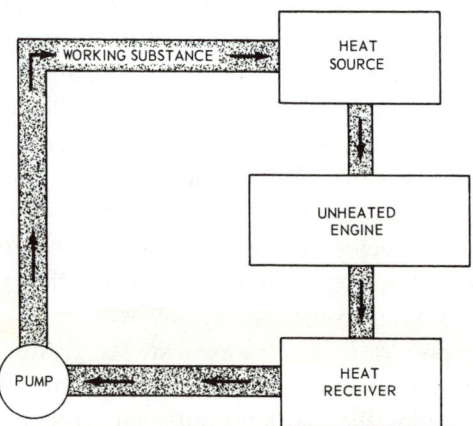

Figure 1–6. *Essential elements of a closed, unheated-engine cycle.*

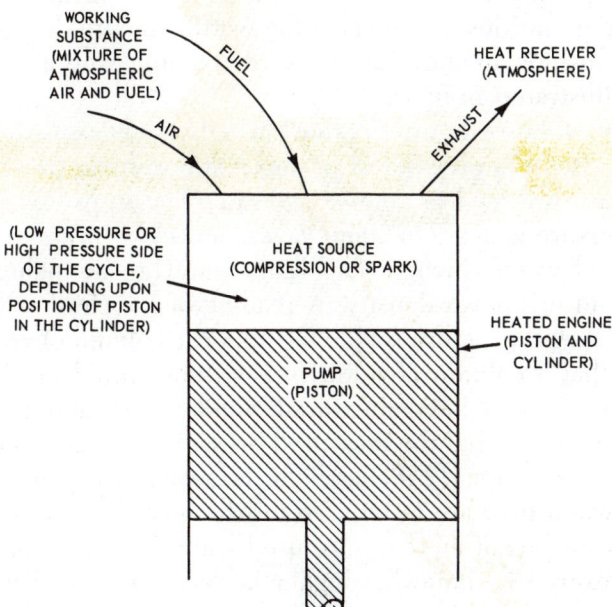

Figure 1-7. Essential elements of an open, heated-engine cycle.

the system. Thus, in terms of function, the piston-and-cylinder arrangement may be thought of as including the heat source, the engine, and the pump. An open, heated-engine cycle might therefore be represented as shown in figure 1-7.

Pressure Definitions

Pressure, like temperature, is one of the basic engineering variables, and one that must frequently be measured on board ship. Let us consider certain definitions that are important in any discussion of pressure measurement.

Pressure is defined as *force per unit area*. We usually think of pressure as being exerted *on* a substance, but it is important to remember that pressure is also exerted *by* the substance. In the case of vapors and gases, the pressure exerted by the fluid arises from the fact that the molecules of the fluid are in continual motion and are constantly impinging upon the walls of the container. Thus the pressure exerted by a vapor or a gas depends upon the number, mass, and velocity of the molecules.

A great deal of confusion arises from the fact that the zero point on most pressure gages represents atmospheric pressure, whereas absolute pressure is required for some engineering calculations. To clarify the numerous meanings of the word pressure, let us examine the relationships among gage pressure, vacuum, and absolute pressure (as illustrated in figure 1–8).

Gage pressure is the pressure actually shown on the dial of a gage that registers pressure at or above atmospheric pressure. Gage pressure is expressed in *pounds per square inch* (psi) on most shipboard pressure gages. On some gages, however, pressure is shown in inches of water, inches of mercury (in. Hg), or inches of some other liquid or known density. A reading of 1 inch of water means that the exerted pressure is able to support a column of water 1 inch high, or that a column of water in a U-tube would be displaced 1 inch by the pressure being measured. Similarly, a reading of 12 inches of mercury means that the measured pressure is sufficient to support a column of mercury 12 inches high. Gages are calibrated in inches of water when they are to be used for the measurement of very low pressures; inches of mercury are used when the range of pressures to be measured is somewhat higher because mercury has a considerably greater density than water.

It should be emphasized that an ordinary pressure gage reading of zero does *not* mean that there is no pressure in an absolute sense. It merely means that there is no pressure in excess of atmospheric pressure.

Atmospheric pressure is the pressure exerted by the weight of the atmosphere. At sea level, the *average* pressure of the atmosphere is sufficient to hold a column of mercury at a height of 29.92 (usually

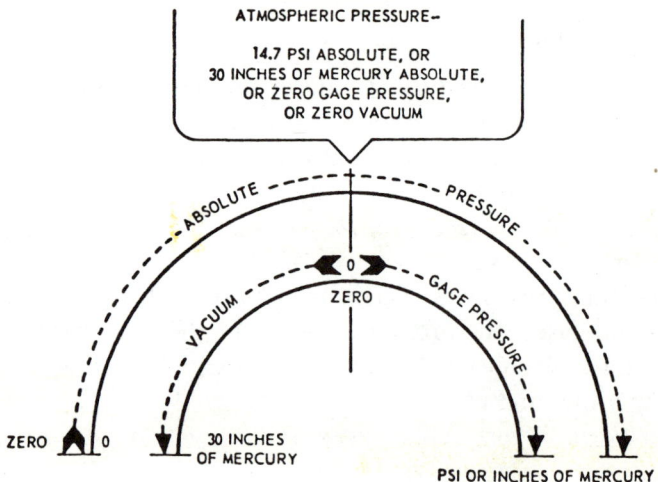

Figure 1–8. *Relationships among gage pressure, atmospheric pressure, vacuum, and absolute pressure.*

rounded off to 30) inches of mercury. Since a column of mercury 1 inch high exerts a pressure of 0.49 pound per square inch, a 30-inch column of mercury exerts a pressure equal to 30 × 0.49, or 14.7 psi. Because we are dealing now in the region below gage pressure, we must say that the average atmospheric pressure at sea level is 14.7 pounds per square inch *absolute*. (It is normally but not necessarily zero on the ordinary pressure gage.)

Notice, however, that the figure of 14.7 psia represents the average atmospheric pressure at sea level; it is therefore referred to as standard atmospheric pressure, and does not always represent the actual pressure being exerted by the atmosphere at the moment the gage is being read.

Barometric pressure is the term used to describe the actual atmospheric pressure that exists at any given moment. Barometric pressure is usually measured by a simple mercury column. In most engineering calculations no significant error is introduced by using the standard atmospheric pressure of 14.7 psia as though it were the actual barometric pressure. For some purposes, however, it is important to know the actual barometric pressure and to know how much it differs from the standard.

A space in which the pressure is less than atmospheric pressure is said to be under *vacuum*. The amount of vacuum is expressed in terms of the difference between the absolute pressure in the space and the pressure in the atmosphere; it is usually expressed in inches of mercury. Vacuum gage scales are marked from 0 to 30 inches of mercury. When a vacuum gage scale reads zero, the pressure in the space is the same as atmospheric pressure—in other words, there is no vacuum. A vacuum gage reading of 29.92 inches of mercury would indicate a perfect vacuum. In actual practice, it is impossible to obtain a perfect vacuum even under laboratory conditions. The highest vacuum readings obtained on shipboard machinery are practically never more than 29 inches of mercury and are usually a good deal less than that.

Absolute pressure is atmospheric pressure plus gage pressure or minus vacuum. For example, a gage pressure of 300 psig is an absolute pressure of 314.7 psia (300 + 14.7). If the measured vacuum is 10 inches of mercury vacuum, the absolute pressure in the space is approximately 20 inches of mercury absolute. It is important to note that the amount of pressure in a space under vacuum can only be expressed in terms of absolute pressure.

A little reflection will indicate that all pressure calculations would be easier and probably more accurate if we used only absolute pressure. To take just one example, consider the following conversation:

QUESTION: What's the pressure in that space?
ANSWER: Twenty inches of mercury.

If we convert the "20 inches of mercury" answer to pounds per square inch absolute, there are three possible interpretations of the answer. Did the individual answering mean 20 inches of mercury absolute? Then the pressure in the space is 9.8 psia. Did he mean 20 inches of mercury vacuum? Then the pressure in the space is 4.9 psia. Or did he mean 20 inches of mercury gage? If so, the pressure in the space is 24.5 psia. In each situation, then, it is necessary to establish clearly just what kind of pressure we are talking about, unless it is immediately clear in the context of the discussion.

Pressure, Temperature, and Volume Relationships *

The energy transformation of primary interest in the shipboard steam plant is the transformation from heat to work. To see how the transformation occurs, we need to consider the pressure, temperature, and volume relationships that hold true for gases. These relationships may be summarized as follows:

1. When the temperature is held constant, increasing the pressure on a gas causes a proportional decrease in volume. Decreasing the pressure causes a proportional increase in volume.
2. When the pressure is held constant, increasing the temperature of a gas causes a proportional increase in volume. Decreasing the temperature causes a proportional decrease in volume.
3. When the volume is held constant, increasing the temperature of a gas causes a proportional increase in pressure. Decreasing the temperature causes a proportional decrease in pressure.

Suppose we have a boiler in which steam has just begun to form. With the steam stop valves still closed, the volume of the steam remains constant while the temperature and the pressure are both increasing. When operating pressure is reached, and the steam stop valves are opened, the high pressure of the steam causes the steam to flow to the turbines. The pressure of the steam thus provides the potential for doing work; the actual conversion of heat to work is done in the steam turbines. The change in the internal energy of the steam (as indicated by changes in pressure and temperature) between the boiler and the condenser is an indication of the amount of heat that has been converted to work in the turbines.

*For a discussion of the gas laws see Arthur Beiser, *Physics* (2nd ed., Menlo Park, Calif.: Benjamin/Cummings Publishing Co., Inc., 1978).

When heat is transferred to a liquid, the average velocity of the molecules is increased, and the amount of internal kinetic energy stored in the liquid is increased. As the average velocity of the molecules increases, some molecules which are at or near the surface of the liquid momentarily achieve unusually high velocities, and some of them escape from the liquid and enter the space above, where they exist in the vapor state. As more and more of the molecules escape and come into the vapor state, the probability increases that some of the vapor molecules will momentarily have unusually low velocities; these molecules will be captured by the liquid. As a result of this exchange of molecules between the liquid and the vapor, a condition of equilibrium is reached, and an equilibrium pressure is established. The equilibrium pressure depends upon the molecular structure of the fluid and upon its temperature. For any given fluid, therefore, there is a definite relationship between the temperature and the pressure at which a liquid and its vapor may exist in equilibrium contact with each other.

The pressure and the temperature that are related in the manner just decribed are known as the saturation pressure and the saturation temperature. Thus, for any specified pressure there is a corresponding temperature of vaporization known as the saturation temperature; and for any specified temperature there is a corresponding saturation pressure.

A liquid that is under any specified pressure and at the saturation temperature for that particular pressure is called a saturated liquid. A liquid that is at any temperature below its saturation temperature is said to be a subcooled liquid. For example, the saturation temperature that corresponds to atmospheric pressure (14.7 psia) is 212°F for water. Hence, water at 212°F and standard atmospheric pressure is said to be a saturated liquid. Water flowing in a river or standing in a pond is also under atmospheric pressure, but it is at a much lower temperature; hence, this water is said to be subcooled.

A vapor that is under any specified pressure and at the saturation temperature corresponding to that pressure is said to be a saturated vapor. Thus, water at 14.7 psia and 212°F produces a vapor known as saturated steam. As previously noted, it is impossible to raise the temperature of a vapor above the temperature of its liquid as long as the two are in contact. If the vapor is drawn off into a separate container, however, and additional heat is supplied to the vapor, the temperature of the vapor is raised. A vapor that has been raised to a temperature above its saturation temperature is called a superheated vapor, and the vessel or container in which the saturated steam is superheated is called a superheater. The elementary boiler

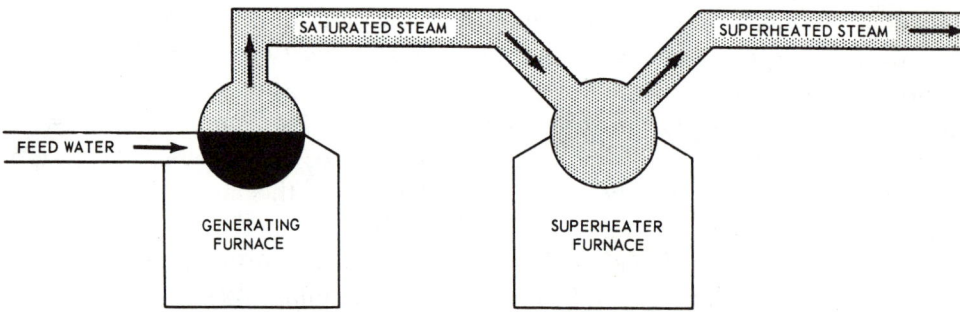

Figure 1–9. Elementary boiler and superheater.

and superheater illustrated in figure 1–9 show the general principle of generating and superheating steam. Practically all naval propulsion boilers have superheaters for superheating the saturated steam generated in the generating sections of the boiler; the steam is then called superheated steam. The amount by which the temperature of a superheated vapor exceeds the temperature of a saturated vapor at the same pressure is known as the degree of superheat. For example, if saturated steam at a pressure of 1275 psi and an approximate saturation temperature of 576°F is superheated to 975°F, the degree of superheat is 399°F.

2

The Main Steam Cycle

Ships can be powered in a variety of ways. Sails were first; then came steam. Now, try to imagine how many sails would be needed to move the 70,000-ton-plus super carrier *John F. Kennedy* at full speed.

Steam is one of the most dependable and effective means of powering a ship; therefore, steam is used to power the majority of our Navy's fighting ships. This chapter will cover the steam cycle and its relationship to a ship's propulsion plant. The illustration used in this chapter is representative of a ship's propulsion plant in general.

This chapter is mainly based on the typical FF 1052 class fast frigate main engineering plant. It should be noted, however, that the Navy employs various other types of steam plants ranging from 600 psi, 700 psi, 1200 psi to nuclear reactors.

Refer to figure 2–1 and locate each component as it as discussed. The main steam cycle has four phases: *generation, expansion, condensation,* and *feed*. Each phase will be described in the order listed.

Function and Purpose of the Boiler and the Components of the Generation Phase of the Main Steam Cycle

The steam-generation phase of the steam cycle takes place in the *boiler* where fuel, which has chemical energy, is burned in the boiler *furnace* to produce *heat*, or thermal energy.

The boiler is the largest and heaviest component of a steam propulsion plant, as illustrated in figure 2–1.

Fuel is burned in the furnace of a boiler to produce heat (about 3000°F), which is transferred by convection and radiation to the steam generating tubes of the boiler. This heat is then transferred by conduction through the metal tube walls to the water in the tubes. When the water is hot enough, "bubbles" start to form and rise through the water in the tubes. These "bubbles" are gas—a gas that we call *steam*.

When steam is being generated, it collects within a large cylindrical container, which is part of the boiler, called a *steam drum*. Continuous steam generation causes the pressure inside the drum to rise, which brings us to an important point: At sea level, with an atmospheric pressure of 14.7 psia, water boils and changes state into steam when it is heated to a temperature of 212°F. In a 1200-psig boiler, the water in the steam drum won't boil until is it heated to a temperature of 567°F! Why do we operate at a higher pressure? So that more heat energy can be transferred from the burning fuel to the water in the boiler!

Steam that collects in the steam drum is called saturated steam because it has the same temperature as the water with which it is in contact. Although saturated steam is sometimes referred to as "wet" steam, it is at least 99.75 percent steam; so its moisture content is very small.

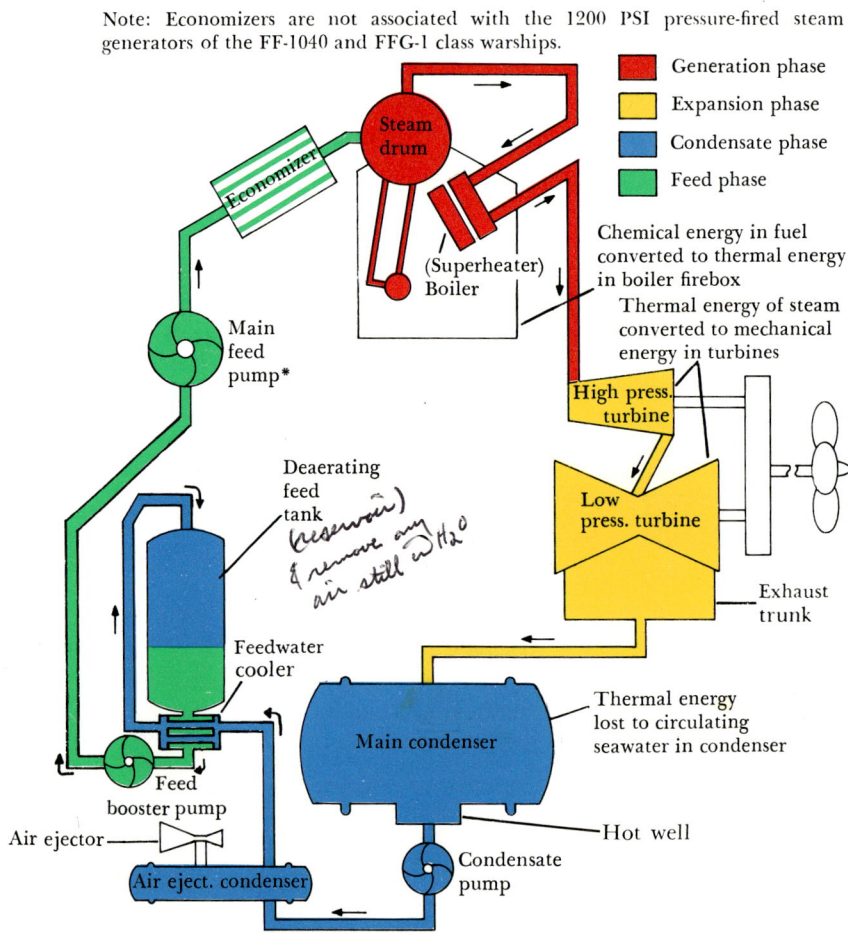

Figure 2–1. *Basic elements of a 1200 psi steam cycle.*

Unfortunately, even saturated steam doesn't contain enough thermal energy to cause the turbines of the main engines, turbogenerators, and main feed pumps to operate at their best efficiency, except in nuclear plants which only operate saturated. The required increase in thermal energy is produced by causing saturated steam to leave the steam drum and pass through the tubes of the superheater section of the boiler.

The *superheater* is a heat exchanger that uses heat from the furnace of a boiler to raise the thermal energy of the steam, which results in an increase in the *temperature* of the steam. Steam is considered to be "superheated" when it is *not* in contact with the water in the steam drum, and it has a temperature that is *greater* than the temperature of the saturated steam (even as little as 1° F). (Note: The superheater causes an increase in steam temperature only; there is *no* increase in the steam pressure.)

Remember what you learned about the properties of heat? Recall that when Btu's of heat are added to or removed from a substance, and there is a change in temperature with no change of state, it is called sensible heat. On the other hand, when Btu's of heat are added to or removed from a substance, and there is a change of state with no change in temperature, it is called latent heat.

In addition to increasing the thermal energy and temperature of saturated steam, the superheater also makes the steam passing through it "drier." Superheated steam is preferred to saturated steam because even the low moisture content of saturated steam can cause the blading in the turbines of main engines, turbogenerators, and main feed pumps to erode. Erosion, or the wearing away of the blades in a steam turbine, is caused by the friction of "heavy" water molecules in the steam rubbing against the blades as the steam passes through the turbine. These same water molecules also carry chemicals that leave deposits of scale on the blades, which could upset the dynamic balance of the turbine rotor. (Note: The above discussion is not dealing with carryover.*)

Function and Purpose of the High- and Low-Pressure Turbines and the Components of the Expansion Phase of the Main Steam Cycle

To provide steam with enough thermal energy for efficient operation of the main engines, turbogenerators, and main feed pumps,

---

*Carryover is the result of boiler water being entrained with the steam because of foaming (failure of the steam bubbles to break) or priming (boiler water carried out with the steam because of malfunctioning steam drum internals, high water level, etc.).

the superheater of a 1200-psig boiler heats the steam to a temperature of approximately 950°F! This high-pressure, high-temperature steam is carried from the superheater outlet of the boiler to the *high pressure (HP) turbine* of the main engine, through the main steam system, which must be able to withstand extremely high pressure and temperature.

As steam from the main steam system enters the HP turbine, the expansion phase of the steam cycle begins. Superheated steam passing through the turbine expands within the turbine, where thermal energy is converted to mechanical energy to turn the rotor of the turbine.

Although the pressure of the steam leaving the HP turbine is much less than it was when the steam entered, the steam still contains a considerable amount of thermal energy. To take advantage of this, the steam is allowed to expand even more as it passes through the blading of the *low pressure (LP) turbine*, which is connected to the *exhaust* of the HP turbine. (It should be noted that nuclear plants use a single-case turbine.)

Function and Purpose of the Main Condenser and the Components in the Condensation Phase of the Main Steam Cycle

Exhaust steam from the LP turbine flows into the *main condenser*, where the condensation phase of the steam cycle begins. The main condenser is a sealed container—a huge, indirect-type, shell-and-tube heat exchanger that contains miles of tubing through which cool seawater flows. As steam from the exhaust of the LP turbine passes over these tubes, latent heat flows out of the steam into the seawater, which carries the heat away with it on its way out of the main condenser. When a sufficient amount of latent heat has been removed from the steam, it changes state and turns back into water, called *condensate*.

This brings up an important point. Since the exhaust steam from the LP turbine has already expanded and is at a very low pressure, how can it flow into the main condenser? The answer to this question is that it can do so because the pressure inside the main condenser is even lower! Here's why: When the steam, having expanded to a large volume, passes over the colder tubes of the condenser, it condenses (changes state from steam to condensate or water). A given amount of water occupies much less space (volume) than the same amount of steam. Since the main condenser is a sealed container, the pressure has to drop, owing to the change in volume (space) occupied by the steam; and a vacuum is created. To help "create" this vacuum, *there must be a constant supply of cool circulating water flowing through the tubes.* This very "high" vacuum, between 26 and 29 in. Hg, means that the absolute pressure in the main condenser, during steady operation of the plant, is less than 1 psia!

As the exhaust steam turns into condensate, it flows down into a collecting tank at the bottom of the main condenser, called a *hotwell.* From this point the condensate is moved into the main condensate system, where it must pass through (a) the main condensate pump and (b) the air ejector condenser.

The *main condensate pump* is a one- or two-stage centrifugal pump. This electrically driven pump moves condensate from the hotwell of the main condenser, through the main air ejector condenser, to the deaerating feed tank.

The *air ejector,* that discharges into the main air ejector condenser, is a two-stage jet pump supplied with steam from the 150-psi auxiliary steam system. The suction side of the first stage of this pump is connected to the main condenser for the purpose of removing air that may collect or leak into the condenser.

In "starting up" the plant, the air ejector establishes and helps to maintain, along with the condensing steam, a "high" vacuum of approximately 29 in. Hg in the main condenser.

Before it enters the deaerating feed tank, the condensate must pass through the *air ejector condenser.* This shell-and-tube heat exchanger has two major functions: (1) it removes latent heat from steam being discharged from the air ejector; and (2) it transfers heat from the steam being discharged from the air ejector to preheat the condensate before it enters the deaerating feed tank.

After the condensate leaves the air ejector condenser, it passes through another heat exchanger, called a *feedwater cooler.* Then it flows into the upper section of the *deaerating feed tank,* where the condensate is heated and any free oxygen held in solution is removed. The condensate, thus heated and oxygen-free, falls to the lower section of the deaerating feed tank. This heated, oxygen-free *feedwater* is then stored until needed in the feed system. Because feedwater by definition must be oxygen-free, the lower section of the deaerating feed tank is the beginning of the feed phase.

The Components of the Feed Phase of the Main Steam Cycle and Their Function

As noted, the lower section of the deaerating feed tank is the beginning of the feed phase of the steam cycle. The deaerating feed tank is a direct heat exchanger which has three functions: (1) it contains apparatus to remove dissolved oxygen entrained in the condensate; (2) it heats the condensate and maintains the proper temperature of feedwater stored in the tank; (3) it is a storage tank, for heated, oxygen-free feedwater. (Note: Oxygen is removed owing to its corrosive effects on boiler watersides at high operating pressures.)

The main feed phase begins with the heated, deaerated feedwater stored in the deaerating feed tank. The other major components in this phase are the *main feed booster pump*, the *main feed pump*, and the *economizer*.

Feedwater must pass from the deaerating feed tank to the suction side of the main feed booster pump, a one- or two-stage double-suction centrifugal pump, which is installed below the deaerating feed tank. The force that moves feedwater into the suction side of the main feed booster pump is due to gravity and steam pressure of about 15 psig inside the deaerating feed tank.

The main feed booster pump discharges into the suction side of the main feed pump, with a normal pressure of 80–100 psig. Although this seems to be a strange way to do it, there is a perfectly good reason for this arrangement. A positive pressure must always be maintained on the suction side of a main feed pump, *to prevent hot feedwater (about 250°F) in the pump casing from "flashing" into steam*! (Note: To "flash" into steam means to "boil.") Here's what can happen. The steam turbine driving the main feed pump will overspeed, then "trip out," because of the sudden loss of load on the pump end as the "heavy" water in the casing is replaced by "light" steam. Now, with no pressure to push water into the boiler, the boiler will soon "run out" of feedwater, and the people in the fireroom will have a "low water in the boiler" casualty on their hands. As you can see, one thing leads to another, which could shut down a steam propulsion plant in a hurry!

The main feed pump itself is a turbine-driven, double-suction, multistage centrifugal pump, one of the more complex units of machinery in a steam propulsion plant.

A multistage centrifugal pump is used because it is one of the few pumps that can deliver feedwater in sufficient amounts (volume) and, at the same time, "develop" enough pressure to force the water into the boiler against the pressure in the steam drum.

From the discharge side of the main feed pump, boiler feedwater must pass through an economizer. At its name implies, the economizer improves the efficiency of the steam plant; it uses thermal energy normally wasted by "going up the smoke pipe," to preheat the feedwater before it passes into the steam drum of the boiler.

An economizer is a heat exchanger, simply a nest of tubes located between the furnace of the boiler and the smoke pipe. By this means, hot gases, created by the combustion of fuel in the furnace, must pass over the tubes so heat can be transferred from the hot furnace gases to the feedwater. Normally, the temperature of the feedwater at the outlet of the economizer will be about 140°F higher than the inlet temperature.

3

Overview of Oil-Fired Marine Boilers

In the conventional steam turbine propulsion plant, the steam generator (boiler) (figure 3–1) is the source or high-temperature region of the thermodynamic cycle. The steam that is generated in the boiler is led to the propulsion turbines and other machinery and heat exchangers where its thermal energy is converted into mechanical energy which drives the ship and provides power for vital services.

In essence, a boiler is merely a container in which water can be boiled and steam generated. When a boiler is designed to produce a large amount of steam, a large heat-transfer surface must be provided. In most modern boilers, the steam-generating surface consists of 1000 to 2000 tubes, which provide a maximum amount of heat-transfer surface in a relatively small space. As a rule, the tubes communicate with a steam drum at the top of the boiler and with water drums and headers at the bottom of the boiler. The tubes and part of the drums are enclosed in an insulated casing that has space inside it for a furnace. As we will see presently, a boiler appears to be a fairly complicated piece of equipment when it is considered with all its fittings, piping, and accessories. It may be helpful, therefore, to remember that the basic components of a saturated-steam boiler are merely the tubes in which steam is generated, the drums and headers in which water is contained and steam is collected, and the furnace in which combustion takes place.

Practically all boilers used in propulsion plants are designed to produce either saturated and superheated steam or superheated and desuperheated steam. Steam that is at, or near, the saturation temperature for its pressure is said to be "saturated steam." Very small decreases in temperature or increases in pressure will change this steam back into some moisture and water. Steam that is not at, or near, its saturation temperature and would require large changes in temperature or pressure to change it into water is said to be superheated. Superheated steam is obtained by adding more heat at constant pressure to the saturated steam. To our basic boiler, therefore, we must now add another component: the superheater. The superheater on most boilers consists of header manifolds, usually located at the back or bottom of the boiler, and a number of superheater

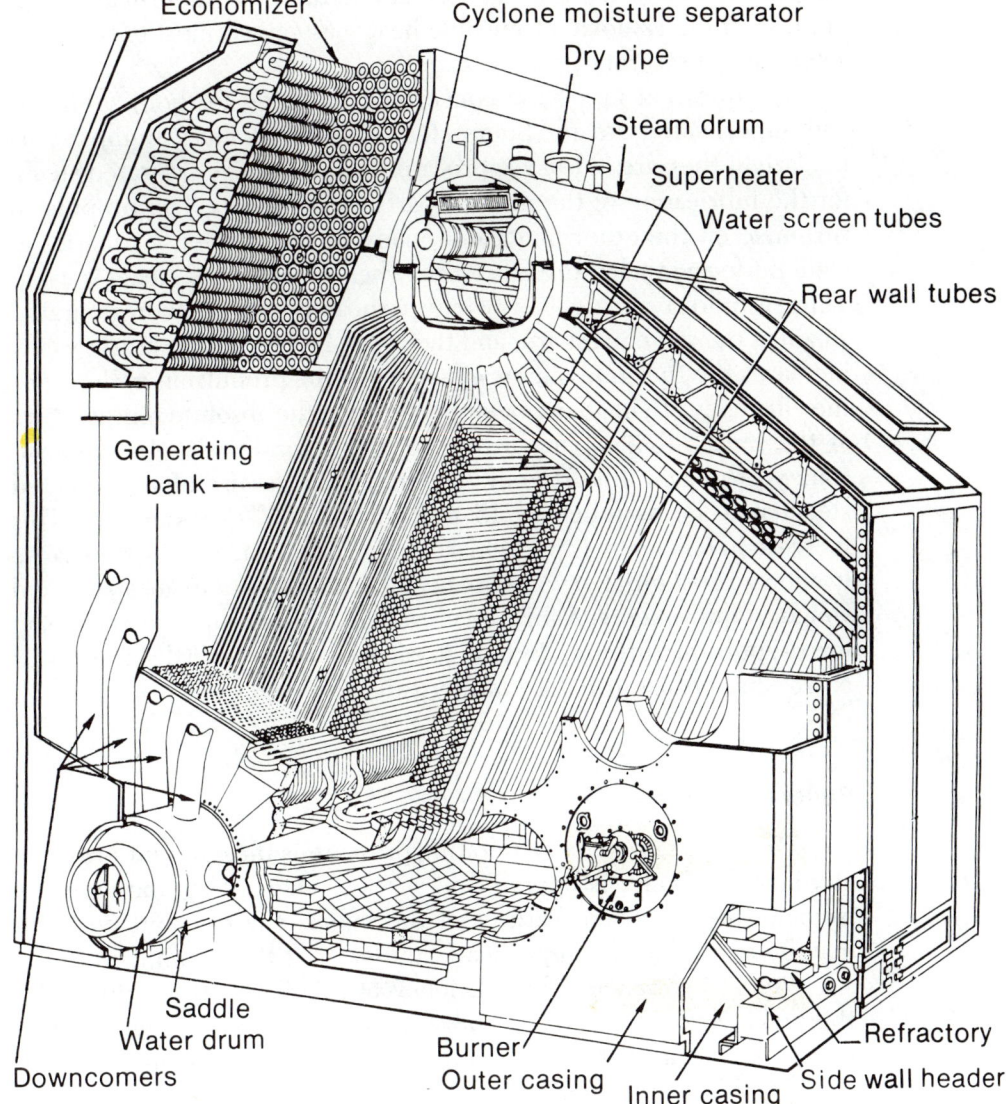

Figure 3-1. Cutaway frontal view of a 1200 psi D-type boiler.

tubes, which are connected to the headers. Saturated steam from the steam drum is led through the superheater; since the steam is now no longer in contact with the water from which it was generated, the steam becomes superheated without any appreciable increase in pressure as additional heat is supplied. In some boilers, there is a separate superheater furnace; in others, the superheater tubes project into the same furnace that is used for the generation of saturated steam.

Desuperheated steam is steam that at one time was superheated, but because of the removal of sensible heat, has gained most of the characteristics of saturated steam.

Some question may arise concerning the need for both saturated and superheated steam. Some steam-driven auxiliary pumps, particularly if they are of the reciprocating type, require saturated steam for the lubrication of the moving parts of the pump. The propulsion turbines, on the other hand, and many turbine-driven auxiliaries as well, perform much more efficiently when superheated steam is used. There is more available energy in superheated steam than in saturated steam at the same pressure, and the use of higher temperatures vastly increases the thermodynamic efficiency of the propulsion cycle, since the efficiency of a heat engine depends upon the absolute temperature at the source (boiler) and at the receiver (condenser). In some instances, the gain in efficiency resulting from the use of superheated steam may be as much as 15 percent for 200° of superheat. This increase in efficiency is particularly important for naval ships because it allows substantial savings in fuel consumption and in space and weight requirements. A further advantage in using superheated steam for propulsion turbines is that it causes relatively little erosion or corrosion because it is free of moisture.

Boiler Definitions

Operating Pressure. Operating pressure is the constant pressure at which the boiler is operated in service. Depending upon various factors, chiefly design features of the boiler, the constant pressure may be carried at the superheater outlet or at the steam drum.

Boiler Efficiency. The efficiency of a boiler is the ratio of the thermal energy that is actually absorbed by the water to the thermal energy that is available in the fuel. Typical boiler efficiencies are in the 80–90 percent range.

Steaming Hours. The term steaming hours is used to indicate all time during which the boiler has fires lighted for raising and generating steam. Steaming hours includes the time when fires are lighted for baking out boiler refractories. Time during which fires are not lighted is *not* included in steaming hours.

Heating Surfaces. The heating surfaces of a modern boiler are formed primarily by the use of a great many metal tubes. The total heating surface of a boiler includes all parts of the boiler that are exposed on one side to the gases of combustion and on the other side to the water and steam being heated. The total heating surface is composed of the generating surface, the superheater surface, and the economizer surface.

The *generating surface* is that part of the total heating surface in which water is being heated and steam is being generated. The generating surface includes the generating tubes, the water wall tubes, and the water screen tubes. Note that the actual transformation from water to steam occurs in these tubes.

The *superheater surface* is that part of the total heating surface in which the steam is superheated after leaving the boiler steam drum.

The *economizer surface* is that part of the total heating surface in which the feedwater is preheated before it enters the generating part of the boiler. Not all boilers have economizers.

Boiler Classification

Although boilers may vary considerably in details of design, most boilers may be classified in terms of a few basic features or characteristics. Some knowledge of these methods of classification provides a useful basis for understanding the design and construction of various types of naval boilers. As you review the following material, refer to figure 3–1 for reference.

Location of Fire and Water Spaces

One basic classification of boilers is made according to the relative location of the fire and water spaces. All boilers used in the propulsion plants of modern naval ships are of the *water-tube type.* In water-tube boilers, the water flows through the tubes and is heated by the gases of combustion that fill the furnace and heat the outside of the metal surfaces of the tubes.

Type of Circulation

Water-tube boilers are further classified according to the cause of water circulation. By this mode of classification, we have *natural circulation* boilers (figure 3–2) and *controlled circulation* boilers.

In natural circulation boilers, the circulation of water depends on the difference between the density of an ascending mixture of hot water and steam and a descending body of relatively cool and steam-free water. The difference in density occurs because the water expands as it is heated and thus becomes less dense. Another way to describe natural circulation is to say that it is caused by convection currents which result from the uneven heating of the water contained in the boiler.

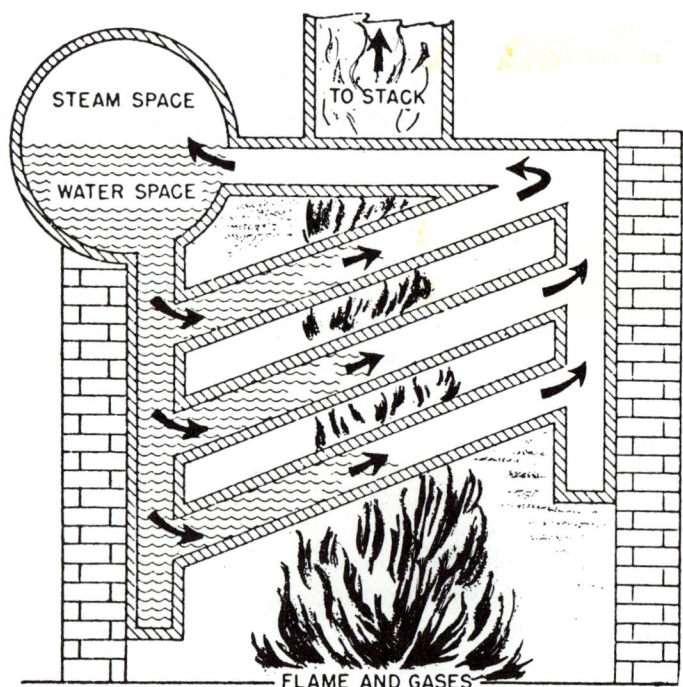

Figure 3–2. Natural circulation (free type).

Most modern naval boilers are designed for *accelerated natural circulation* (figure 3–3). In such boilers, large tubes (3 or more inches in diameter) are installed between the steam drum and the water drums. These large tubes, called downcomers, are located outside the furnace and away from the heat of combustion, thereby serving as pathways for the downward flow of relatively cool water. When a sufficient number of downcomers are installed, all small tubes can be generating tubes, carrying steam and water upward, and all downward flow can be carried by the downcomers. The size and number of the downcomers installed varies from one type of boiler to another, but some are installed on all modern naval boilers.

Controlled circulation boilers are, as their names implies, quite different in design from the boilers that utilize natural circulation. Controlled circulation boilers depend upon pumps, rather than upon natural differences in density, for the circulation of water within the boiler. Because controlled circulation boilers are not limited by the requirement that hot water and steam must be allowed to flow upward while cooler water flows downward, a greater variety of arrangements may be found in controlled circulation boilers.

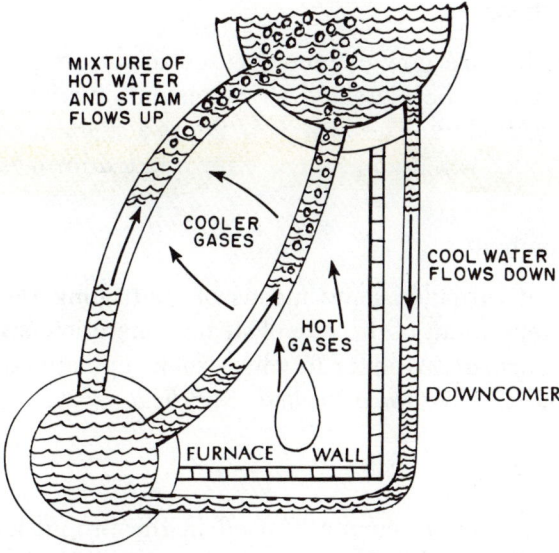

Figure 3-3. Natural circulation (accelerated type).

Arrangement of Steam and Water Spaces

Natural circulation water-tube boilers are classified as drum-type boilers or header-type boilers, depending upon the arrangement of the steam and water spaces. *Drum-type* boilers have one or more water drums (and usually one or more water headers as well). *Header-type* boilers have no water drum; instead, the tubes enter a great many headers which serve the same purpose as water drums.

What is a header, and what is the difference between a header and a drum? The term header is commonly used in engineering to describe any tube, chamber, drum, or similar piece to which a series of tubes or pipes is connected in such a way as to permit the flow of fluid from one tube (or group of tubes) to another. Essentially, a header is a type of manifold. As far as boilers are concerned, the only distinction between a drum and a header is the distinction of size. Drums are larger than headers, but both serve basically the same purpose.

Drum-type boilers are further classified according to the overall shape formed by the steam and water spaces—that is, by the tubes. For example, double-furnace boilers are often called *M-type* boilers because the arrangement of the tubes is roughly M-shaped. Single-furnace boilers, like that shown in figure 3-1, are often called *D-type* boilers because the tubes form a shape that is approximately like the letter D.

Number of Furnaces

All boilers that are now commonly used in the propulsion plants of naval ships may be classified as being either *single-furnace* or *double-furnace* boilers. The D-type boiler is a single-furnace boiler; the M-type boiler is a double-furnace (or divided-furnace) boiler.

Control of Superheat

A boiler that provides some means of controlling the degree of superheat independent of the rate of steam generation is said to have *controlled superheat*. A boiler in which such separate control is not possible is said to have *uncontrolled superheat*.

Type of Superheater

On almost all boilers currently used in the propulsion plants of naval ships, the superheater tubes are protected from radiant heat of the furnace, and the superheater tubes are heated by convection currents rather than by radiation. Hence, the superheaters are referred to as *convection-type superheaters*.

The term *integral* is often used in connection with superheaters to indicate that the superheater is installed as an integral part of the generating-tube bank, rather than in a separate part of the boiler. The single-furnace boiler has an integral superheater, while the double-furnace or M-type boiler does not.

Operating Pressure

For some purposes it is convenient to classify boilers according to operating pressure. Most classifications of this type are approximate rather than exact. The term 600 psi boiler is often applied to various double-furnace and single-furnace boilers with operating pressures ranging around 600 psi.

The term high pressure boiler is at present used rather loosely to identify any boiler that operates at a substantially higher pressure than the so-called 600 psi boilers. In general, we will consider that any boiler that operates at 750 psi or above is a *high pressure* boiler. A good many boilers recently installed on board naval ships operate at approximately 1200 psi; for some purposes, it is convenient to group these boilers together and refer to them as 1200 psi boilers.

As may be seen, boiler classification by operating pressures may vary widely within one group. Also, any classification based on operating pressure may easily become obsolete. What is called a high pressure boiler today might well be called a low pressure boiler tomorrow.

Most propulsion boilers now used by the Navy have essentially the same components: steam and water drums, generating and circulating tubes, superheaters, economizers, fuel oil burners, furnaces, casings, supports, and a number of accessories and fittings required for boiler operation and control. The basic components of boilers are described here. Later we will see how the components are arranged to form various common types of naval propulsion boilers.

Single-Furnace or D-Type Boilers

Single-furnace boilers are the most widely used types in the Navy today. A single-furnace boiler is shown schematically in figure 3–1. This boiler does not have controlled superheat. When the boiler is lit off, both the generating tubes and the superheater tubes are heated. In order to protect the superheater tubes from overheating, all steam generated in the boiler must be led through the superheater. The saturated steam goes from the dry pipe in the steam drum to the superheater inlet; it then goes through the superheater tubes, out the superheater outlet, and into the main steam line.

Saturated steam must go through the superheater (in order to provide a sufficient steam flow to protect the superheater), but must then be desuperheated. Desuperheating is accomplished by passing some of the superheated steam through a desuperheater, which is basically a coil of piping submerged in the water in the water drum. Heat transfer takes place from the steam in the desuperheater to the water in the drum. The desuperheated steam passes out of the desuperheater and into the desuperheated steam line.

On the basis of the classification methods given earlier, we may consider this single-furnace boiler as one that has the following characteristics: (1) It is a water-tube boiler with natural circulation of the accelerated type. (2) It is a drum-type (rather than a header-type) boiler. (3) It has tubes that are arranged roughly in the shape of the letter D—hence it is often called a D-type boiler. (4) It has only one furnace. (5) It does not have controlled superheat. (Note: Most newer fast frigate D-type boilers would be classified as 1200 psi boilers, and the latest oil-fired boilers are in AOs, ASs, ADs, and LHAs, which are 700 psi.)

The basic design of the single-furnace boiler is seeing increased use. Except for experimental boilers, no double-furnace boilers have been installed on board combatant ships since World War II. The newer single-furnace fast frigate boilers operate at approximately 1200 psi. Operating temperature at the superheater outlet is quite commonly 925–975°F for the 1200 psi boilers; this is 75–100°F higher than the operating temperature of the older 600 psi single-furnace boilers and provides more energy to the turbines.

One of the most noticeable differences between the older and the newer single-furnace boilers is the change in furnace design. Higher heat release rates are possible in the newer boilers. Although these newer single-furnace boilers are not the type that we refer to as "pressurized-furnace" boilers, they do often use a slightly higher combustion air pressure than the older single-furnace boilers. The use of higher air pressure causes an increase in the velocity of the combustion gases, and the increased velocity results in a higher rate of heat transfer to the generating tubes. Because of the increased heat release rates, a newer single-furnace boiler is likely to have a water-cooled roof and water-cooled rear walls as well as water-cooled side walls.

4

1200 psi D-type Boiler Water and Steam Side Systems

Boiler Overview

Refer to figure 4–1 for the following discussion:

All feedwater entering the boiler enters the economizer first. The economizer is located in the upper part of the boiler. The function of the economizer is to preheat the feedwater prior to its entering the steam drum. The feedwater is heated by the combustion gases from the furnace passing around the economizer tubes (elements) and out of the stack.

After the feedwater is preheated in the economizer, the water enters the steam drum. The steam drum is located at the uppermost part of the boiler opposite the economizer. Its purpose is to accumulate steam generated in the tubes and also to serve as a reservoir for the large quantity of water required for proper operation of the boiler. It is held in place and supported by boiler tubes.

Downcomers are large tubes used to distribute water to the lower parts of the boiler. The downcomers are located between the inner and outer casings and extend from the steam drum to the headers and water drum. The water drum is located in the lower part of the boiler and acts as the reservoir that supplies water to the main generating tubes. The desuperheater is also located in the water drum.

The generating tubes connect the water drum to the steam drum. The water enters the generating tubes from the water drum, and as the water is heated, it rises in the tubes, forming steam. The steam rises to the steam drum because it is lighter than water. The generating tubes produce the majority of steam generated in the boiler. Boiler water also flows through the downcomers to certain headers, which are the lower rear wall, water screen, and sidewall headers. The headers distribute water to the boiler tubes. Other headers such as the upper rear wall and superheater headers serve as a collection point for steam.

Steam collected in the steam drum is piped to the superheater headers. The steam is then directed through the superheater tubes, making four passes, during which the saturated steam is heated to superheat temperature.

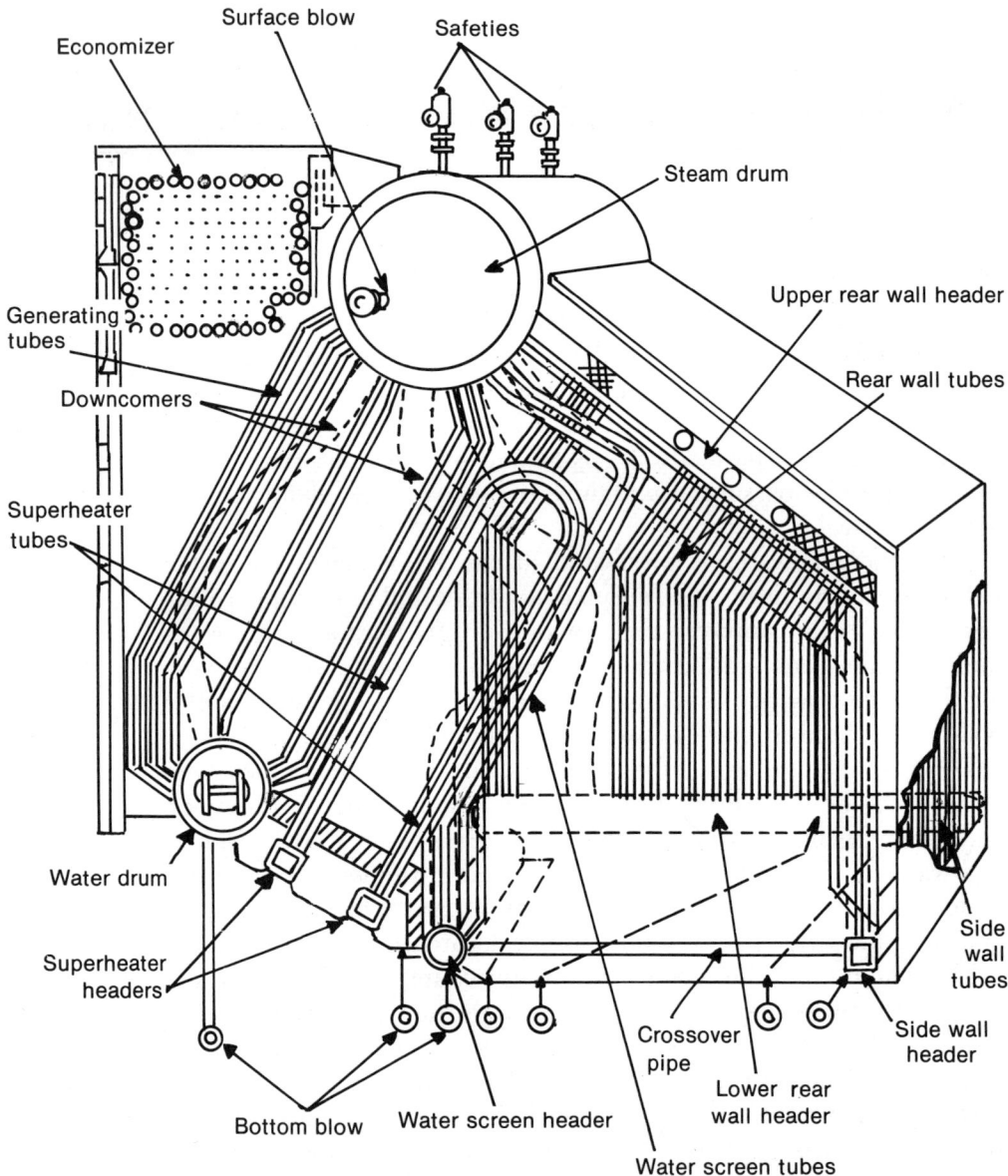

Figure 4–1. Basic tube arrangement of a 1200 psi D-type boiler.

Boiler Components

Safety valves are located on the boiler steam drum and on the superheater outlet. The purpose of these valves is to prevent excessive boiler pressure. There are always some impurities in boiler water, such as dissolved and suspended solids. These impurities are con-

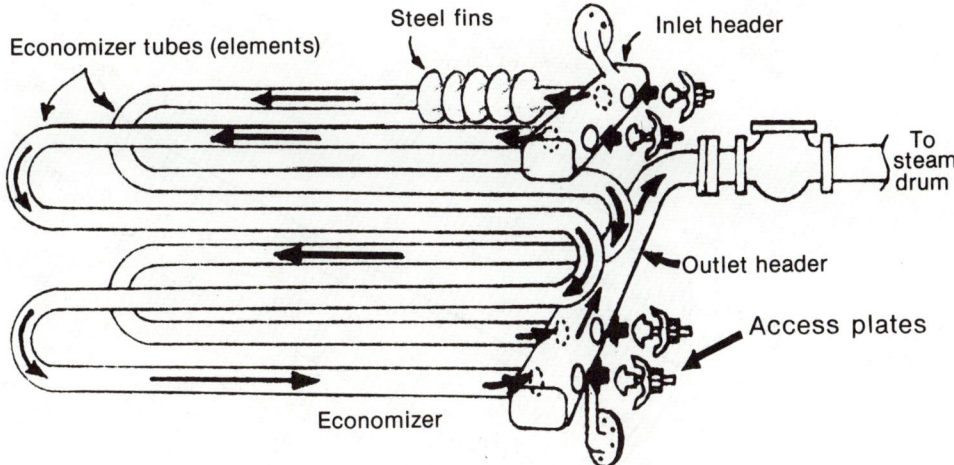

Figure 4–2. Basic parts of an economizer with direction of water flow.

trolled through the use of the surface blow connection and valve, which is installed on the steam drum. Other impurities, which are heavier than water, settle in the water drum and headers. They are controlled by using the bottom blow connections installed on the water drum, side wall, lower rear wall, and screen wall headers, which are attached to the bottom blow valves and piping system.

Economizer

One of the components of the economizer (figure 4–2) is the inlet header. All boiler feedwater entering the economizer goes into the inlet header, which distributes the water evenly to the economizer tubes (elements). The water going through the economizer tubes makes several passes and is preheated by the gases of combustion passing around the outside of the tubes. The outside of the economizer tubes is equipped with steel fins, which are the continuous spiral type. The steel fins give a larger heating surface and aid in the heat transfer from the combustion gases to the feedwater. After the water passes through the tubes, it enters the outlet header, which provides a collection area for the feedwater leaving the economizer tubes.

Located on the inlet header at the highest section is a vent connection. This vent connection allows an outlet for trapped air to be expelled from the economizer. A valve is attached to this connection, but is not shown in this illustration.

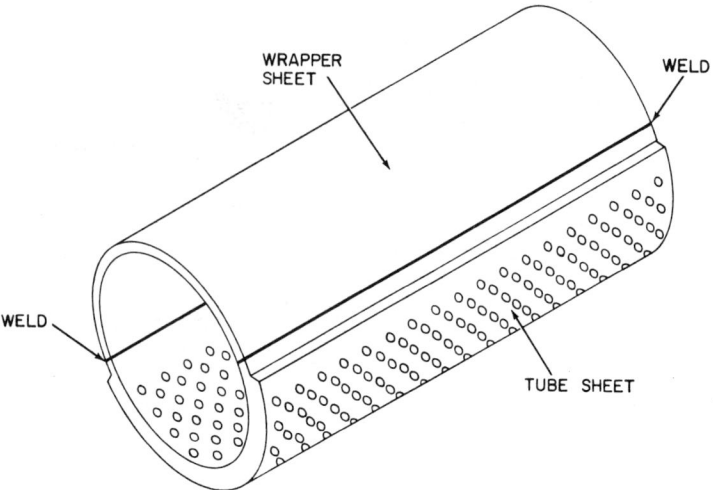

Figure 4–3. Basic design elements of a steam drum.

Steam Drum

The feedwater leaves the economizer and enters the steam drum, which is located at the uppermost part of the boiler. Here all steam generated in the tubes accumulates; it also serves as a reservoir for boiler water. Inside the steam drum there are many internal fittings to aid in distribution of water and in directing the flow of steam and water within the steam drum. The steam drum consists of two sheets of steel (figure 4–3) rolled or bent to the required semicircular shape and then welded together. The upper half is called the wrapper sheet, while the lower half is called the tube sheet. As illustrated, the tube sheet is thicker than the wrapper sheet. The extra thickness is required to ensure adequate strength because the tube sheet has many holes to accommodate the side wall, screen, and generating tube ends.

After the tube sheet and wrapper sheet are welded together, the steam drumheads are welded on each end. One drumhead is made with a large opening, which serves as an access for cleaning and repairs. A manhole plate is designed to fit the opening, as in figure 4–4; the plate is hinged to the steam drum. There is a gasket located between the edge of the drumhead and the lip of the manhole plate. This gasket aids in setting up a seal between the seating surface of the drumhead and the manhole plate. After the gasket is in place and the manhole plate closed against the drumhead, the strongback assemblies are put in place to hold the manhole plate tight to the drumhead.

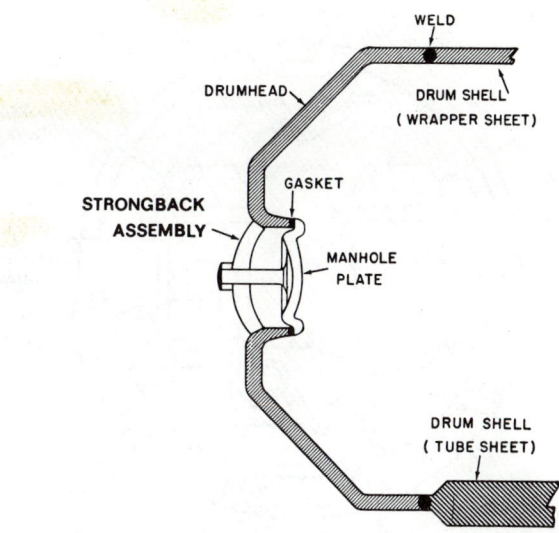

Figure 4–4. Typical elements of a manhole plate in the steam drum.

There are many internal fittings within the steam drum. About 3 inches above the tube sheet in the steam drum are the support plates and panel assemblies (figure 4–5), which are secured to the bottom of the steam drum and direct the flow of steam and water mixture to the steam separators; additionally, they support the weight of the steam separators, feed pipe, and surface blow pipe.

The mixture of steam and water enters the primary separator through the lower support mounted on the panel assemblies. Generally, there are 14 primary steam separators. Inside the primary separator, the mixture of steam and water hits the spinner blades, producing a swirling motion. The heavier water is thrown to the sides where it collects and drops back into the lower half of the drum.

Secondary separators are located on top of and bolted to the 14 primary separators. These secondary separators are made of several corrugated plates in the shape of an inverted basket. The corrugated plates separate the steam and water mixture by creating rapid changes in direction. The steam can adapt to these rapid changes but the heavier water cannot; therefore the water collects and falls back into the lower half of the steam drum.

Steam leaving the steam separators rises to the top of the steam drum where the dry box is located. The dry box is designed to separate any steam and water mixture not separated in the steam separators, and works on the same principle as the separators (rapid change in direction). At each end of the dry box there is a drain hole where

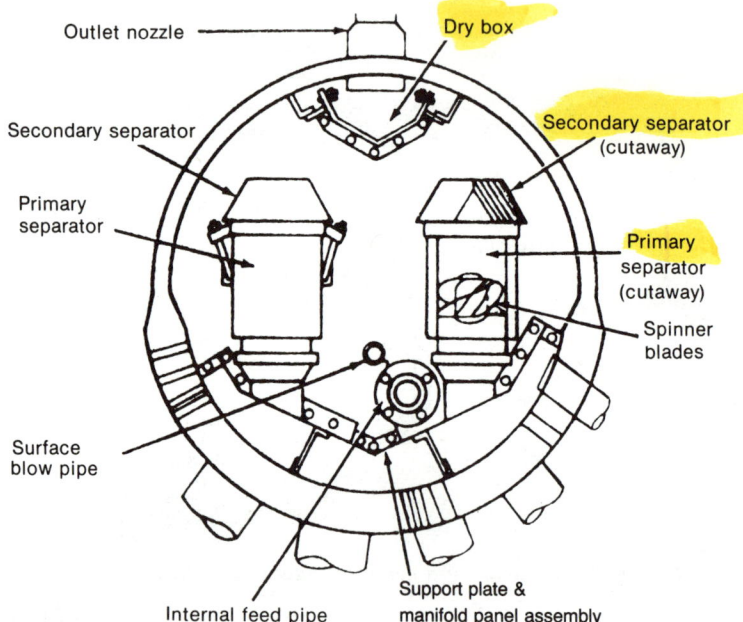

Figure 4–5. Vertical cutaway of the stream drum showing internals.

collected water may drop back into the lower half of the steam drum; steam leaves the dry box through the outlet nozzle (connection) located on top of the steam drum. The feed pipe assembly receives the feedwater from the economizer outlet header and distributes the water evenly within the steam drum. This assembly consists of a nozzle (opening in the drum) that enters the drum through the rear drumhead. Inside the drum the internal feed pipe has holes for the feedwater to disperse within the drum.

The surface blow pipe is located above the internal feed pipe about 4 inches below normal water level. One end of the surface blow pipe is blanked off; the other end is connected to a flanged nozzle on the front drumhead. The surface blow pipe has holes along the entire length of the pipe, which permit removal of impurities that collect in the boiler and float on the surface of the water. (Normal water level is the level at which a boiler is normally operated.)

The internal feed pipe and surface blow pipe have nozzles (connections) located on the drumheads. Other nozzles (connections) that enter the steam drumhead are the drum level transmitter nozzles and the water gage glass nozzles (figure 4–6). The drum level transmitter nozzles provide a connection that is part of the automatic boiler control system, while the water gage glass provides a visual monitoring point for the level of water in the steam drum. Located on the

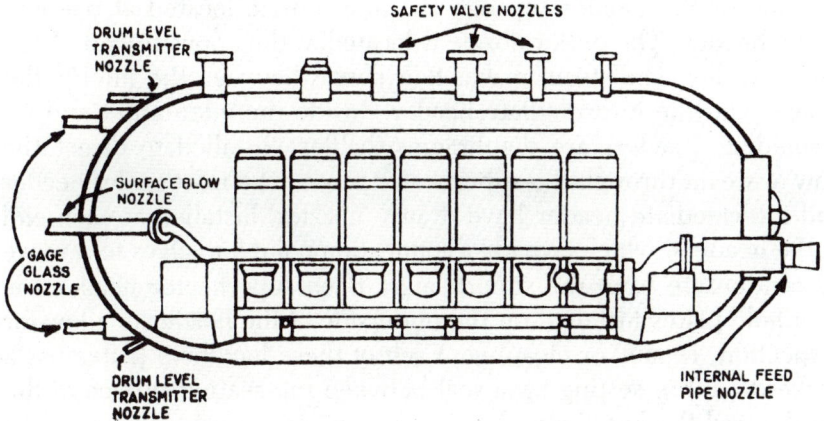

Figure 4–6. Longitudinal cutaway of the steam drum showing internals.

steam drum wrapper sheet are nozzles (connections) for mounting safety valves, which prevent excessive boiler pressure.

Superheater

Saturated steam leaving the steam drum through the steam outlet nozzle is piped to the superheater inlet/outlet header; the saturated steam is heated to superheat temperature. All steam leaving the steam drum goes through the superheater (figure 4–7), and steam enters

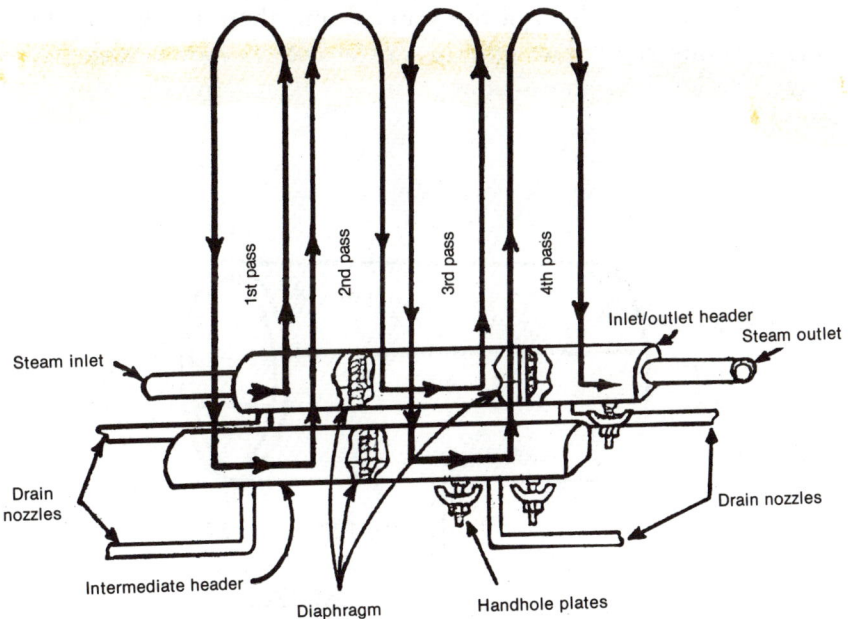

Figure 4–7. Basic elements of a typical superheater with steam flow indicated.

1200 psi D-Type Boiler Water and Steam Side Systems

the inlet/outlet header through the inlet nozzle located at one end of the header. The outlet nozzle is located at the opposite end of the same header. The steam makes four passes between the inlet/outlet header and the intermediate header. Inside the inlet/outlet and intermediate headers are diaphragms (baffles) installed to direct the flow of steam through the superheater tubes. The inlet/outlet header and intermediate header have drains (nozzles) installed at each end of the headers, which provide a connection point for valves to remove all condensate from the superheater. Each superheater header has handhole plates installed on the underside of the header to allow for inspection, repair, or cleaning. Each of these handhole plates has a gasket to aid in setting up a seal between the seating surface of the header and the handhole plate.

Water Drum

The water drum is located at the lower part of the boiler and receives water from the steam drum by way of the downcomers. One purpose of the water drum is to distribute water to all the generating tubes. Construction of the water drum (figure 4–8) is the same as for the steam drum except for the thickness of the tube sheet, which is the top part of the water drum. The holes in the tube sheet are to accommodate the tubes coming down from the steam drum.

On the wrapper sheet of the water drum, there is a bottom blow nozzle (connection). The purpose of the bottom blow connection is to provide for the removal of impurities contained in the boiler water drum.

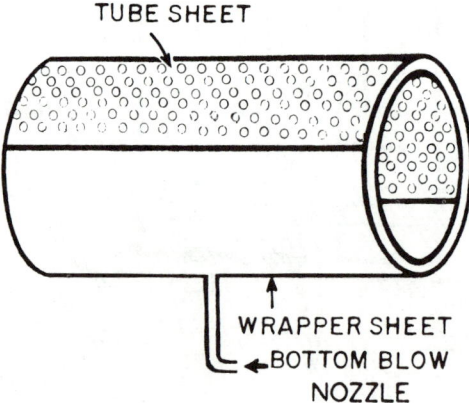

Figure 4–8. Basic design elements of a water drum.

Desuperheater

The desuperheater assembly is located in the water drum. As you have learned, all steam must pass through the superheater, with a large majority of superheated steam being piped to the main engine. Some of the steam leaving the inlet/outlet header of the superheater is piped to the inlet nozzle of the desuperheater where its temperature is reduced to about the same temperature as that of the water in the water drum as it passes through the tubes. This desuperheater steam is used in auxiliary machinery. The water drum has drumheads welded to both ends. One drumhead has a large access opening, to provide for access and inspection or repairs. The desuperheater (figure 4–9) inlet and outlet nozzles (connections) are located in the drumhead that does not contain the manhole plate. The inlet nozzle is on the bottom, and the outlet nozzle is on the top.

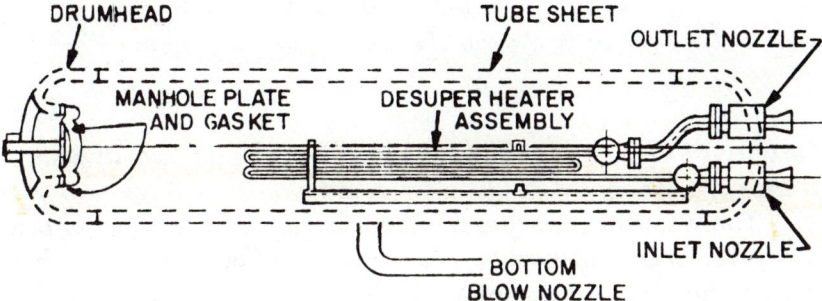

Figure 4–9. Longitudinal view of a typical desuperheater located in the water drum.

Downcomers

Downcomers supply the water drum with water from the steam drum. They also supply the headers (water screen, side wall, and lower rear wall) with water. These headers (figure 4–10) distribute the water received from the downcomers to the side wall tubes, water screen tubes, and rear wall tubes.

Tubes and Headers

The water screen header distributes the water received from the steam drum through the downcomers to the water screen tubes, whose purpose is to prevent radiant heat in the furnace from coming in contact with the superheater tubes. The water screen tubes extend from the header to the steam drum, and tubes from the side wall header (side wall tubes) extend from the header to the steam drum.

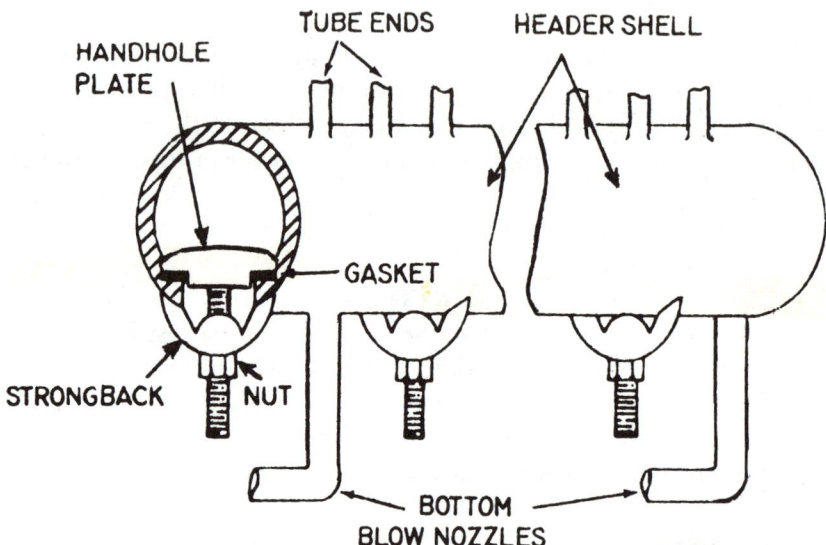

Figure 4–10. Basic design elements of a header.

The purpose of the side wall tubes is to protect the furnace refractories (brickwork) next to the casing located in the side wall from excessive heat. The water passing through the tubes absorbs this excess heat. The lower rear wall header receives water from the steam drum through the downcomers just as the side wall and screen wall headers do; the lower rear wall header distributes the water received to the rear wall tubes. The purpose of the rear wall tubes is to protect the furnace refractories (brickwork) at the rear of the furnace next to the casing from excessive heat. The rear and side wall tubes are referred to as water walls.

The rear wall tubes extend from the lower rear wall header to the upper rear wall header, which is located at and in line with the furnace roof. The upper rear wall header serves as a collection point for water and steam from the rear wall tubes, with circulation from the upper rear wall header back to the steam drum provided by riser pipes. These riser pipes are spaced along the length of the upper rear wall header. The water screen header is connected to the side wall header by a pipe. This connecting pipe is called a crossover pipe; it runs in between the inner and outer casings. The side wall header is also connected to the lower rear wall header in the same manner as with the water screen header.

The headers in the boiler are too small for a man to enter; therefore, handhole plates are installed, which are large enough only for a hand to be inserted. They are constructed like manhole plates but are smaller. Their purpose is to allow for inspection, repair, or cleaning of the inside of the headers. The handhole plate has a gasket strongback and nut just as the manhole plate has.

Located on the underside of the side wall water screen and lower rear wall headers are bottom blow nozzles (connections). There is one bottom blow nozzle at each end of each header. These nozzles are connected to valves in a piping system used for removing heavy solids and sludge from the headers. The upper rear wall header does not have bottom blow nozzles.

5

1200 psi D-Type Boiler Firesides and Support Systems

Boiler Overview

This chapter covers the 1200 psi D-type boiler fireside system (figure 5–1). This boiler does not have controlled superheat. When the boiler is lighted off, combustion takes place in a single furnace, with heat of combustion being transferred to both the generating tubes and the superheater tubes. The purpose of the fireside system is to convert chemical energy (fuel oil) to thermal energy (heat). When this conversion takes place, the thermal energy heats the water and generates steam and superheated steam. The heat and gases of combustion are produced in the furnace, which is enclosed by the side wall, rear wall, and screen tubes. Burner assemblies that are mounted on the boiler front provide for the proper mixing of the fuel and air and their entry into the furnace. The function of the economizer is to preheat the feedwater before it goes into the steam drum, which is done by combustion gases from the furnace passing around the economizer tubes and out the smoke pipe. The economizer is located in the upper part of the boiler at the base of the smoke pipe. Most of the tubes in a boiler are generating or circulating tubes and consist of the generating tubes, water wall tubes, screen tubes, and downcomers. Generating tubes generate most of the saturated steam. Water wall and screen tubes act as generating tubes at high firing rates. Downcomers connect the steam drum with the headers and the water drum. These are the largest tubes in the boiler, and are located between the inner and outer air casings. The purpose of downcomers is to ensure water circulation from the steam drum to the water drum and headers. Downcomers are never used to generate steam, but only for circulation of water to the water drum and headers.

The temperature of the saturated steam is raised in the superheater tubes. Raising the temperature of the steam this way adds thermal energy to the steam, resulting in greater efficiency and economy of the engineering plant. The function of the uptake is to conduct the combustion gases from the furnace to the stack after they leave the economizer. The inner and outer casings of this boiler are constructed of steel panels bolted together. The purpose of the two casings is to

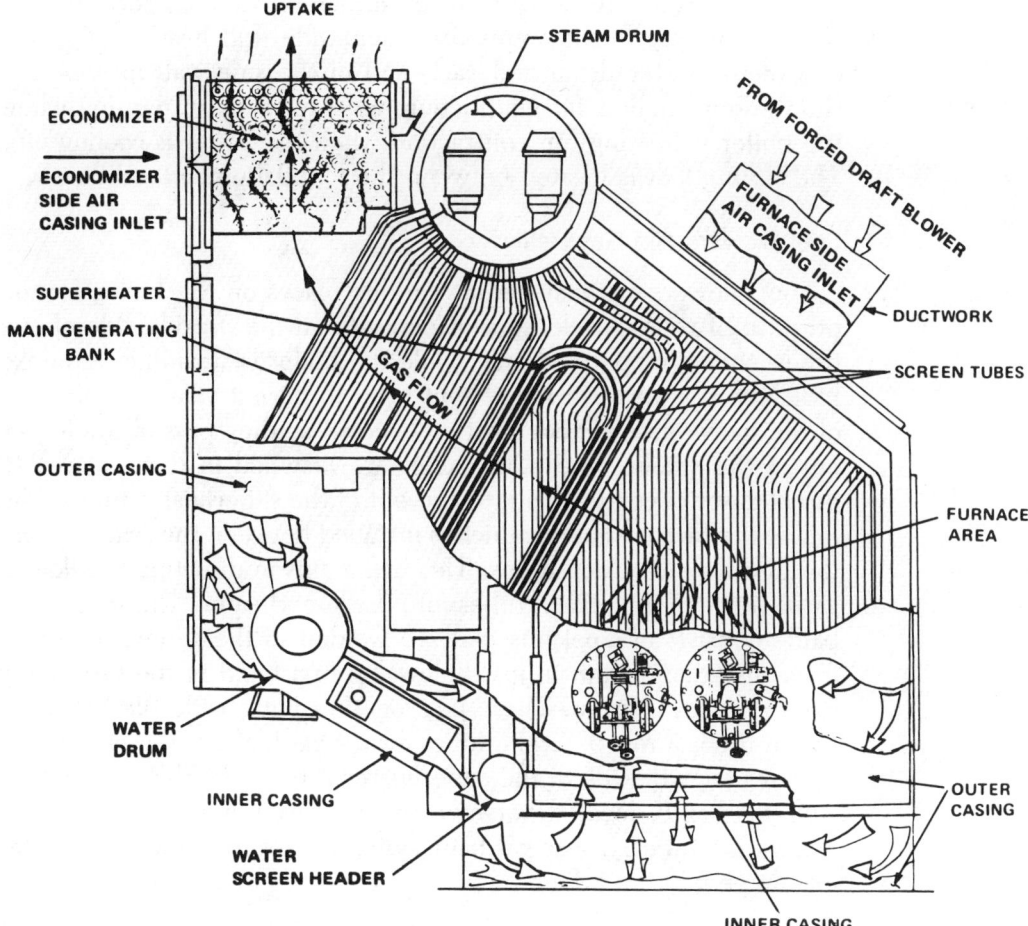

Figure 5–1. Detailed cutaway of a 1200 psi D-type boiler.

allow air from the blower to flow between both casings, around the boiler, and into the furnace. Access panels are designed to allow entry for inspection and/or repairs, and are constructed of steel panels that bolt onto the inner and outer casings.

Boiler Components

The foundation arrangement of a boiler provides support at several locations and is welded to the hull of the ship. The support arrangement provides support at each end of the water drum and at each end of the headers. These supports are constructed of heavy structural steel. On top of the foundation are saddles, which are constructed flat on the bottom and rounded or curved on top to fit the shape of the water drum and headers. The saddles provide support and are welded directly to the water drum and side wall headers. On one end of the water drum and each end of the side wall header is a sliding foot to allow for contraction and expansion (expansion when the boiler is heating up, contraction when the boiler is cooling off). The sliding foot is located between the foundation and saddle.

Soot Blowers and Gas Baffles

Soot blowers are installed at various places on the boiler for the purpose of removing loose soot deposits on the fireside side of the tubes; they are operated only while the boiler is steaming. A smoke indicator is a mirror arrangement used to permit visual observation of the heavy gases of combustion (smoke) as they pass on their way out of the smoke pipe. Gas baffles are installed in three places to direct gases of combusion into and out of the superheater tubes. The first is the furnace baffle, which is installed between the rear wall and the first row of screen tubes. This baffle aids in directing the flow of gases through the screen tubes into the superheater. The second gas baffle consists of steel fins that are welded to the generating tubes from the steam drum to just below the first bend in the tube away from the steam drum. These fins together with baffle tile fitted between the last row of screen tubes form a gas baffle to direct the flow of gases over the top of the superheater tubes. The third is the rear wall gas baffle, located between the rear wall and the first row of generating tubes. These gas baffles direct gases from the superheater on the generating tubes.

Refractories and Burners

The function of the boiler refractories (brickwork) is to confine the heat of the furnace and to protect the casings and structural parts from overheating. In order to have higher heat release rates, the rear wall and side wall refractories are protected by tubes. The rear wall and side wall tubes also act as generating tubes at high firing rates. Screen tubes separate the furnace from the superheater tubes and block the superheater tubes, preventing overheating of the tubes from direct radiant heat. The inner row of screen tubes is used to support the superheater tubes; these tubes are called superheater support tubes. Specially shaped alignment tile is fitted between the inner row of screen tubes and the superheater tubes.

In the front wall of the boiler, there are burner openings to accommodate the burners (air register and atomizer assembly). The burner openings allow for admission of air and oil for combustion. In the furnace refractory, there are expansion joints to provide space for refractories to expand without causing damage to the refractory or boiler linings. The boiler has four Ce-wallsend burners. The burners utilize steam to atomize the fuel oil. (Atomization is the process of breaking down a solid flow of fuel oil into tiny droplets to obtain more efficient combustion.) The Ce-wallsend burner has two types of atomization, steam and mechanical. During normal operation, the burners use steam to atomize the fuel oil, but when steam is not available, purely mechanical sprayer plates are used. In these, the fuel oil pressure alone is the force for atomizing the fuel oil.

The fuel oil burner assembly controls and directs the amount of combustion air to be mixed with the fuel oil to ensure proper combustion. The observation and light-off port of the burner has a twofold purpose: when fires are being lighted, the port is opened to insert the lighting-off torch; after fires are lighted, it is used to observe the flame for proper combustion. The control block has two separate connections, one for steam and one for oil. It contains stopcocks that are used to start and stop the flow of fuel oil and atomizing steam into the burner assembly. Attached to the stopcocks in the control blocks are levers that allow the operator to start or stop the flow of fuel oil and atomizing steam to the atomizer.

On the right side of the burner next to the control block is the sleeve operating lever. This lever positions the sliding sleeve (open or shut) which permits the admission of combustion air into the burner register. The burner (figure 5–2) has a converging section to increase velocity and decrease pressure to the throat. After the throat is the diverging section or discharge, which converts velocity to pressure. Included in this burner is a Venturi assembly, which directs air to the furnace area for combustion at the correct pressure and velocity at all firing rates. It also gives a minimum air turbulence with only slight friction losses to the combustion air flow.

Guide bars give proper alignment for the sliding sleeve when the operator is opening or closing the sleeve operating lever. The air diffuser assembly gives combustion air a whirling motion, intermixing the air with the mist of fuel oil leaving the atomizing tip. This is done to stabilize the flame near the atomizer tip and helps determine the shape of the flame.

Ships in the Navy use Diesel Fuel Marine (DFM). The function of the fuel oil atomizer assembly (figure 5–3) is to deliver fuel oil and steam from the burner head (inlet block of the air register to the furnace). The steam breaks the fuel oil into a fine mist to obtain complete combustion. When steam assist is used, the fuel oil and steam enter the inlet block; both travel down the atomizer in separate

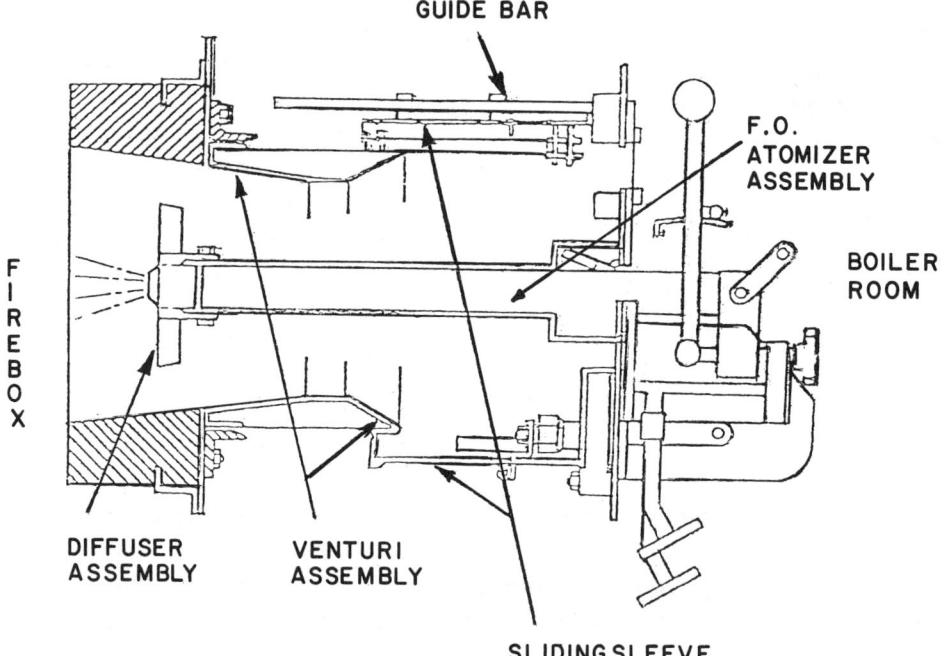

Figure 5–2. Fuel oil burner assembly.

tubes. The two separate tubes of the atomizer lead to the atomizer tip, which is screwed onto the end of the atomizer. In the atomizer tip, the fuel oil and steam mix at the internal mix sprayer plate to form a fine mist for combustion. The last component part of the atomizer assembly is the cap nut; it screws onto the atomizer, providing a tight seal with the front part of the sprayer plate. The space between the cap nut and sprayer plate is the location where the

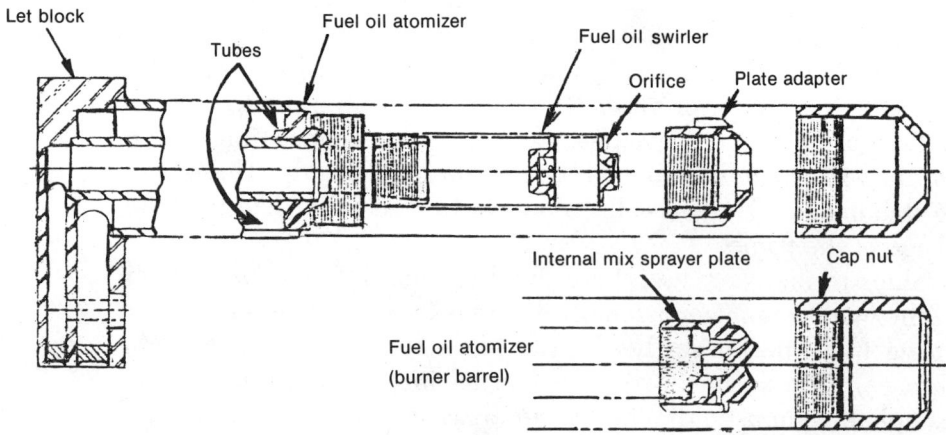

Figure 5–3. Fuel oil atomizer (burner barrel).

atomizing steam enters the sprayer plate. Steam is not always available for lighting off; therefore straight mechanical atomization is used, and different components are used in the tip, permitting the use of fuel oil only.

In the case of mechanical atomization, a swirler and orifice plate replace the internal mix sprayer plate. The fuel oil enters the fuel oil swirler, which gives it a swirling motion; then it goes to the orifice plate. In the orifice plate, the fuel oil continues its swirling motion and comes out in a foglike spray.

The fuel oil swirler and orifice plate are positioned back to back (flat surface to flat surface), then placed in the adapter. Then the adapter is screwed on the end of the atomizer. The adapter closes off the steam tube of the atomizer, preventing oil from backing up into the steam side of the atomizer. The cap nut, which is placed over the adapter and screwed onto the end of the atomizer, helps secure the adapter to the atomizer and also closes off the steam tube.

Smoke Indicator

Installed in the uptake of each boiler is a smoke indicator, which permits the operator to observe the gases of combustion as they leave

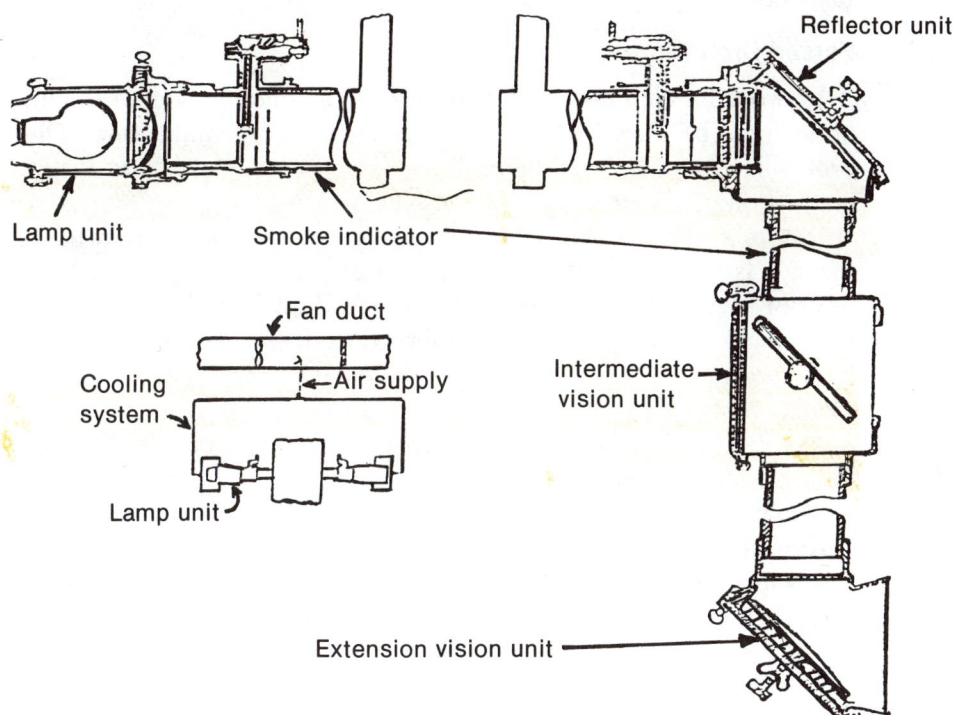

Figure 5–4. Cutaway of a typical smoke indicator.

the furnace and enter the smoke pipe on their way to the atmosphere (figure 5-4).

The lamp unit projects a beam of light across the uptake where it is picked up by the reflector unit mirror. The reflector unit is an adjustable mirror that reflects the light beam down to the intermediate and extension vision units. The intermediate vision unit is installed so that the console operator may observe the exit gases. This unit is different from most, in that the reflecting glass, called 50–50 type, not only reflects light but also allows the light beam to pass through and down to the extension vision unit, an adjustable mirror. This unit is installed in a convenient location at eye level for the burnerman to observe exit gases. The smoke indicator has a cooling system, which originates at the forced draft blower duct and pipes air into the lamp unit and the reflector unit. This system serves to cool the lens and vision glass in these units, and also keeps soot out of the units, aiding in keeping them clean. The boiler has viewports in the casing. They are like small portholes and allow the operator to inspect the casing visually for leakage of fuel oil or water without having to dismantel it. Along with viewports, there are inspection lamps, which give sufficient lighting inside the casing for the visual inspection.

Forced Draft Blowers

In order for combustion to take place in a boiler, an ample amount of air must be mixed with the fuel to assure total combustion. Theoretically, 14 pounds of air must be mixed with 1 pound of fuel to assure complete combustion. However, in actual practice, more than 14 pounds of air is required. Supplying combustion air to a boiler furnace is the purpose of the forced draft blowers.

Prior to a detailed study of the forced draft blowers, you should have some knowledge of how *all* forced draft blowers are classified. Forced draft blowers are generaly classified according to three main characteristics: (1) the type of power source, (2) the relative position of the blower drive shaft, and (3) the direction in which the air flows through the blower fan.

The power source for a forced draft blower is usually either a steam turbine or an electric motor. Most modern naval ships have both types installed on each boiler. Two or more steam-turbine-powered forced draft blowers are used to supply combustion air during the normal operation of the boilers, and one small motor-powered blower is installed for use in lighting off a boiler if high-pressure steam is not available. However, some noncombat ships have only motor-driven blowers.

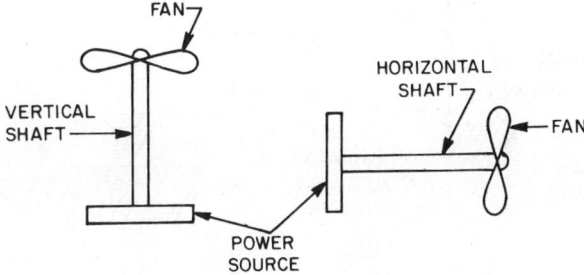

Figure 5–5. Second means of basic forced draft blower classification.

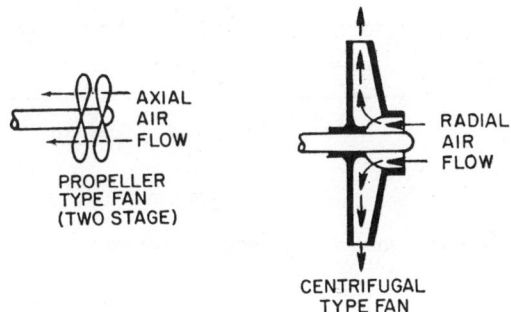

Figure 5–6. Third means of basic forced draft blower classification.

The second classification of forced draft blowers (figure 5–5) is by the relative position of the blower drive shaft; that is, the shaft is either in the vertical or the horizontal position. Both types are used on board modern combat ships.

The last classification characteristic of forced draft blowers is the direction in which the air flows through the blower fan (figure 5–6). This characteristic is determined by the type of blower fan utilized. On forced draft blowers with a propeller-type fan, the air flow is axial (parallel to the blower shaft); on blowers with centrifugal fans, the air flow is radial (tangent to the blower shaft). Most modern forced draft blowers have propeller-type fans.

Although forced draft blowers on board modern combat ships are highly sophisticated machines, you may have guessed that their basic design is similar to that of the common house fan.

As you have already learned in this chapter, the power source for the main forced draft blowers installed in the 1200 psi ships is steam turbines. The supply steam for these turbines comes from the 1200 psi auxiliary steam system.

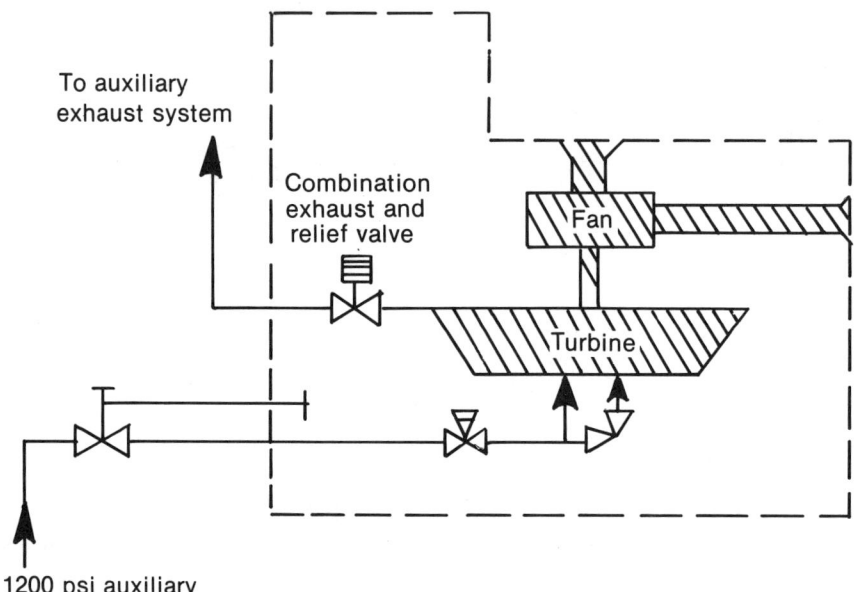

Figure 5–7. Mock-up of a forced draft blower steam supply and exhaust systems.

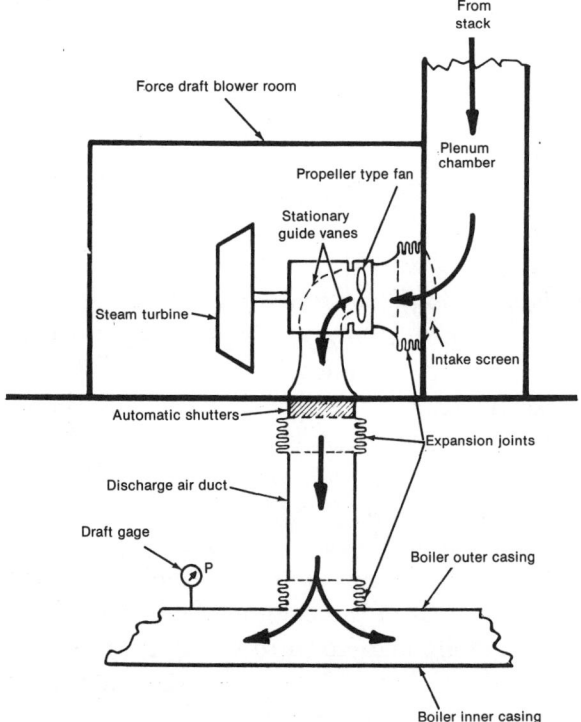

Figure 5–8. Cutaway of a typical forced draft blower, room, and air flow.

62 Introduction to Naval Engineering

In figure 5–7 the line labeled "To auxiliary exhaust system" allows for the removal of "used" (exhaust) steam from the blower turbine after the thermal energy of the steam has been converted to mechanical energy (work) by the forced draft blower turbine.

Figure 5–8 shows the forced draft blower and the ducting needed to guide the flow of air from the atmosphere to the boiler.

The flow of supply air for the forced draft blower starts at an opening in the side of the stack. This opening directs the flow of air to the forced draft blower fan through a relatively large air duct called a plenum chamber.

A mesh screen is installed on the intake side of the forced draft blower fan, which prevents foreign objects (paper, rags, etc.) from entering and damaging the blower fan.

After passing through the intake screen, the air flows through one of several expansion joints installed in the forced draft blower air ducting. All of the expansion joints serve the same purpose, to relieve stress at points of contact between the ship structure and machinery by "bending with the wind." This structural stress is even more pronounced on board ship because as you know, or will soon find out, ships at sea have a tendency to "rock and roll." This rocking causes the ship's structure to bend and twist, thus playing havoc with rigid connections.

The turbine drive forced draft blowers in the ships have two-stage propeller-type fans. "Two-stage" means there are two propeller-type fans mounted back to back on the main shaft (See figure 5–9.) The first-stage fan discharges to the second, and the second-stage discharges to the air discharge outlet at the base of the forced draft blower. By staging fans, higher discharge air pressures are possible than if only one fan of similar size was used.

The air flow from a propeller-type fan is axial (parallel to the blower shaft). The stationary guide vanes installed on the discharge side of the fan change the direction of the air flow with the least possible resistance.

Installed on each boiler are two turbine-driven forced draft blowers. When only one forced draft blower is in operation, automatic

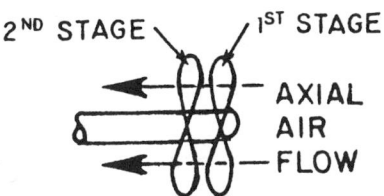

Figure 5–9. *Drawing of a two-stage propeller-type fan forced draft blower.*

shutters prevent air flow in the reverse direction through the idle blower. The automatic shutters will close any time the air pressure on the outlet side is equal to or greater than the pressure on the inlet side of the shutters.

After the air passes through the automatic shutters, it is directed to the boiler through the forced draft blower discharge air duct. Notice in figure 5–8 that the discharge air duct has an expansion joint at each end.

The air enters into the boiler in an area between the inner and outer boiler casings. These boiler casings direct the air to the boiler furnace, where it is mixed with the fuel.

The lube oil sump stores lubricating oil ready for use in the forced draft blower lube oil system.

The lubricating oil system for the turbine-driven forced draft blower is pressure-fed by two simple gear-type pumps. One lube oil pump is geared to the turbine shaft and will pump oil any time the forced draft blower is rotating. The discharge pressure of this turbine-driven lube oil pump is directly proportional to the speed of the forced draft blower, and for this reason it does not provide enough oil pressure when the blower is operated at low speeds (1500 rpm or less). The electric-motor-driven lube oil pump provides lubrication before the forced draft blower starts to operate, while the blower is operating at low speeds, and immediately after the blower is shut down.

Before a forced draft blower is started, the electric lube oil pump must be manually started by a switch located in the blower room. (The turbine-driven blower *cannot* be started until lube oil pressure is established.) When the blower is started, the electric pump supplies the oil needed to lubricate the blower at lower speeds. As the blower speed is increased, the discharge pressure of the turbine-driven lube oil pump will also be increased to a point where the electric lube oil pump is no longer needed. At this time, the electric lube oil pump will automatically shut down and remain off until the speed of the forced draft blower is again brought below the effective operating range of the turbine-driven lube oil pump. After the blower has been secured, the electric pump will be allowed to continue to run to cool the blower until the operators manually turn the pump off.

As the forced draft blower bearings are lubricated, heat of friction is conducted from the bearings to the lubricating oil. The purpose of the lube oil cooler is to reduce the temperature of the lube oil before the oil passes through the main shaft bearings. As the lube oil leaves the cooler, its temperature should be about 125°F.

The cooling medium for the lube oil cooler on board ship is sea-

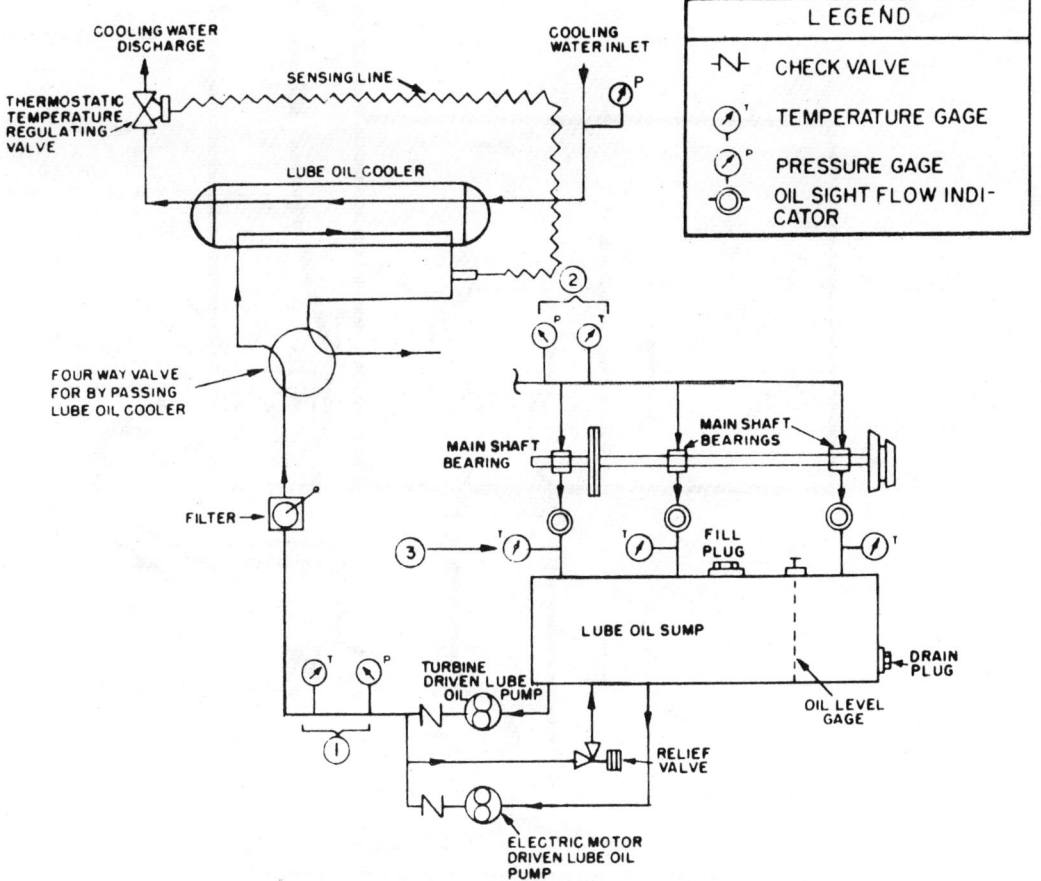

Figure 5–10. Forced draft blower lube oil system, schematic drawing.

water; that is, seawater is directed through the inside of the tubes as lube oil flows on the outside of the tubes. Notice in the drawing (figure 5–10) that the flow of lube oil and seawater is in opposite directions. This method is used in many heat exchangers because it encourages greater heat transfer through the tube walls.

Locate the thermostatic temperature regulating valve and sensing line in the drawing. The sensing line measures the cooler outlet oil temperature and causes the regulating valve to increase or decrease the flow of seawater from the cooler, in this way maintaining the lube oil at the desired temperature (about 125°F).

Normally a cold propulsion plant cannot be lighted off unless the boilers are supplied with air for combustion, and the steam-turbine-driven forced blowers are useless when 1200 psi auxiliary steam is not available. However, in an emergency the plant can be lighted-

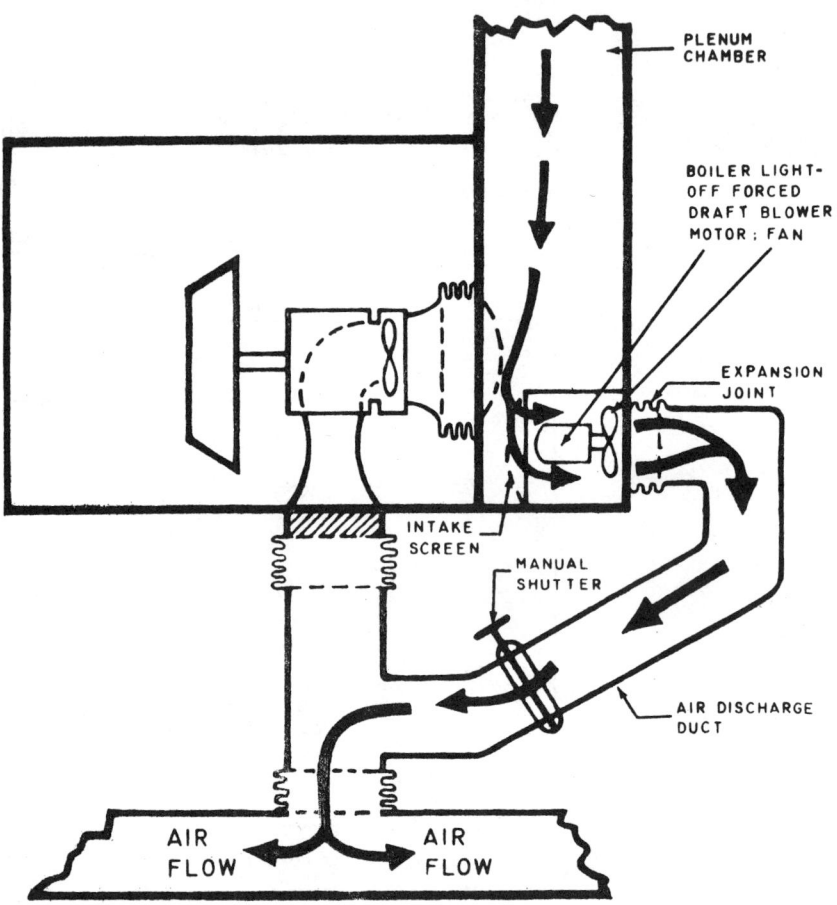

Figure 5–11. Light-off forced draft blower location and air flow.

off, but it must be in an extreme emergency and under very careful guidance.

How is combustion air supplied in a boiler lighting-off situation? This task is accomplished by the boiler light-off forced draft blower (figure 5–11). Each boiler in the propulsion plant has one boiler light-off forced draft blower. It is motor-driven and therefore *not* dependent on a supply of steam. Under normal conditions, the source of electrical power for these forced draft blowers may be obtained from a shore-based power plant; however, in case of an emergency, electrical power can be acquired from the ship's emergency generators (diesel-engine-driven generators).

Locate the boiler light-off forced draft blower in figure 5–11. Notice that both the blower motor and the fan are installed inside the plenum chamber. The control switch (on and off) for this blower is located

66   Introduction to Naval Engineering

on the upper level of the fireroom, at a point where it is easily accessible to the operating personnel.

Like the turbine-driven forced draft blower, the boiler light-off blower has a mesh screen on the air intake to protect the fan from large objects (paper, rags, etc.) that may have passed into the plenum chamber.

An expansion joint is installed on the outlet side of the boiler light-off forced draft blower to absorb shock and relieve stress between the ship's structure and the blower air ducting.

After the air passes through the shutter, its flow path is the same as it is for the steam-turbine-driven forced draft blower.

After the air passes through the expansion joint, it flows into the air discharge duct (see figure). Installed in this duct is a manually operated shutter that must be opened any time the boiler light-off forced draft blower is in use and shut at all other times. The shutter is operated by a handwheel that is easily accessible to fireroom personnel.

Fuel Oil Service System (figure 5–12)

In order that steam be produced for the propulsion plant, fuel oil must be supplied to the boiler for combustion. This fuel oil must be available for delivery to the boilers at the proper pressure and temperature for proper atomization and combustion in the boiler furnaces. The fuel to be burned in the boilers is stored in two fuel oil service tanks, located in the fireroom. Two electric-motor-driven main fuel oil service pumps are used to pump fuel oil from the service tanks to the boilers. The main fuel oil service pump discharge pressure is maintained at a pressure of 350 psi. In addition to the main fuel oil service pumps, there is a port-use fuel oil service pump. This pump is electric-motor-driven but has a lower capacity than the main fuel oil service pumps. The port-use fuel oil pump is used during light-off and during auxiliary steaming.

Connected to the discharge piping of each main fuel oil service pump downstream of the pump discharge valves are the fuel oil accumulators. There are four bladder-type nitrogen-charged accumulators which have a capacity of 10 gallons of fuel oil each. In the event of a loss of fuel oil pressure, the fuel oil is forced out of the accumulator into the fuel oil system, preventing a loss of fires in the boilers for a period of time (approximately 30 seconds).

Fuel oil service pump discharge pressure is maintained at 350 psi by the use of the fuel oil unloading valves, which are located downstream of the pump discharges. The proper fuel oil pressure is main-

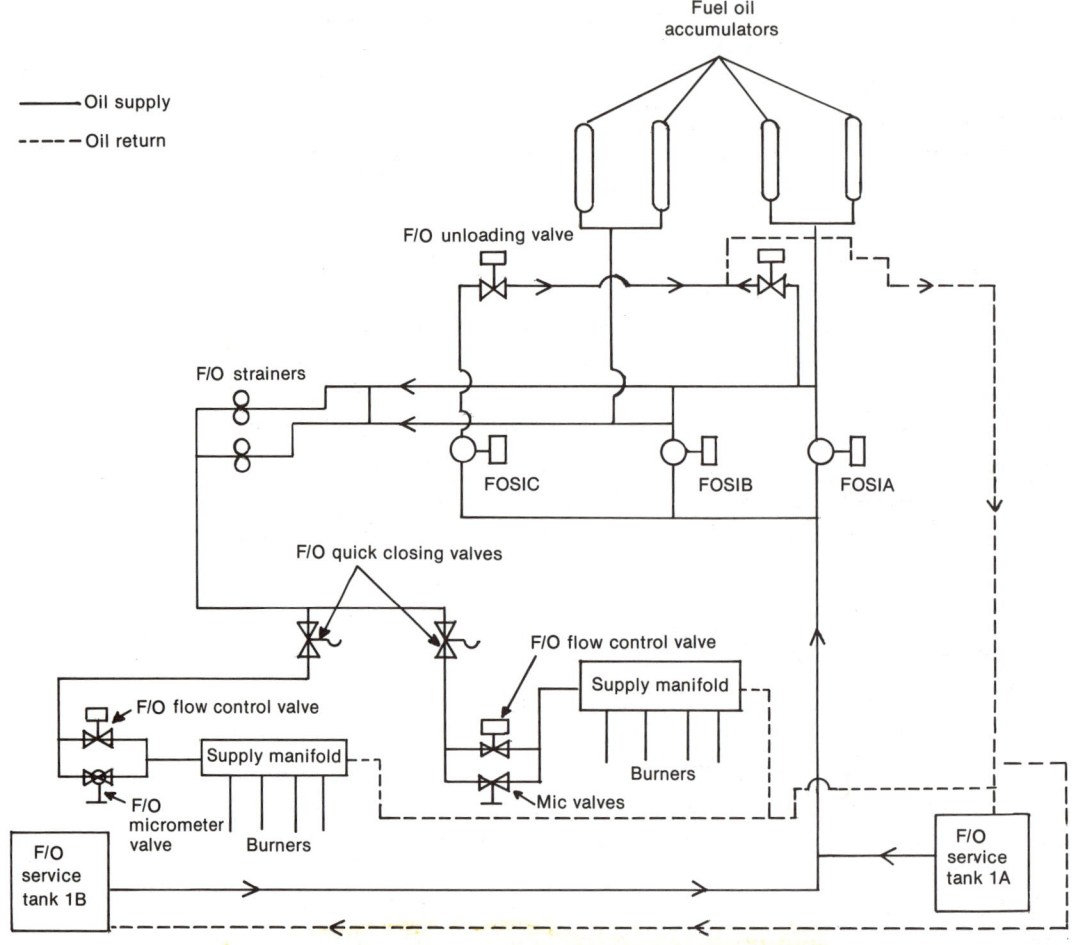

Figure 5-12. Schematic drawing of the fuel oil service system.

tained by the unloading valves, which recirculate excess oil back to the fuel oil service tank that is on suction.

Installed in the fuel oil system after the fuel oil heaters, which are used mainly with low-viscosity fuels, are duplex strainers. These strainers are used to collect particles of foreign matter (dirt, sediment, etc.) that might otherwise interfere with atomization of the fuel oil. The strainers are installed with a valve arrangement that allows removal of one basket for cleaning while the other basket is in use.

During an emergency situation, the boiler operator must have a means of securing fuel oil to the boilers. For this reason, the fuel oil quick-closing valves are installed in the fuel oil supply lines upstream of the burner supply headers of boilers 1A and 1B. The fuel oil quick-closing valves may be operated from the lower level of the boiler fronts or the fireroom control station.

With constant pressure from the fuel oil service pumps, there must be a way of varying the fuel oil pressure at the burners to allow for the varying firing rates. This is accomplished by the use of the fuel oil flow control valves, which are components of the automatic boiler control system. These valves operate by an air control signal that will open, close, or regulate them. They can also be operated manually by the use of their handwheel.

A secondary means of controlling the fuel oil pressure to the burners is by the fuel oil micrometer valves. These valves are manually operated and bypass the fuel oil flow control valve. They are used during system start-up and when the control valves do not function properly.

Each boiler has a fuel oil supply manifold which is used to distribute the fuel oil to the burners (fuel oil atomizers). Each supply manifold has a stop valve and a two-way valve for recirculation of the fuel oil. Normally, during lighting off, the fuel oil is recirculated to the service tank on suction. After the fires are lit, the recirculating valves are secured.

Installed in the lines from the supply manifold to each burner (fuel oil atomizer) are atomizer root valves. They are used to start or stop the flow of fuel oil to each burner.

As was stated, the two-way valve recirculates fuel oil to the service tank on suction. It also can be positioned to send fuel oil to the contaminated oil storage tank. This would be done if the fuel oil system should become contaminated with water.

Steam used for atomization of the fuel oil comes from the 150 psi system to the atomizing steam supply system.

Coming off the atomizing steam supply header are four atomizing steam cutout valves, one for each burner. They allow the operator to start or stop the flow of atomizing steam to each burner.

Fuel oil is stored in tanks located fore and aft in the ship's bottom. The fuel, taken in on the 01 level, is discharged to the fill main and then to these tanks. There are two electric fuel oil transfer pumps (located in auxiliary machinery room #1) which take a suction on the storage tanks and transfer the fuel to the service tanks for use in the service system.

Soot Blowers

Soot collects on boiler tubes as a result of combustion, and soot is an insulator. One-fifth of an inch of soot has an insulating quality equal to that of an inch of asbestos. Therefore we must remove these

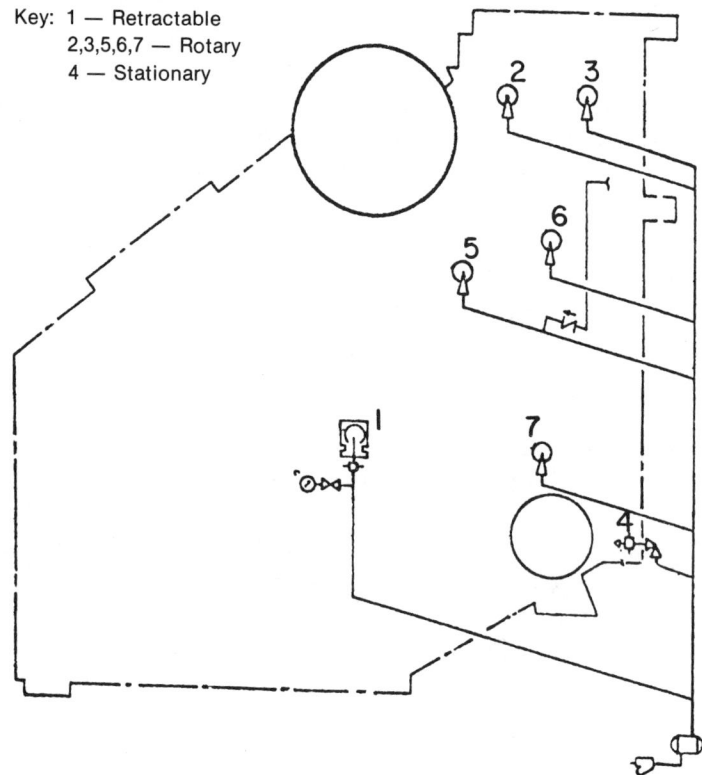

Figure 5–13. Location of soot blowers of a 1200 psi D-type boiler.

soot deposits to have an efficiently operating boiler. Soot blowers (figure 5–13) are used for this purpose. The process of using soot blowers is called blowing tubes.

There are three basic types of soot blowers: rotary, retractable, and stationary (figure 5–14). Soot blowers Nos. 2, 3, 5, 6, and 7 (figure 5–13) are manually operated rotary-type soot blowers, which are used to remove soot from the economizer and generating tube bank. Soot blower No. 4 is a manually operated stationary-type soot blower, used to remove soot from the lower section of the generating tube bank behind the water drum. Soot blower No. 1 is a retractable-type soot blower; it is air-motor-driven and is used to remove soot from the superheater tubes.

Located inside the boiler is the soot blower element. It is a long pipe with nozzle outlets, which projects into the tube banks of the boiler. Steam is evenly directed over the boiler tubes by the soot blower element.

The retractable soot blower is used to remove soot from the su-

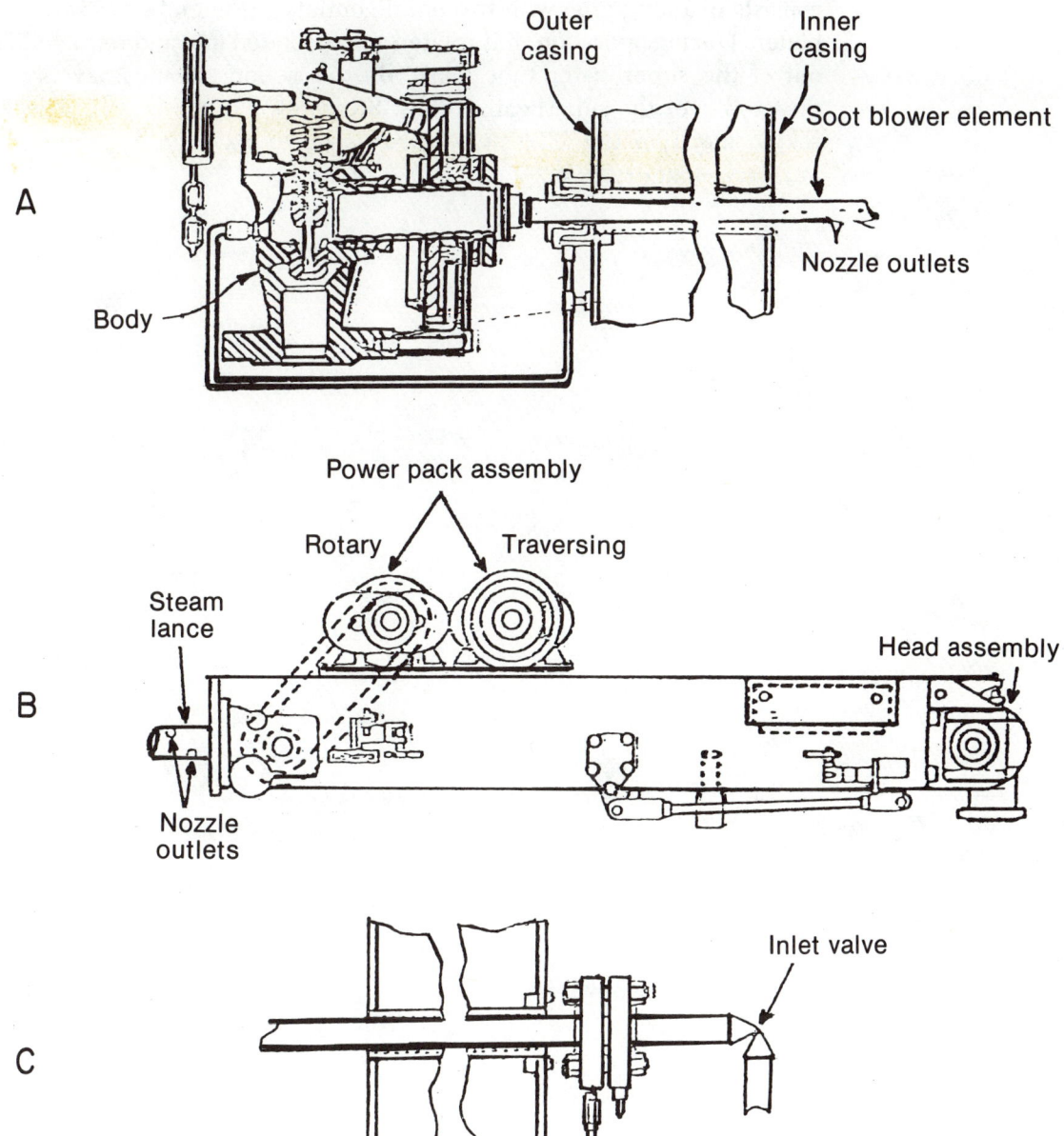

Figure 5-14. Three basic types of soot blowers: (A) rotary, (B) retractable, (C) stationary.

perheater tubes. Because of the high temperatures attained in the superheater, it is necessary for the steam lance (element) to remain outside the boiler except when actually being used. The steam lance consists of a long pipe with two nozzle outlets at the end nearest the boiler. During operation, it is rotated and extended into and retracted out of the superheater tube bank. By this action, steam is evenly directed over the superheater tubes. Steam used in the soot blowers may come from the 1200 psi auxiliary steam system or the 600 psi auxiliary steam system.

6

Main Propulsion Turbines, Reduction Gear, Shafting, and Auxiliary Turbines

Steam Flow

As we begin the study of steam turbines, we have reached that part of the thermodynamic cycle in which the actual conversion of thermal energy to mechanical energy takes place.

We know by simple observation of pressures and temperatures that the steam leaving a turbine has far less thermal energy than it had when it entered the turbine. By observation, again, we know that work is performed as the steam passes through the turbine, the work being evidenced by the turning of a shaft and the movement of the ship through the water. Since we know that energy can be transformed but can be neither created nor destroyed, the decrease of thermal energy and the appearance of work cannot be regarded as separate events. Rather, we must infer that thermal energy has been transformed into work—that is, mechanical energy in transition.

Disregarding such irreversible energy losses as those caused by friction and by heat flow to objects outside the system, it can be shown that two energy transformations are involved. First, there is the thermodynamic process by which thermal energy is transformed into mechanical kinetic energy as the steam flows through one or more nozzles. Second, there is the mechanical process by which mechanical kinetic energy is transformed into work as the steam impinges upon projecting blades of the turbine, thereby turning the turbine rotor.

In order to understand the process by which thermal energy is converted into mechanical kinetic energy, we must have some understanding of the process that takes place as steam flows through a nozzle. The second energy transformation, from kinetic energy to work, is best understood by considering some basic principles of turbine design.

Nozzles

Essentially, the nozzle is shaped in such a way as to cause an increase in the velocity of the steam as it expands from a high-pressure area to a low-pressure area. The nozzle also serves to direct the steam so that it will flow in the right direction to impinge upon the turbine blades.

If a quantity of steam is confined in a closed vessel, it exerts a pressure on the walls of the container. If this container is connected by means of a restricted channel, such as a pipe or nozzle, to a region of lower pressure, the steam will flow through the channel and expand to a greater volume, depending upon the lower pressure. Velocity of flow, in feet per second, will depend upon the difference in pressure between the high- and low-pressure regions.

At a given pressure and temperature, steam contains a certain amount of thermal energy, which increases with increase of pressure or temperature, and vice versa. The flow of steam through a restricted channel results, therefore, in a decrease of thermal energy. This decrease is equivalent to its gain in kinetic energy. In other words, the thermal energy is converted to kinetic energy, by causing the steam to flow from a high- to a low-pressure region.

Nozzles take many forms, but in principle all are similar. They consist of an entering section, a throat, and a mouth. Figure 6–1 shows a typical convergent nozzle.

When the pressure into which the nozzle exhausts is relatively low, there is a tendency for the steam to expand rapidly in all directions, causing turbulence. In order to reduce or control this turbulent expansion, a section of gradually increasing cross-sectional area is added to the throat, with the result that the steam will emerge in a smooth stream. We then have the nozzle shown in figure 6–2, which is called a convergent-divergent nozzle.

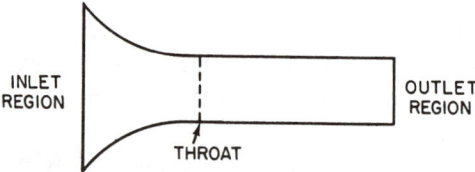

Figure 6–1. A typical convergent nozzle.

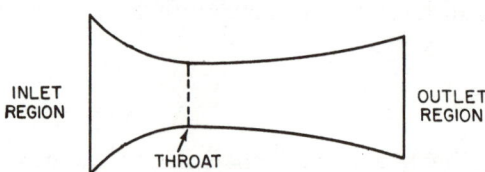

Figure 6–2. A typical convergent-divergent nozzle.

Turbine Design

In essence, a turbine may be thought of as a bladed wheel or rotor that turns when a jet of steam from the nozzle impinges upon the blades. The basic parts of a turbine are the rotor, which has blades projecting radially from its periphery; a casing, in which the rotor revolves; and nozzles, through which the steam is expanded and directed. As we have seen, the conversion of thermal energy to mechanical kinetic energy occurs in the nozzles. The second energy conversion—the conversion of kinetic energy to work—occurs on the blades.

The basic distinction to be made between types of turbines has to do with the manner in which the steam causes the turbine rotor to move. When the rotor is primarily moved by a direct push or "impulse" from the steam impinging upon the blades, the turbine is said to be an *impulse* turbine. When the rotor is moved primarily by the force of reaction, the turbine is said to be a *reaction* turbine.

Although the distinction between impulse turbines and reaction turbines is a useful one, and one that is followed in this text, it should not be considered as an absolute distinction in real turbines. An impulse turbine utilizes both the impulse of the steam jet and, to a lesser extent, the reactive force that results when the curved blades cause the steam to change direction. A reaction turbine is moved primarily by reactive force, but some motion of the rotor is caused by the impact of the steam against the blades because the arrangement of the reaction blading serves to act like a nozzle for the steam entering the next stage.

If an obstruction is placed in the path of the flowing steam, the steam will exert a force or "impulse" on the obstruction. If the obstruction is so arranged that it can move, work will be done upon it by the jet stream, and this amount of work will be equal to the product of the force times the distance through which the force acts.

Let us suppose that the obstruction referred to above is a vane, or blade, mounted upon the rim of a wheel that is free to rotate. If the wheel carries a series of such blades around its entire periphery, so that as each rotates out of the path of the steam jet, another takes its place, there will be a continuous rotational force applied to the wheel, and it will continue to turn as long as the steam jet is maintained. (See figure 6–3.) This device constitutes, in principle, one of the simplest forms of steam turbine. Since its operation depends upon the impulse of the steam jet, it is known as an impulse turbine.

The pressure and velocity changes that occur in the nozzle and in the blades of an impulse turbine are shown in figure 6–4. The steam

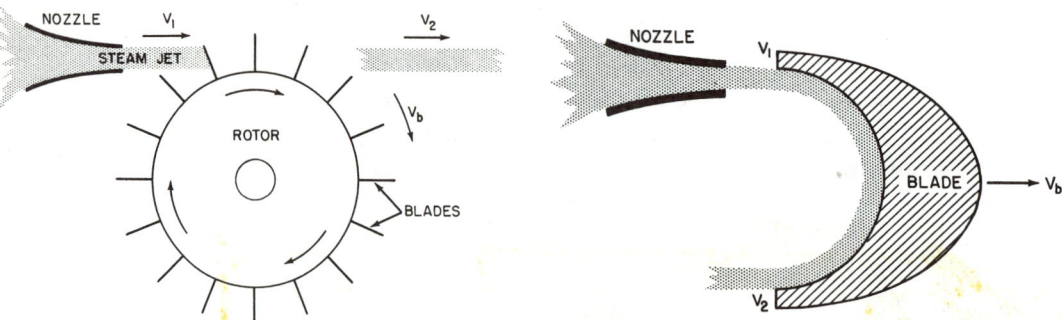

Figure 6–3. Simple steam impulse turbine.

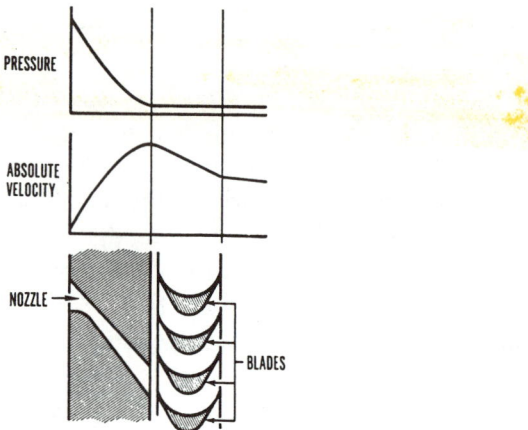

Figure 6–4. P–V relationship as steam passes through a nozzle and onto moving blades.

expands through stationary nozzles only and so loses pressure but gains velocity. In the moving blades, the steam loses velocity, but the pressure remains constant.

We have seen that steam in expanding through a nozzle, acquires velocity and, hence, kinetic energy. As a result it is capable of exerting a force against any object placed in its path. The force is, of course, present whether it is or is not exerted. Now, we know that for every force, or action, there exists an equal and opposite reaction. For example, when the end of a hose is held in the hand, a "kick-back" or reaction is evidenced by the effort required to hold the hose when water is flowing from it. If we attach a number of nozzles to the rim of a wheel, which is free to rotate, in such a way that the jets from all these nozzles will flow in a direction tangent to the periphery of the wheel, and if we connect the nozzles by means of tubes to a hollow shaft on which the wheel is mounted and connect this hollow shaft to a source of steam supply at relatively high pressure, the flow of steam through the nozzles will produce a reactive force which will cause the wheel to rotate in a direction opposite to the direction of steam flow from the nozzles. This simple device, which is exemplified in practice by the common garden sprinkler, constitutes, in principle, a second type of turbine which is known as the reaction turbine, since its operation depends upon the reaction of steam expanding through nozzles.

In actual turbines, the nozzles in our example are replaced by blading formed in such a way that the passages between adjacent blades take the shape of nozzles. Because these blades are mounted on the revolving rotor, they are called moving blades. Fixed or stationary blades of the same shape as the moving blades are fastened to the casing in which the rotor revolves; these fixed blades are installed between successive rows of the moving blades. The fixed blades guide the steam into the moving blade system, and they are also shaped and mounted in such a way as to provide nozzle-shaped spaces between the blades.

Now let us take time to compare the basic difference between impulse and reaction-turbine blading. No matter what the number of fixed and moving blade rows in an impulse turbine, the pressure remains the same throughout the blading of each stage.

In the reaction turbine, on the other hand, the steam pressure decreases in every row of fixed and moving blades (figure 6–5). There are no nozzles in the reaction turbine; the fixed blades serve the same purpose as the nozzles of an impulse turbine. The low pressure turbine is an example of a reaction turbine.

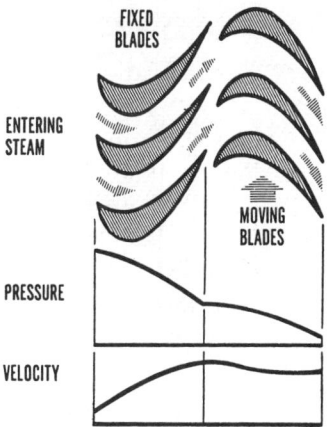

Figure 6–5. P–V *relationship as steam passes through fixed blades and onto moving blades.*

Classification by Staging and Compounding

Thus far in this chapter, we have more or less assumed that an impulse turbine has one set of nozzles and one row of blading on the rotor, and that a reaction turbine has one row of fixed blades and one row of moving blades. In reality, however, propulsion steam turbines are not this simple. Instead, they use several rows of blading, arranged in various ways. In modern naval ships, the amount of available energy per pound of steam is so great that there is no practical way of utilizing the major portion of it in one row of blades. When several rows of blades are used, the steam passes through one row after another, and each row uses part of the energy of the steam. Figure 6–6 is a summation of the following basic staging in P–V diagram form.

Impulse Stage

In an impulse turbine, a stage is defined as one set of nozzles and the succeeding row or rows of moving and fixed blades. Because the only place a pressure drop occurs in an impulse turbine is in the nozzles, another way of defining an impulse stage is to say that it

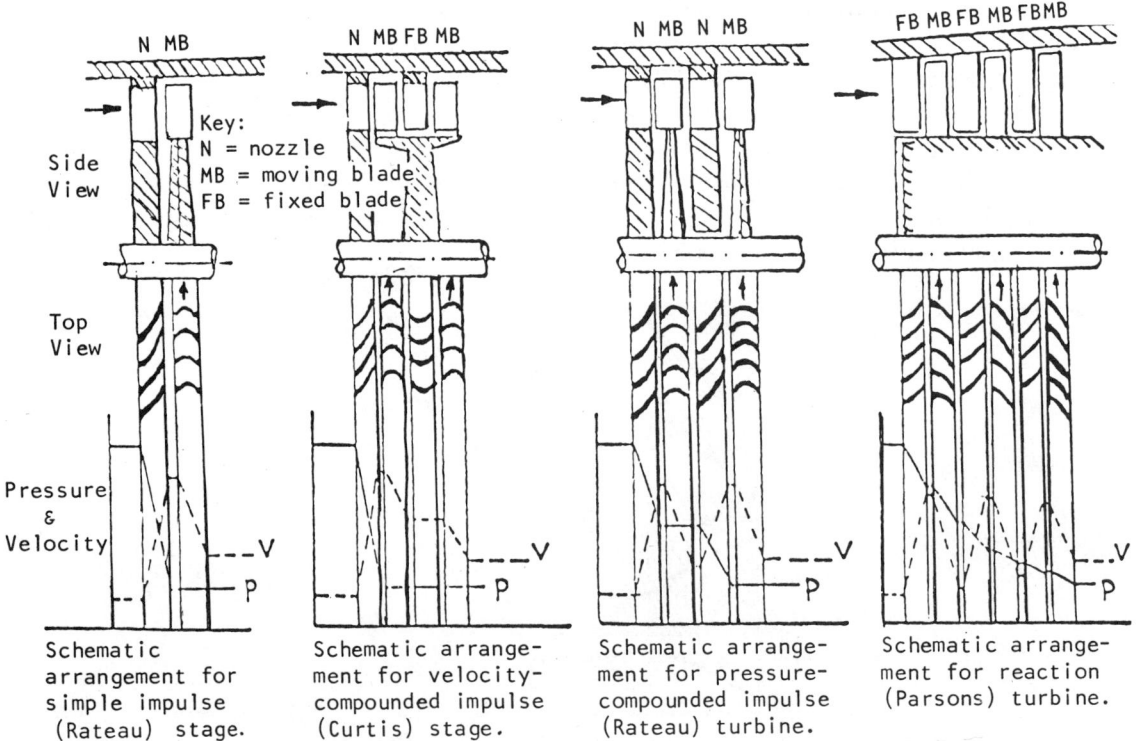

Figure 6–6. P–V *relationship summary for various types of turbines.*

includes the nozzles and blading in which only one pressure drop takes place. A simple impulse stage is often called a *Rateau stage*. Turbines consisting of a single Rateau stage are not used as propulsion turbines but are frequently used to drive small auxiliary units.

Reaction Stage

In reaction turbines, one row of fixed blades and its succeeding row of moving blades are considered to constitute one stage. Since the fixed blades in a reaction turbine are comparable to the nozzles in an impulse turbine, this definition of a reaction stage may seem very similar to the definition of an impulse stage. However, there is this important difference: a reaction stage includes two pressure drops, whereas an impulse stage includes only one.

Velocity-Compounded Impulse Turbine

One way of increasing the efficiency of an impulse turbine is by velocity-compounding—that is, by adding one or more rows of moving blades to the rotor. This type of turbine is called velocity-compounded because the residual velocity of the steam leaving the first row of moving blades is utilized in the third row. The fixed blades, which are fastened to the casing rather than to the rotor, serve to direct the steam from one row of moving blades to another.

A velocity-compounded impulse turbine has only one pressure drop and has, by definition, only one stage. This type of velocity-compounded impulse stage, which is usually called a *Curtis stage*, is shown in figure 6–7.

Pressure-Compounded Impulse Turbine

Another way to increase the efficiency of an impulse turbine is to

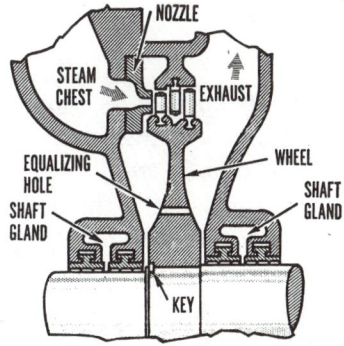

Figure 6–7. Cutaway of a Curtis stage velocity-compounded impulse turbine.

Main Propulsion Turbines, Reduction Gear, Shafting, and Auxiliary Turbines

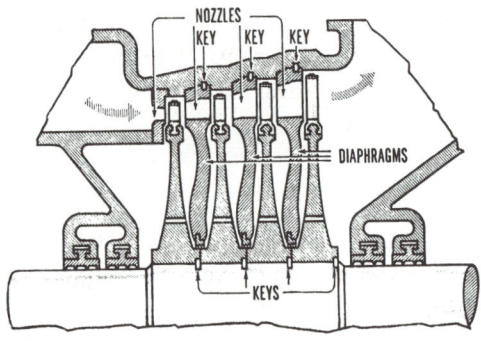

Figure 6–8. Cutaway of a pressure-compounded impulse turbine with four stages.

arrange two or more simple impulse stages in one casing. The casing is internally divided by nozzle diaphragms. The steam leaving the first stage is expanded again through the first nozzle diaphragm, to the second stage; from the second nozzle diaphragm, to the third stage; and so on. This type of turbine is known as a pressure-compounded turbine because a pressure drop occurs in each stage. Figure 6–8 shows a pressure-compounded impulse turbine with four stages. A pressure-compounded impulse turbine is frequently called a *Rateau turbine* because it is essentially a series of simple impulse (Rateau) stages arranged in a sequence in one casing.

Velocity-Pressure-Compounded Impulse Turbine

An impulse turbine that consists of one velocity-compounded (Curtis) stage followed by a series of pressure-compounded (Rateau) stages is generally referred to as a velocity-pressure-compounded impulse turbine. Turbines of this type are commonly used in the propulsion plants of naval ships.

Pressure-Compounded Reaction Turbine

Because the ideal blade speed in a reaction turbine is so high in relation to the velocity of the entering steam, all reaction turbines are pressure-compounded—that is, they are so arranged that the pressure drop from inlet to exhaust is divided into many steps by means of alternate rows of fixed and moving blades. The pressure drop in each set of fixed and moving blades (i.e., in each stage) is therefore small, thus causing a lowered steam velocity in all stages and consequently a lowered ideal blade velocity for the turbine as a whole (figure 6–9).

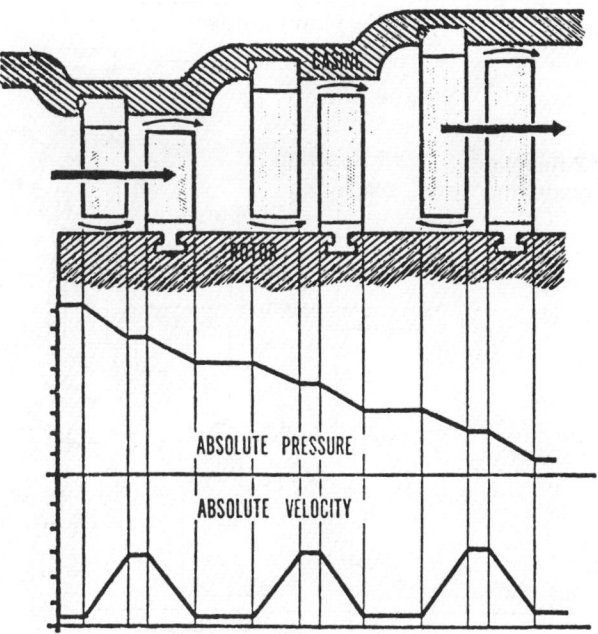

Figure 6–9. Diagrammatic arrangement of reaction blading showing absolute velocity and pressure relationships, and tip leakage.

Combination Impulse and Reaction Turbine

A combination impulse and reaction turbine employs a velocity-compounded impulse (Curtis) stage at the high-pressure end of the turbine, followed by reaction blading (see figure 6–10 on the following page). The impulse blading effects large pressure and temperature drops in the beginning, with a high initial utilization of thermal energy. The reaction blading is more efficient at the low-pressure end of the turbine. Hence, the combination impulse and reaction turbine is a highly efficient machine that utilizes the advantages of both impulse and reaction blading. Combination impulse and reaction turbines are very commonly used as propulsion turbines.

Classification by Mode of Steam Flow

Direction of Steam Flow

The direction of steam flow through a turbine may be axial, radial, or helical. In general, the direction of flow is determined by the relative positions of nozzles, diaphragms, moving blades, and fixed blades.

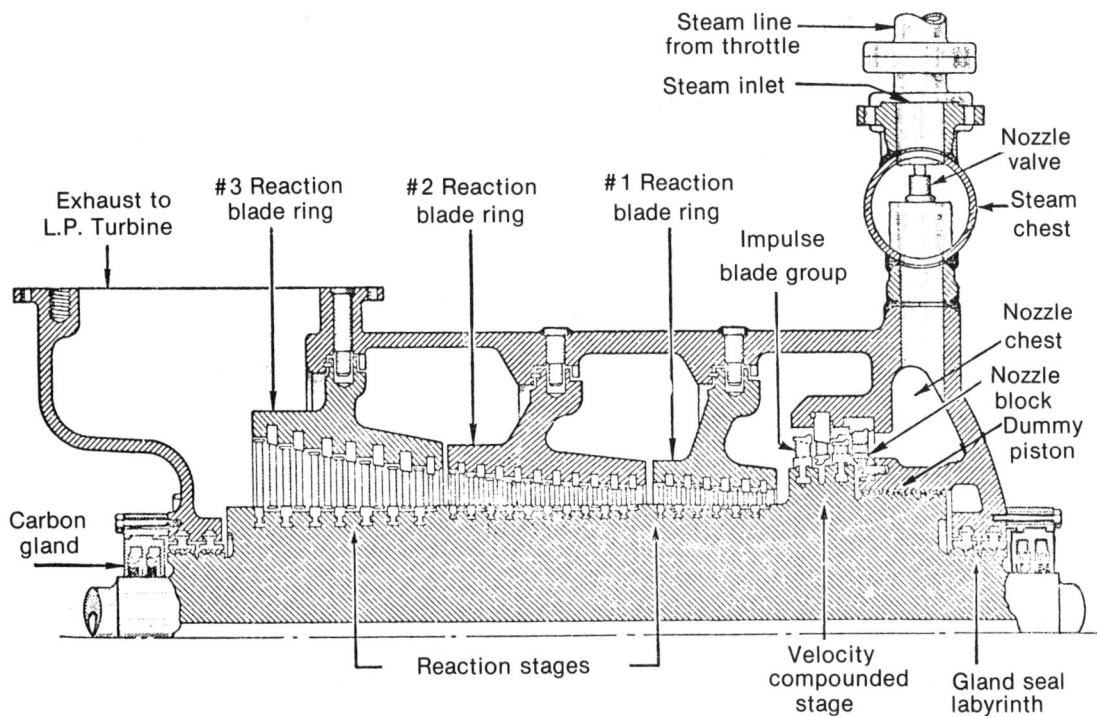

Figure 6–10. Combination, velocity-compounded impulse and pressure-compounded reaction turbine.

Most turbines are of the axial-flow type—that is, the steam flows in a direction approximately parallel to the long axis of the turbine shaft. As we have seen, the blades in an axial-flow turbine project outward from the periphery of the rotor.

Repetition of Steam Flow

Turbines are classified as single-entry turbines or reentry turbines, depending on the number of times the steam enters the blades. If the steam passes through the blades only once, the turbine is called a single-entry turbine. All multistage turbines are of the single-entry type.

Reentry turbines are those in which the steam passes through the blades more than once. Reentry turbines are used to drive some pumps and forced draft blowers but are not used as propulsion units.

Division of Steam Flow

Turbines are classified as single-flow or double-flow, depending upon whether the steam flows in one direction or two. In a single-flow turbine, the steam enters at the inlet or throttle end, flows once through the blading in a more or less axial direction, and emerges at the exhaust end of the turbine. A double-flow turbine consists essentially of two single-flow units mounted on one shaft, in the same casing. The steam enters at the center, between the two units, and flows from the center toward each end of the shaft. The main advantages of the double-flow arrangement are: (1) the blades can be shorter than they would have to be in a single-flow turbine of equal capacity, and (2) axial thrust is avoided by having the steam flow in opposite directions. This second point applies primarily to reaction turbines, since impulse turbines develop relatively little axial thrust in any case.

A double-flow reaction turbine of this type, used as the low pressure turbine in some propulsion plants, is shown in figure 6–11.

Typical Propulsion Turbine Installation

The typical propulsion unit consists of a high pressure turbine, a low pressure turbine, and an astern turbine.

The high pressure turbine would be classified as an axial-flow, single-entry, single-flow, velocity-pressure-compounded, impulse turbine.

The low pressure turbine would be classified as an axial-flow, single-entry, double-flow, pressure-compounded, reaction turbine.

The astern turbine consists of two Curtis stages located at either end of the low pressure turbine. The astern turbine shares a common shaft and turbine casing with the low pressure turbine, but it does not operate at the same time as the low pressure turbine.

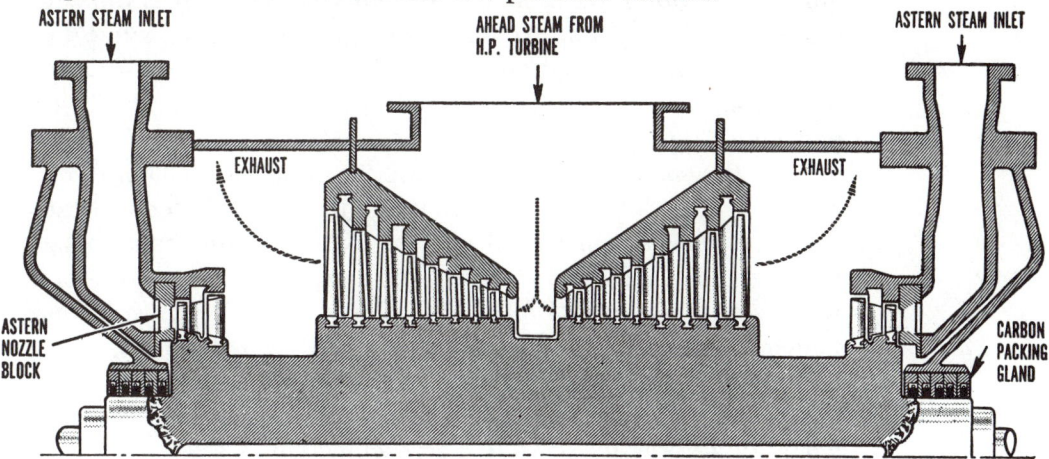

Figure 6–11. Cutaway of a double-flow reaction turbine used as the low pressure turbine.

Gland Seal and Gland Exhaust Systems

Gland sealing steam is supplied to the shaft glands of propulsion turbines and turbogenerator turbines to seal the shaft glands against two kinds of leakage: (1) air leakage into the turbine casings, and (2) steam leakage out of the turbine casings. These two kinds of leakage may seem contradictory; however, each kind of leakage could occur under some operating conditions if the shaft glands were not sealed.

Pressures in the gland seal system are low, ranging from about ¾ psi to 2 psig, depending upon the conditions of operation. Gland exhaust piping carries the steam and air from the turbine shaft glands to the gland exhaust condenser, where the steam is condensed and returned to the condensate system.

On board most ships, gland sealing steam is supplied from the auxiliary exhaust system, although in some ships it is supplied from the 150 psi auxiliary steam system. In either case, the steam is supplied through reducing valves or reducing stations. Figure 6–12 illustrates a typical gland seal and gland exhaust system for propulsion turbines on board a destroyer.

Lube Oil System

The primary purposes of a lubricating system are lubrication and cooling. This is accomplished in two ways: (1) reducing the heat buildup caused by friction, and (2) removing the heat that is generated during that process.

Lubrication reduces friction between moving parts by substituting fluid friction for solid friction. Without lubrication, it is difficult to move a hundred-pound weight across a rough surface; with lubrication, and with proper attention to the design of bearing surfaces, it is possible to move a million-pound load with a motor that is small enough to be held in the hand. By reducing friction, lubrication reduces the amount of energy required to perform mechanical actions and also reduces the amount of energy that is dissipated as heat.

Lubrication is a matter of vital importance throughout the shipboard engineering plant. Moving surfaces must be steadily supplied with the proper kind of lubricants. These lubricants must be maintained at specified standards of purity, and designed pressures and temperatures must be maintained in the lubrication systems. Without adequate lubrication, a good many units of shipboard machinery would quite literally grind to a screeching halt.

Fluid lubrication is based on actual separation of surfaces so that no metal-to-metal contact occurs. As long as the lubricant film remains

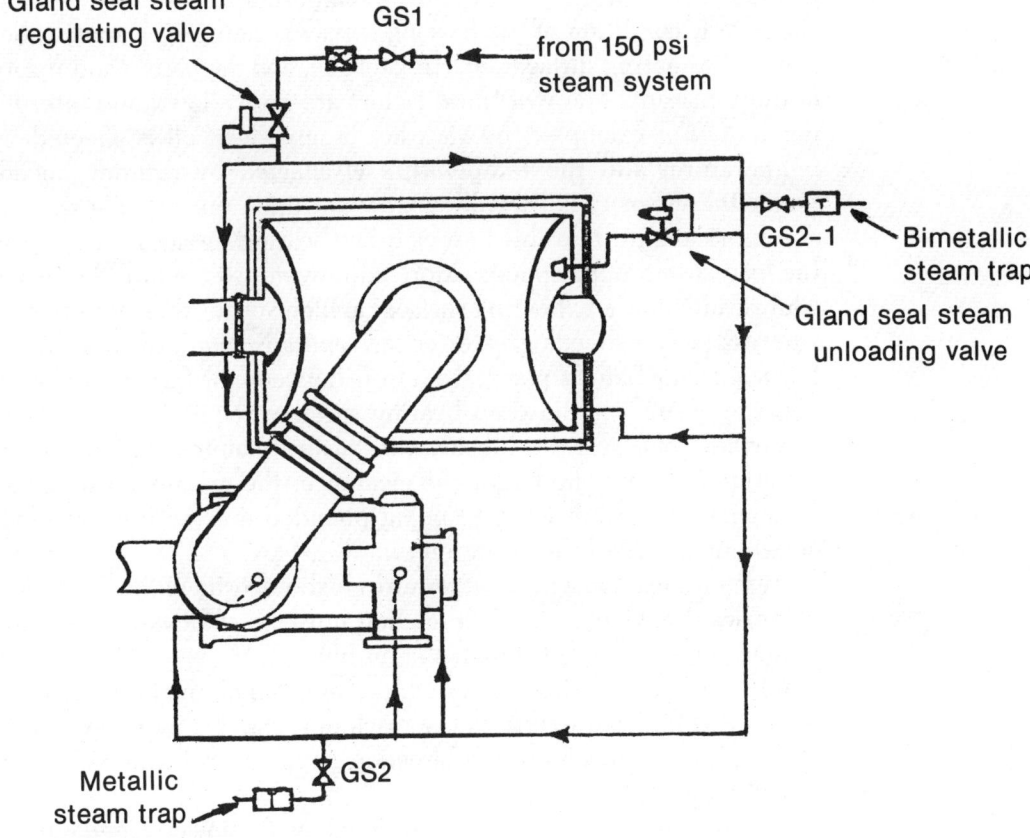

Figure 6–12. Schematic diagram of a typical gland seal steam system.

unbroken, sliding friction and rolling friction are replaced by fluid friction.

In any process involving friction, some power is consumed, and some heat is produced. Overcoming sliding friction consumes the greatest amount of power and produces the greatest amount of heat. Overcoming rolling friction consumes less power and produces less heat. Overcoming fluid friction consumes the least power and produces the least amount of heat.

A presently accepted theory of lubrication is based on the action of fluid films of oil between two surfaces, one or both of which are in motion. When a rotating journal is set in motion, the relationship of the journal to the bearing lining is such that a wedge of oil is formed between two boundary films, thus continuously preventing contact between the two metal surfaces.

A number of factors determine the efficiency of oil film lubrication, including such things as pressure, temperature, viscosity, speed, alignment, condition of the bearing surfaces, running clearances between the bearing surfaces, starting torque, and the nature and purity of the lubricant. Many of these factors are interrelated and interdependent. For example, the viscosity of any given oil is affected by temperature, and the temperature is affected by running speed; hence the viscosity is partially dependent upon the running speed.

A lubricant must be able to stick to the bearing surfaces and support the load at operating speeds. More adhesiveness is required to make a lubricant adhere to bearing surfaces at high speeds than at low ones, whereas at low speeds greater cohesiveness is required to keep the lubricant from being squeezed out from between the bearing surfaces.

Large clearances between bearing surfaces require high viscosity and cohesiveness in the lubricant to ensure maintenance of the lubricating oil film. The larger the clearance, the greater must be the resistance of the lubricant to being pounded out, with consequent destruction of the lubricating oil film.

High unit load on a bearing requires high viscosity of the lubricant. A lubricant subjected to high loading must be sufficiently cohesive to hold together and maintain the oil film.

A forced-feed lubrication system is one in which the lubricating oil is supplied under pressure to the machinery moving parts. With rare exceptions, all machinery lubrication in modern naval vessels is forced-feed.

On auxiliary turbines (pumps, forced draft blowers, generators, etc.), a small attached pump provides the lubricating oil pressure for the turbine bearings and reduction gears, where installed. After providing lubrication, the oil passes down into a sump or oil reservoir, which must be drained periodically, just like the crankcase on a car, because it has no purification system of its own.

The lubrication system for the main engine (propulsion turbines) bearings and main reduction gears includes: lube oil pumps, a main sump, a cooler, filters and strainers, a centrifugal purifier, and a settling tank (figure 6–13). On board most ships, each lube oil service system includes three positive-displacement lube oil service pumps: a shaft-driven pump and two motor-driven pumps. The shaft-driven pump, attached to and driven by either the propulsion shaft or the quill shaft of the reduction gear, is used as the regular lube oil service pump when the shaft is turning fast enough so that the pump can supply the required lube oil pressure.

Lubricating oils may be kept in service for long periods of time, provided the purity of the oils is maintained at the required standard. The simple fact is that lubricating oil does not wear out, although it can become unfit for use when it is robbed of its lubricating properties by the presence of water, sand, sludge, fine metallic particles, acid, and other contaminants. Proper care of lubricating oil requires, then, that the oil be kept as free from contamination as possible and that, once contaminated, the oil be purified before it is used again.

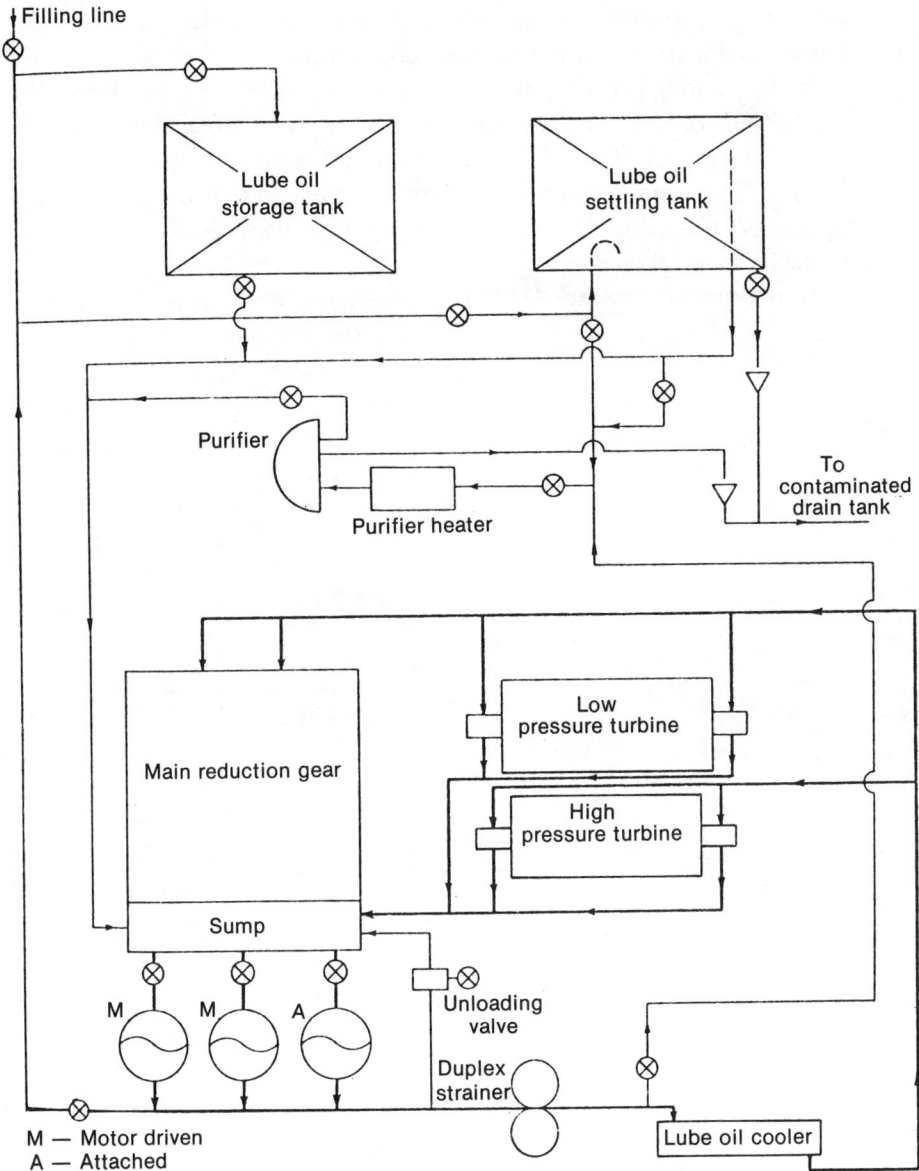

Figure 6–13. Schematic diagram of a typical main lube oil system.

Strainers or filters are used in many lubricating systems to prevent the passage of grit, scale, dirt, and other foreign matter (figure 6–14). The use of strainers and filters, however, does not solve the problem of water contamination of lubricating oil. Even a very small amount of water in lubricating oil can be extremely damaging to machinery, piping, valves, and other equipment; water in the oil causes widespread pitting and corrosion, and, by increasing frictional resistance, can cause the oil film to break down prematurely. Thus, lubricating oil piping is generally arranged to permit two methods of purification: batch purification and continuous purification.

In the batch process, the lubricating oil is transferred from the sump to a settling tank by means of a purifier or a transfer pump. In the settling tank, the oil is heated and allowed to settle for several hours. Water and other impurities are removed from the settling tank, and the oil is then centrifuged and returned to the sump from which it was taken.

In the continuous purification process, a centrifugal purifier (figure

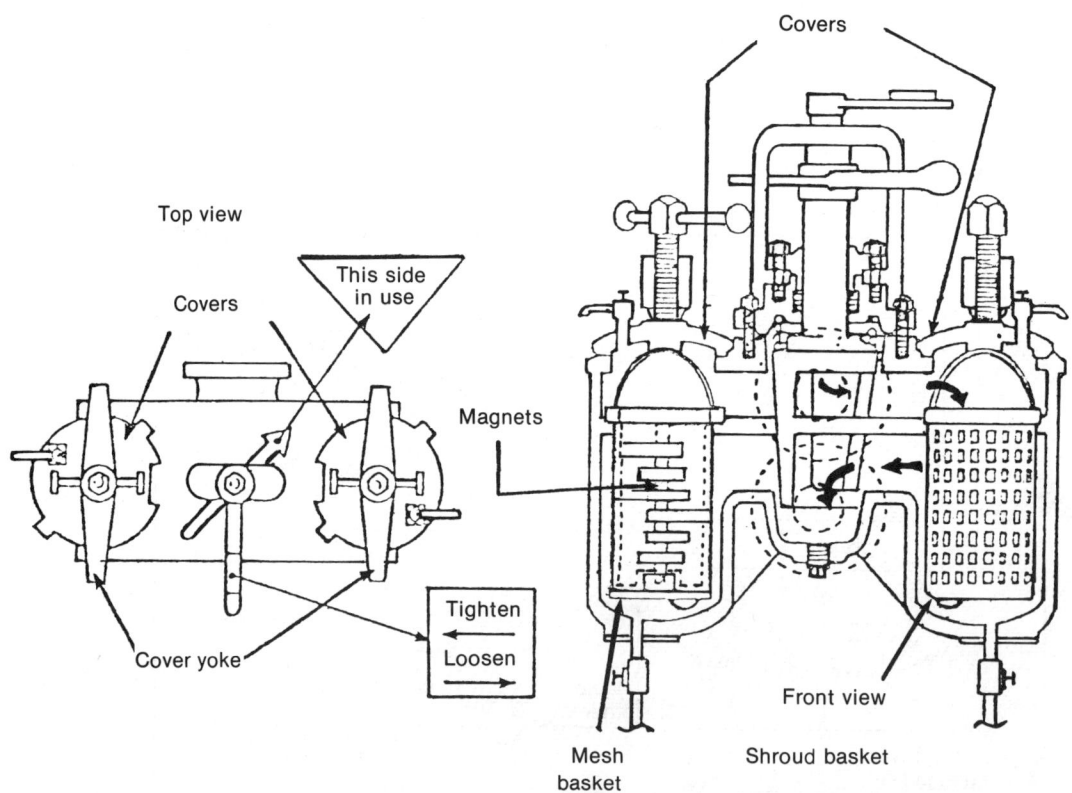

Figure 6–14. *Cutaway of a typical main lube oil duplex strainer.*

6–15) takes suction from a sump tank and, after purifying the oil, discharges it back to the same sump. The centrifugal purifier is essentially a bowl or hollow cylindrical container that is rotated at high speed while contaminated oil is forced through and rotated with the container. The centrifugal force imposed on the oil by the high rotational speed of the container causes suspended foreign matter and water to separate from the oil. Continuous purification is done while a ship is under way, whereas batch purification is usually done when the ship is in port.

Auxiliary Steam Turbines

The preceding chapters have described the principles of operation and the major construction features of the various types of turbines, with the major emphasis on the design of large turbines such as those

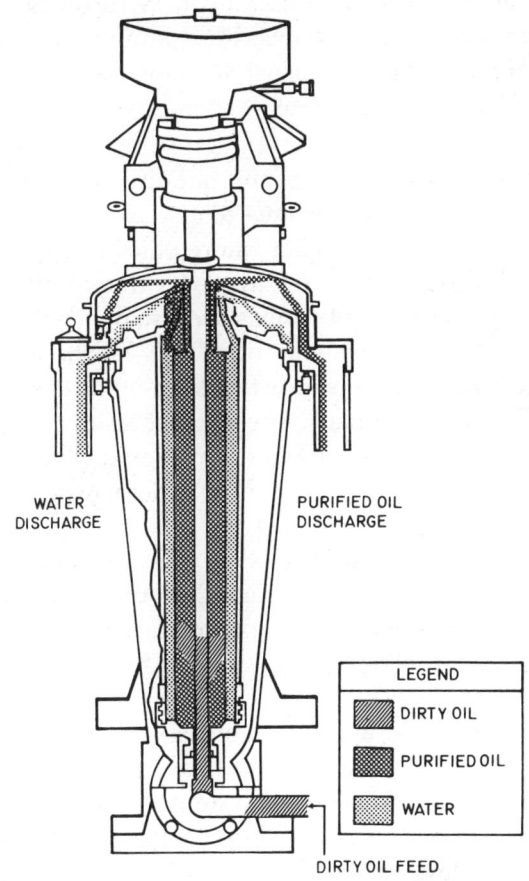

Figure 6–15. Cutaway of a typical main lube purifier with oil/water flow indicated.

used for ship propulsion. In addition to these large units, there are, in all naval engineering plants, a considerable number of small turbines which are employed for driving auxiliary machinery such as generators, forced draft blowers, pumps, and air compressors.

Characteristics

Until about 1950, many generator turbines were designed and installed in such a way that they could be operated on steam from either superheated or saturated steam lines at full boiler pressure. Most of the other auxiliary turbines at this time operated on saturated steam at full boiler pressure. During the early 1950s, a few ships were built in which all auxiliary turbines were designed to operate on steam at full superheat and full boiler pressure. In most oil-fired ships built since 1953, steam at full superheat and full boiler presure is supplied to the auxiliary turbines for generators, main feed pumps, and forced draft blowers; the other auxiliary turbines on these ships usually operate on steam at reduced temperature and pressure. The generator turbines usually exhaust to their own separate auxiliary condensers. Most other auxiliary turbines exhaust to the auxiliary exhaust system.

Many auxiliary turbines are of the impulse type. Reduction gears are used with most auxiliary turbines to increase efficiency by allowing the turbine to operate at a higher, more efficient rpm and the pump at lower rpm's. Since space requirements frequently demand relatively small units, auxiliary turbines are usually designed with comparatively few stages, often only one. This means a large pressure drop must occur in each stage, creating high-velocity steam. To obtain maximum efficiency, the blade speed must also be high. With auxiliary turbines, such as those used as ship's service turbine generators, reduction gears serve to reconcile the conflicting speed requirements of the driving and the driven shaft.

Most auxiliary turbines are axial-flow units, which are quite similar (except for size and number of stages) to the axial-flow propulsion turbines. However, some auxiliary turbines are designed for helical flow and some for radial flow—types of flow never used for propulsion turbines.

Auxiliary turbines designed to naval specifications have pressure lubrication systems to lubricate the bearings, reduction gears, and governors. A pressure lubrication system requires a lube oil pump. As a rule, the lube oil pumps used for auxiliary turbines are attached to simple gear pumps driven by the turbine itself instead of being separate pump units.

The lubricating oil passes through an oil sight flow indicator, a filter, and an oil cooler. Oil is then piped to the bearings on the turbine shaft, to the governor, and to the worm gear on the pump shaft. The bearings and gear on the oil pump and governor shaft are lubricated by oil that drains from the governor and passes back into the oil reservoir. A relief valve is built into the gear casing. This valve serves to protect the system against the development of excessive pressures.

Speed Control Devices

The principles involved in the control of auxiliary turbines are basically the same as those for main propulsion turbines. However, the great variety of uses to which auxiliary turbines are put makes it necessary to have different types of control apparatus. From the point of view of service requirements, auxiliary turbines may be divided into two classes: constant-speed and variable-speed. Turbo-generators and turbine-driven air compressors are of the first class, while forced draft blowers and most pumps are of the second class.

The constant-speed governor, sometimes called the speed-regulating governor, is used on constant-speed machinery to maintain a constant turbine speed regardless of the load on the turbine. Constant-speed governors are used primarily on generator turbines and on air compressor turbines. With an increased load on the generator, the turbine tends to slow down. Since the governor is driven by the turbine shaft, through reduction gears, the governor also slows. Centrifugal weights on the governor move inward as the speed decreases, causing the nozzle valves to open and admit additional steam to the turbines. A reverse process occurs when the load on the generator decreases.

Many turbine-drive pumps are fitted with constant-pressure pump governors. The function of a constant-pressure pump governor is to maintain a constant pump discharge pressure under conditions of varying flow. The governor, which is installed in the steam line to the pump, controls the pump discharge pressure by controlling the amount of steam admitted to the driving turbine, hence the speed of the turbine.

Safety Devices

Safety devices used on auxiliary turbines include speed-limiting governors and several kinds of trips. Safety devices differ from speed control devices in that they have no control over the turbine under normal operating conditions. It is only when some abnormal condition occurs that the safety devices come into use to stop the unit or to limit its speed.

The speed-limiting governor is essentially a safety device for variable-speed units. It allows the turbine to operate under all conditions from no-load to overload, up to the speed for which the governor is set, but it does not allow operation in excess of 107 percent of rated speed. It is important to note that this type of governor is adjusted to the maximum operating speed of the turbine and therefore has no control over the admission of steam until the upper limit of safe operating speed is reached. The governor has no effect on the speed of the turbine at speeds below about 95 percent of rated speed.

Overspeed trips are used on turbines that have constant-speed governors. The overspeed trip shuts off the supply of steam to the turbine and thus stops the unit when a predetermined turbine speed has been reached. Overspeed trips are usually set to trip out at 110 percent of normal operating speed.

Back-pressure trips are installed on turbogenerators to protect the turbine by closing the throttle automatically when the back pressure (exhaust pressure) becomes too high. Emergency hand trips are installed on turbogenerators to provide a means for closing the throttle quickly, by hand, in case of damage to either the turbine or the generator.

Low oil pressure alarms are installed on generator turbines, in particular, to warn operating personnel when the lubricating oil pressure becomes dangerously low. When the oil pressure drops below normal, an audible alarm is sounded. Other devices may actually trip the throttle closed (similar to the overspeed trip) if the oil pressure drops too low.

Auxiliary Exhaust System

Main propulsion turbines exhaust into the main condenser. Similarly the turbogenerator, being a sizable unit, exhausts into its own condenser (but also may exhaust into the main condenser). It is impracticable, however, to provide an individual condenser for each small steam turbine that drives an auxiliary pump. Consequently, a piping circuit is provided to collect the discharged steam from all of the auxiliary turbines in a propulsion plant. This circuit is called the auxiliary exhaust system. Pressure in this system is maintained at 15 psig by an automatic valve arrangement.

Steam in the auxiliary exhaust system at approximately 15 psig retains considerable thermal energy, which is utilized in two important ways. First, the steam is used in the deaerating feed tank (DFT), to heat the condensate and to purge it of dissolved air. A major portion of the auxiliary exhaust steam is used in this manner. This portion combines with the feedwater in the heating and deaerating process and so returns to the boiler. Second, a smaller portion is used to provide a part of the thermal energy required to operate the distilling plant, in which fresh water is obtained from seawater. This portion is condensed, and finds its way back to the boiler through the low pressure drain system.

When more steam is delivered to the auxiliary exhaust system than is required for these two functions, the pressure in the system may exceed 15 psig. This excess pressure causes the automatic unloading valve to open, allowing steam to flow to the main condenser. This valve closes when the pressure falls back to 15 psig. The steam that flows to the main condenser loses its thermal energy to the circulating seawater and is condensed. Note that the thermal energy in the portions of the auxiliary exhaust steam used in the DFT and distilling plant performs useful functions. However, the steam unloaded to the condenser loses its thermal energy to the sea. Therefore, it is desirable to control the supply of auxiliary exhaust steam discharged into the system to balance the sum of the requirements of the DFT and the distilling plant. This is done by the proper choice of steam-driven pumps and electric-driven pumps in use under various plant load conditions.

Under some load conditions, there may be insufficient steam in the auxiliary exhaust system to maintain the required pressure, even when no electric pumps are in use. In such circumstances provision is made for augmenting the pressure by admitting steam directly to the auxiliary exhaust steam system through a pressure-reducing valve called the augmentation valve.

Transmission of Power and Thrust

The ability to move through the water and the ability to control the direction of movement are among the most fundamental of all ship requirements. Ship propulsion is achieved through the conversion, transmision, and utilization of energy in a sequence of events that includes the development of power in a prime mover (turbine), the transmission of power to the propellers, the development of thrust on the working surfaces of the propeller blades, and the transmission of thrust to the ship's structure in such a way as to move the ship through the water.

Reduction Gears

On most main propulsion plants, reduction gears serve two purposes. The first, and most important, is to allow both the prime mover (the turbines) and the propeller to operate at their most efficient speed. Higher turbine rotor speeds lead to greater energy conversion per stage and therefore more powerful turbines for a given size. Mechanical energy developed by the turbine, however, must be connected to thrust delivered by the propellers; and propellers work efficiently at relatively lower speeds. The coupling of the higher turbine speed with the desired lower propeller speed is accomplished by use of mechanical reduction gears.

The second purpose or function of reduction gears is to combine the output of two or more shafts into one output shaft (i.e., combining the output of the low pressure turbine with the output of the high pressure turbine). The output of the two shafts is transferred to a larger gear called the bull gear, which drives the main propulsion shaft.

Most main reduction gearing in current ships make use of double helical gears. Reduction gears are classified by the number of steps used to bring about speed reduction and by the general arrangement of the gearings. A single reduction gear consists of a small pinion gear, which is driven by the turbine shaft, and a large main gear (or bull gear), which is driven by the pinion. The ratio of speed reduction is proportional to the diameter of the pinion and the bull gear. In a 2 to 1 single reduction gear, for example, the diameter of the driven gear is twice that of the driving pinion. In a 10 to 1 single reduction gear, the diameter of the driven gear is ten times that of the pinion.

A double reduction gear consists of first reduction (or high speed) pinions, first reduction gears, second reduction (or low speed) pinions, and a second reduction (or bull) gear. Speed reduction is accomplished in two steps. The turbine shaft drives the first reduction pinions, which drive the first reduction gears. Each first reduction gear is mounted on a common shaft with a second reduction pinion, so that they turn at the same speed. The second reduction pinion drives the second reduction gear, which drives the propeller shaft.

A typical double reduction gear installation for a destroyer is shown in figure 6–16. The high pressure turbine and the low pressure turbine are connected to the propeller shaft through a locked train double reduction gear. First reduction pinions are connected by flexible couplings to the turbines. Each of the first reduction gears drives a second reduction (low speed) pinion. These four pinions drive the second reduction gear (bull gear), which is attached to the propeller shaft.

Propulsion turbine shafts are connected to the reduction gears by flexible couplings, which are designed to take care of very slight misalignment between the two units.

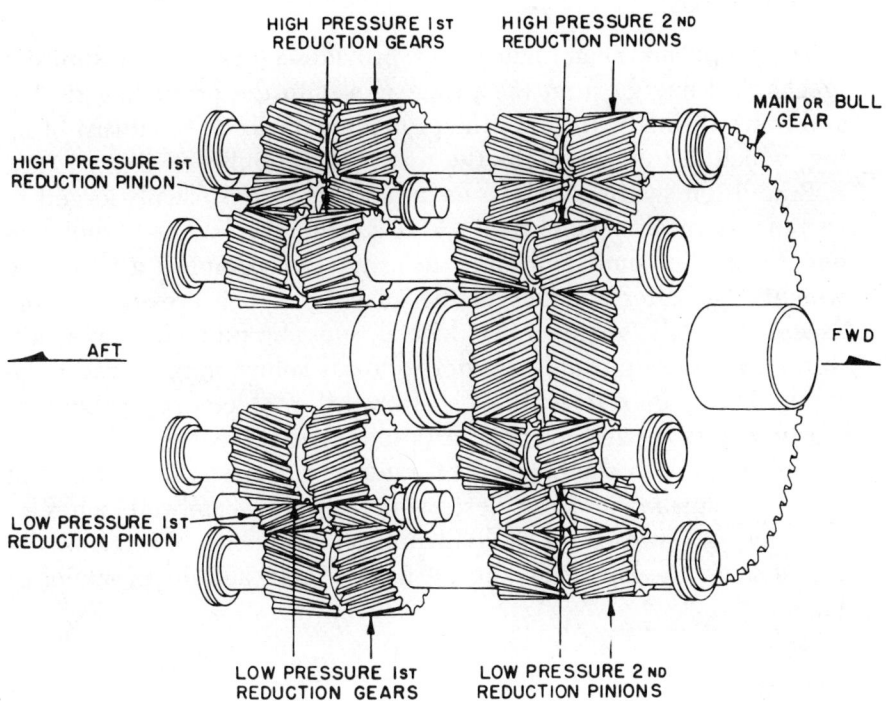

Figure 6–16. A typical double reduction gear installation.

Shaft Turning Gear

It is often necessary to turn over the main propulsion shaft and the turbines without the use of steam. Some such occasions are: (1) immediately after securing the turbine, to assure even cooling of the rotor to prevent warping; (2) in port at regular intervals, to ensure maintenance of the oil film in the turbine and gear bearings; (3) when starting a cold plant before admitting steam and for at least 15 minutes after admitting steam, to provide even heating and expansion of the rotor; and (4) during maintenance, to position certain components such as propellers, gears, or rotors for repair work.

To provide this capacity, an electric-motor-powered gear train called a shaft turning gear or jacking gear is installed on the reduction gear. The jacking gear is arranged so as to engage the high pressure first reduction pinion shaft inside the reduction gear casing. In addition to driving the shaft without steam, the jacking gear can be used to lock the propeller shaft in a certain position if necessary.

Propulsion Shaft

The propulsion shaft (figure 6–17) provides a means of transmitting mechanical energy from the prime mover to the propelling device and transmitting thrust from the propelling device to the thrust bearing, which in turn transmits the thrust to the hull.

Propulsion shafting for vessels in naval service is usually forged in sections from alloy steel ingots and is generally hollow-bored from end to end in combatant ships in order to accomplish a saving of weight. The after sections of the shafting, which are inaccessible except when the vessel is dry-docked, must be protected from saltwater corrosion (pitting) in order to avoid failure from corrosion fatigue. These after sections of shafting are protected by rubber or plastic paints applied over the exposed areas of steel shafts.

The propulsion shafting, which ranges in diameter from 15 to 21 inches for small twin-screw destroyers to approximately 30 inches for large four-screw carriers, is divided into four functional sections: the thrust shaft, the line shaft, the stern tube shaft, and the propeller or tail shaft.

Bearings

Generally speaking, the term bearing may be used to designate anything that supports a moving element. In this section we will discuss only those bearings that support or confine the propulsion

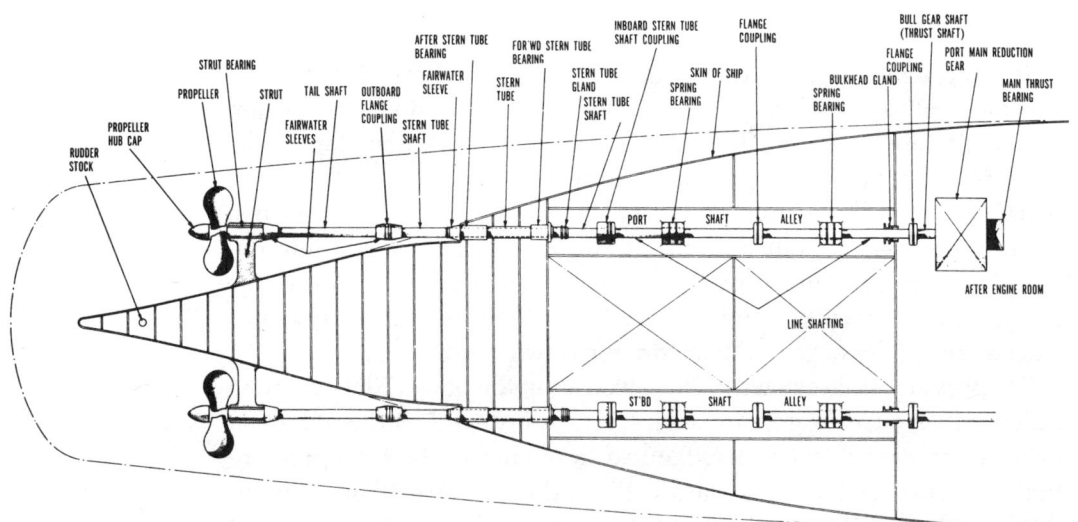

Figure 6–17. A typical propulsion shaft arrangement.

shaft. Classified by function they would either be thrust bearings, which position the shaft longitudinally and absorb any force parallel to the axis, or radial bearings, which support and position the shaft by absorbing the radial load.

Main Thrust Bearing

The main thrust bearing (figure 6–18), which is usually located in the reduction gear casing, serves to absorb the axial thrust transmitted through the propulsion shaft from the propeller. As the propulsion shaft turns, and the propeller pushes the water away from the ship's hull, the propulsion shaft has a tendency to move forward. The main thrust bearing absorbs this thrust and transmits it to the ship's hull, causing the hull to move through the water.

Line Shaft Bearings

Line shaft bearings or spring bearings (figure 6–19) are journal bearings that support the shaft inside the hull. The length of the shaft determines the number of spring bearings. The bearing is lined with

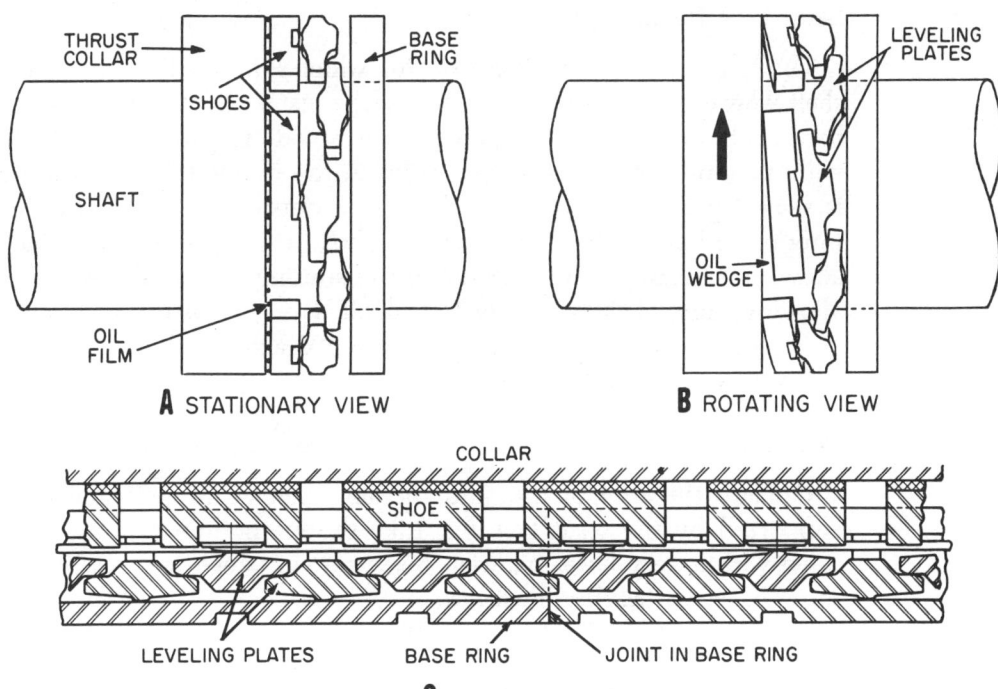

Figure 6–18. (A) Stationary, (B) rotating, and (C) extended view of the main thrust bearing. (Courtesy Kingsburg Machine Tools Corp.)

Main Propulsion Turbines, Reduction Gear, Shafting, and Auxiliary Turbines **97**

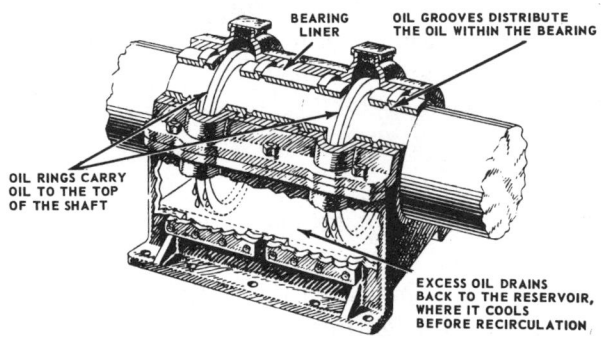

Figure 6–19. Line shaft bearing.

a nonferrous metal alloy called babbit and is split into upper and lower halves to facilitate removal. Lubrication is accomplished by a ring or chain on the turning shaft dipping into an oil reservoir. An abnormal rise in temperature is an indication of trouble, and prompt corrective measures must be taken, often including slowing the ship down; otherwise, the bearing will wipe (the babbit will melt), and the shaft must be stopped.

Stern Tube Bearing

The stern tube bearing (figure 6–20), which supports the propulsion shaft where it pierces the hull, is basically just a pipe lined with very hard rubber or lignum vitae, a very hard wood. The stern tube shaft, with a copper or bronze jacket, rides in the bearing. The lubricant is seawater leaking along the shaft from outside the hull. On the interior end of the bearing is a stuffing box, which prevents the shaft alley from flooding. The packing in the stuffing box is lubricated by allowing some of the seawater to leak off. Every hour the shaft alley is checked to ensure that there is sufficient leakoff, and to determine when it needs to be pumped.

Strut Bearing

Outside the hull of the ship the tail shaft or propeller shaft is supported by the strut bearing. Like the stern tube bearing, it is lined with a very hard material. Seawater, in which it is totally immersed, serves as the lubricant.

Propellers

The propelling device imparts velocity to a column of water and moves it in a direction opposite to the direction in which it is desired

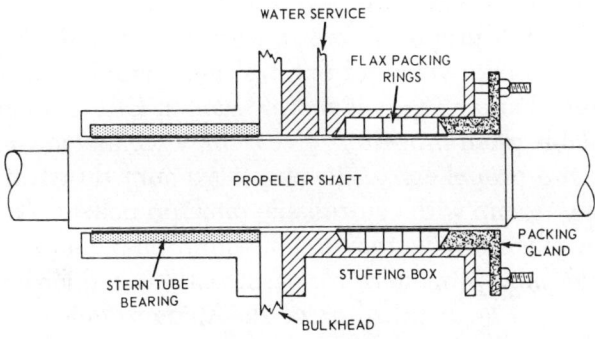

Figure 6–20. Stern tube stuffing box and gland arrangement.

to move the ship. A reactive force (thrust) is thereby developed against the velocity-imparting device; and this thrust, when transmitted to the ship's structure, causes the ship to move through the water. In essence, then, we may think of propelling devices as pumps which are designed to move a column of water to build up a reactive force sufficient to move the ship.

The propelling device most commonly used for naval ships is the screw propeller, so called because it advances through the water in somewhat the same way that a screw advances through wood, or like a bolt when it is screwed into a nut. With the screw propeller, as with a screw, the axial distance advanced with each complete revolution is known as the pitch. Propeller blades are placed at such an angle as to advance with each revolution; this angle is called the pitch angle.

A screw propeller consists of a hub and several (usually three or four) blades spaced at equal angles about the axis. When the blades are integral with the hub, the propeller is known as a solid propeller. When the blades are separately cast and secured to the hub by means of studs and nuts, the propeller is referred to as a built-up propeller.

All propellers may be further classified as having constant pitch or variable pitch. In a constant pitch propeller, the pitch of each radius is the same. On a variable pitch propeller, the pitch at each radius may vary. Solid propellers of the variable pitch type are the most commonly used for naval ships. The radius is measured as a distance from the hub.

Screw propellers may be broadly classified as fixed pitch propellers or controllable pitch propellers. The pitch of a fixed propeller cannot be altered during operation; the pitch of a controllable pitch propeller can be changed continuously subject to bridge or engineroom control.

Most propellers in naval use are of the fixed pitch type, but some controllable pitch propellers are in service on board LSTs, gas-turbine-powered destroyers and frigates, and various smaller combatants. (Note: Chapter 12 gives more detail on CRP systems.)

Controllable pitch propellers give a ship excellent maneuverability and allow the propellers to develop maximum thrust at any given engine rpm. A ship with controllable pitch propellers requires much less distance for stopping than a ship with fixed pitch propellers. The controllable pitch propellers are particularly useful for landing ships because they make it possible for the ships to hover offshore and because they make it easier for the ships to retract and turn away from the beach.

Propellers are classified as being right-hand or left-hand propellers, depending upon the direction of rotation. When viewed from astern, with the ship moving ahead, a right-hand propeller rotates in a clockwise direction, and a left-hand propeller rotates in a counterclockwise direction. The great majority of single-screw ships have right-hand propellers. Multiple-screw ships have right-hand propellers to starboard and left-hand propellers to port. Reversing the direction of rotation of a propeller reverses the direction of thrust and consequently reverses the direction of the ship's movement.

One of the major problems encountered with propellers is known as cavitation. Cavitation is the formation of a vacuum around a propeller that is revolving at a speed above a certain critical value (which varies, depending upon the size, number, and shape of the propeller blades). The speed at which cavitation begins to occur is different in different types of ships; the turbulence increases in proportion to the propeller rpm. Specifically, a propeller rotating at a high speed will develop a stream velocity that creates a low pressure. This low pressure is less than the vaporization point of the water, and from each blade tip there appears to develop a spiral of bubbles. The water boils at the low pressure points. As the vapor bubbles of cavitation move into regions where the pressure is higher, the bubbles collapse rapidly and produce a high-pitched noise.

The net result of cavitation is to produce: (1) a high-level underwater noise; (2) erosion of propeller blades; (3) vibration with subsequent blade failure from metallic fatigue; and (4) overall loss in propeller efficiency, requiring a proportionate increase in power for a given speed.

In naval warfare, the movements of surface ships and submarines can be plotted by sonar bearings on propeller noise. Because of the high static water pressure at submarine operational depths, cavitation sets in when a submarine is operating at a much higher rpm than it uses when near the surface. For obvious reasons, a submarine that is under attack will immediately dive deep so that it can use high propeller rpm with the least amount of noise.

A certain amount of vibration is always present on board ship. Propeller vibration, however, may also be caused by a fouled blade or by seaweed. If a propeller strikes a submerged object, the blades may be nicked, causing the propeller to vibrate excessively.

7

Main Condensate System

This section deals with construction and operation of the main condenser, which is the major heat transfer apparatus found in the condensate system. Besides the main condenser, this section discusses the function and operation of the main circulating pump and the main condensate pump.

Vacuum

A space in which the pressure is less than atmospheric pressure is said to be under vacuum. The amount of vacuum is expressed in terms of the difference between the pressure in the space and the existing atmospheric pressure. Vacuum is normally measured in inches of mercury—that is, the number of inches a column of mercury in a U-tube would be displaced by a pressure equal to the difference between the pressure in the vacuum space and the existing atmospheric pressure.

Vacuum gage scales are marked from 0 to 30. When a vacuum gage reads zero, the pressure in the space is the same as the existing atmospheric pressure—in other words, there is no vacuum. A vacuum gage reading of 30 inches of mercury would indicate a nearly perfect vacuum. In actual practice, it is impossible to obtain a perfect vacuum, and the highest vacuum gage readings are seldom over 29 inches of mercury. (See figure 1–8 on p. 22.)

Main Condenser

The main condenser (figure 7–1) is a cross flow heat exchanger in which exhaust steam from the propulsion turbines is condensed as it comes in contact with tubes through which cool seawater is flowing. The main condenser is the heat receiver of the thermodynamic cycle—that is, it is the low-temperature heat sink to which some heat must be rejected. The main condenser is also the means by which feedwater is recovered and returned to the feed system. If we imagine a shipboard propulsion plant in which there is no main condenser and the turbines exhaust to atmosphere, and if we consider the vast

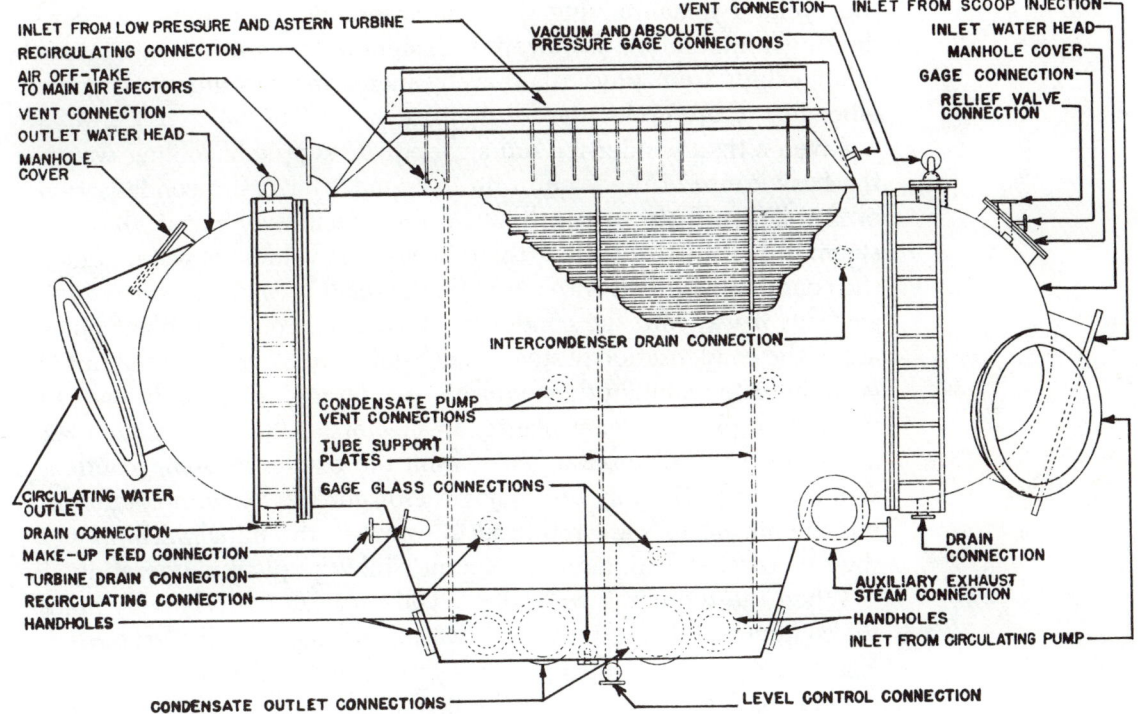

Figure 7-1. Drawing of a main condenser.

quantities of fresh water that would be required to support even one boiler generating 150,000 pounds of steam, hence 150,000 pounds of water or almost 18,000 gallons per hour, it is immediately apparent that the main condenser serves a vital function in recovering feedwater.

The main condenser is maintained under a vacuum of approximately 25 to 28.5 inches of mercury. The designed vacuum varies according to the design of the turbine installation and according to such operational factors as the load on the condenser, the temperature of the outside seawater, and the tightness of the condenser.

It is often said that an engine can do a greater amount of useful work if it exhausts to a low-pressure space than if it exhausts against a high pressure. This statement is undeniably true, but for the condensing steam power plant it may be somewhat misleading because of its emphasis on pressure. The pressure is important because it determines the temperature at which the steam condenses. An increase in the temperature difference between the source (boiler) and the receiver (condenser) increases the thermodynamic efficiency of the cycle. By maintaining the condenser under vacuum, the condensing temperature is lowered (remember pressure–temperature relationships from physics), thereby increasing the thermodynamic efficiency of the cycle.

Given a tight condenser and an adequate supply of cooling water, the basic cause of the vacuum in the condenser is the condensation of the steam. This is true because the specific volume (ft^3/lbm) of steam is enormously greater than the specific volume of water. Since the condenser is filled with air when the plant is cold, and since some air finds it way into the condenser during the course of plant operation, the condensation of steam is not sufficient to establish the initial vacuum or to maintain the required vacuum under all conditions. In modern shipboard steam plants, air ejectors are used to remove air and other noncondensable gases from the condenser. The condensation of steam is thus the major reason for the vacuum condition, but the air ejectors are required to establish the initial vacuum and then to assist in maintaining vacuum while the plant is operating.

When the temperature of the outside seawater is relatively high, the condenser tubes are relatively warm, and heat transfer is retarded. For this reason, a ship operating in warm tropical waters cannot develop as high a vacuum in the condenser as the same ship could develop when operating in colder waters.

In any main condenser, there are two separate paths of flow. The first is the steam vapor-condensate circuit, in which the exhaust steam enters the condenser at the top of the shell and is condensed as it comes in contact with the outer surfaces of the condenser tubes. (Refer to figure 7–2.) The condensate then falls to the bottom of the condenser, drains into a space called the hotwell, and is removed by the condensate pump. Air and other noncondensable gases that enter with the exhaust steam or that otherwise find their way into the condenser are drawn off by the air ejector through the air ejector suction opening in the shell of the condenser, above the condensate level. (See figure 7–3.)

The second circuit is the circulating water circuit. During normal ahead operation, a scoop injection system provides the flow of seawater through the main condenser. The scoop, as its name implies, uses the forward movement of the ship through the water to "scoop" up the seawater and, by large pipes, direct it through the tubes of the condenser. When it comes out of the tubes on the other side of the condenser, it is directed over the side of the ship, thus carrying away the latent heat of condensation it picks up from the condensing steam. A steam-turbine-driven or electric-driven main circulating pump provides positive circulation of seawater through the condenser

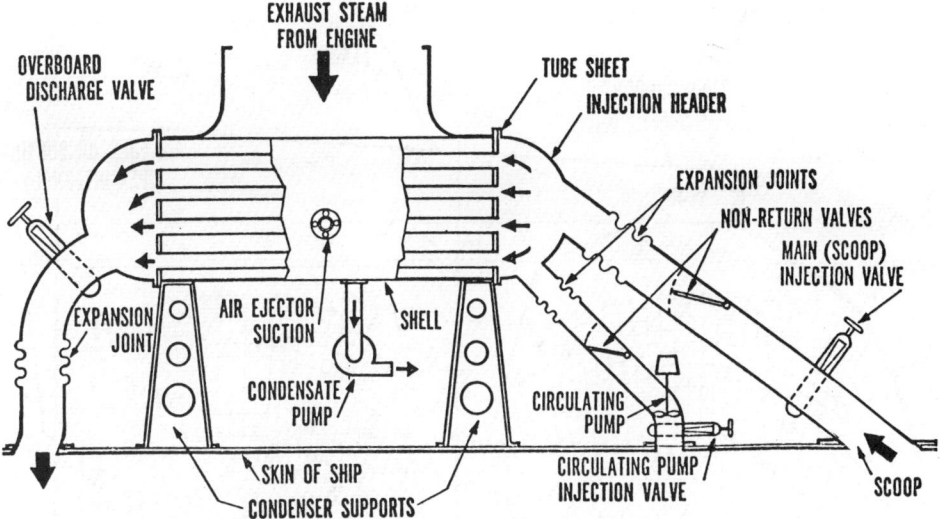

Figure 7–2. Drawing displaying steam, seawater, and condensate flow in a main condenser.

at times when the scoop injection system is not effective—when the ship is stopped, backing down, or moving ahead at very slow speeds.

Condensate Depression

The difference between the temperature of the condensate discharge and the condensing temperature corresponding to the vacuum maintained at the exhaust steam inlet to the condenser is called condensate depression or subcooling. The amount of condensate depression that exists at any time can be calculated by finding the temperature difference between the condenser inlet and the bottom of the condenser, the hotwell. The temperature in the hotwell is the lowest temperature reached by the working substance in the steam cycle.

One measure of the efficiency of design and operation of any condenser is its ability to maintain condensate depression at a reasonably low value under all normal conditions of operation; this temperature difference is normally maintained between 2 and 5°F. The amount of condensate depression is regulated by controlling the flow of circulating water passing through the condenser, which is done by throttling the overboard discharge valve on the main circulating line. For any given sea injection temperature, increasing the flow rate of circulating water will increase the amount of condensate depression; decreasing the flow rate will decrease condensate depression. Excessive condensate depression (above 5°F) decreases the operating

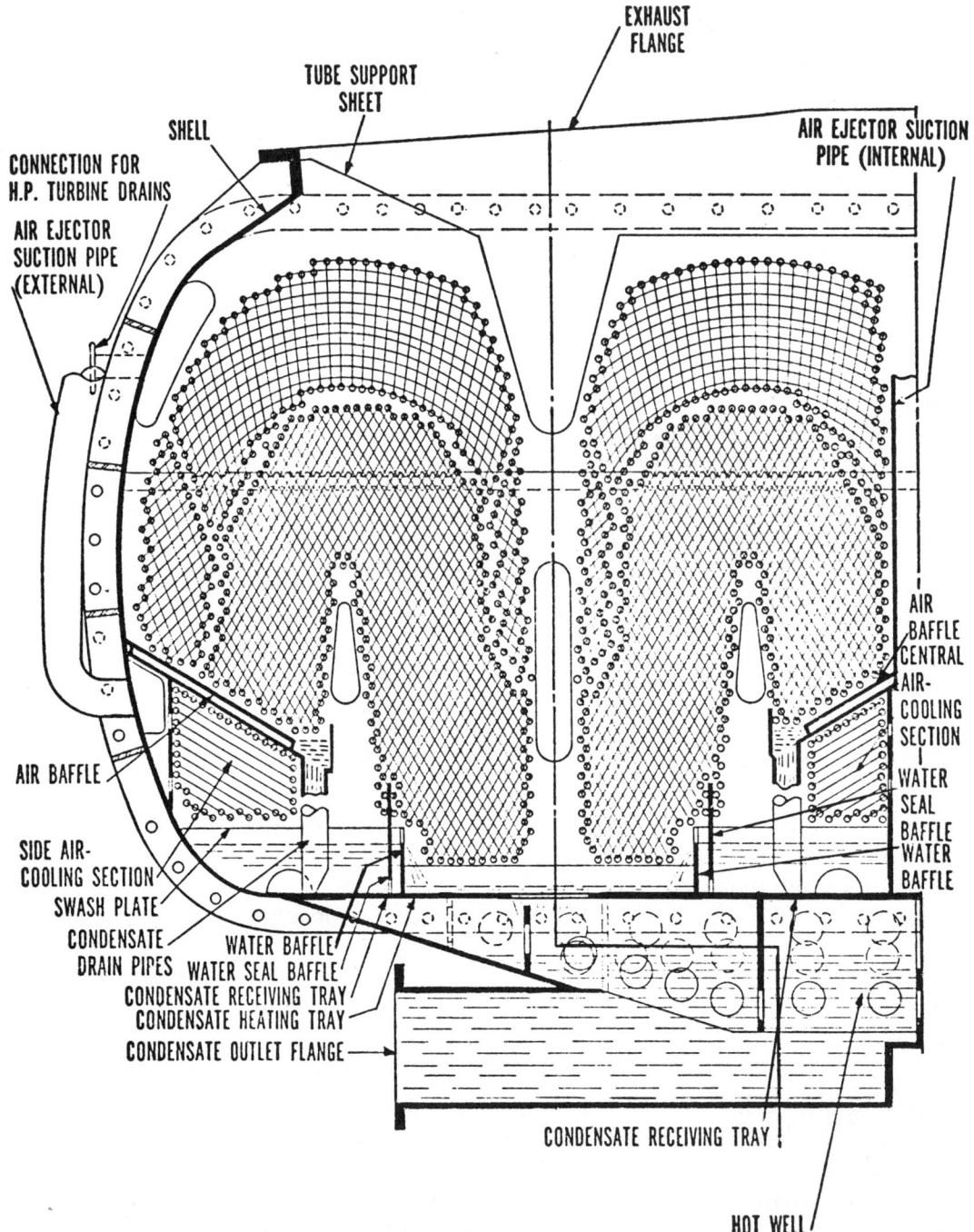

Figure 7–3. Main condenser cross section.

efficiency of the plant because the excessively subcooled condensate must be reheated in the feed system, with a consequent expenditure of steam and waste energy. Excessive condensate depression also allows an increased absorption of air by the condensate, and this air must later be removed in order to prevent oxygen corrosion of piping and boilers.

Main Circulating Pump

The main circulating pump provides a flow of water through the condenser at times when scoop injection is inadequate—as for example, when a ship is stopped, backing down, or moving ahead at a very low speed. A large swing-check valve or nonreturn valve, which prevents backflow of water, is installed in the main injection line; another one is installed in the main circulating pump discharge line. When water is flowing through the main injection line into the condenser, the nonreturn valve in the main circulating pump discharge line prevents the backflow of water through the line.

The main circulating pump can be either a steam or an electric-driven propeller-type pump. Propeller-type pumps are capable of pumping large volumes of water at low pressures. Considering that the main condenser is operating in a vacuum, it is desirable that only low-pressure seawater be pumped through the cooling tubes. High pressure could easily result in ruptured cooling tubes in the condenser.

In addition to providing cooling water to the condenser, most main circulating pumps are provided with a bilge suction line to pump water overboard in case of serious flooding of the engineroom. The main circulating pump generally constitutes the largest potential capacity for pumping engineroom bilges in an emergency.

Main Condensate Pump

The purpose of the main condensate pump (figure 7–4) is to pump water from the condenser hotwell through the main air ejector condensers to the DFT. It is normally an electric-driven, centrifugal-type pump. The lowest pressure reached in the steam cycle occurs at the inlet to the condensate pump. Two condensate pumps are usually installed for each main condenser.

Air Ejector Assemblies

The primary function of air ejectors is to remove air and other noncondensable gases from the condenser.

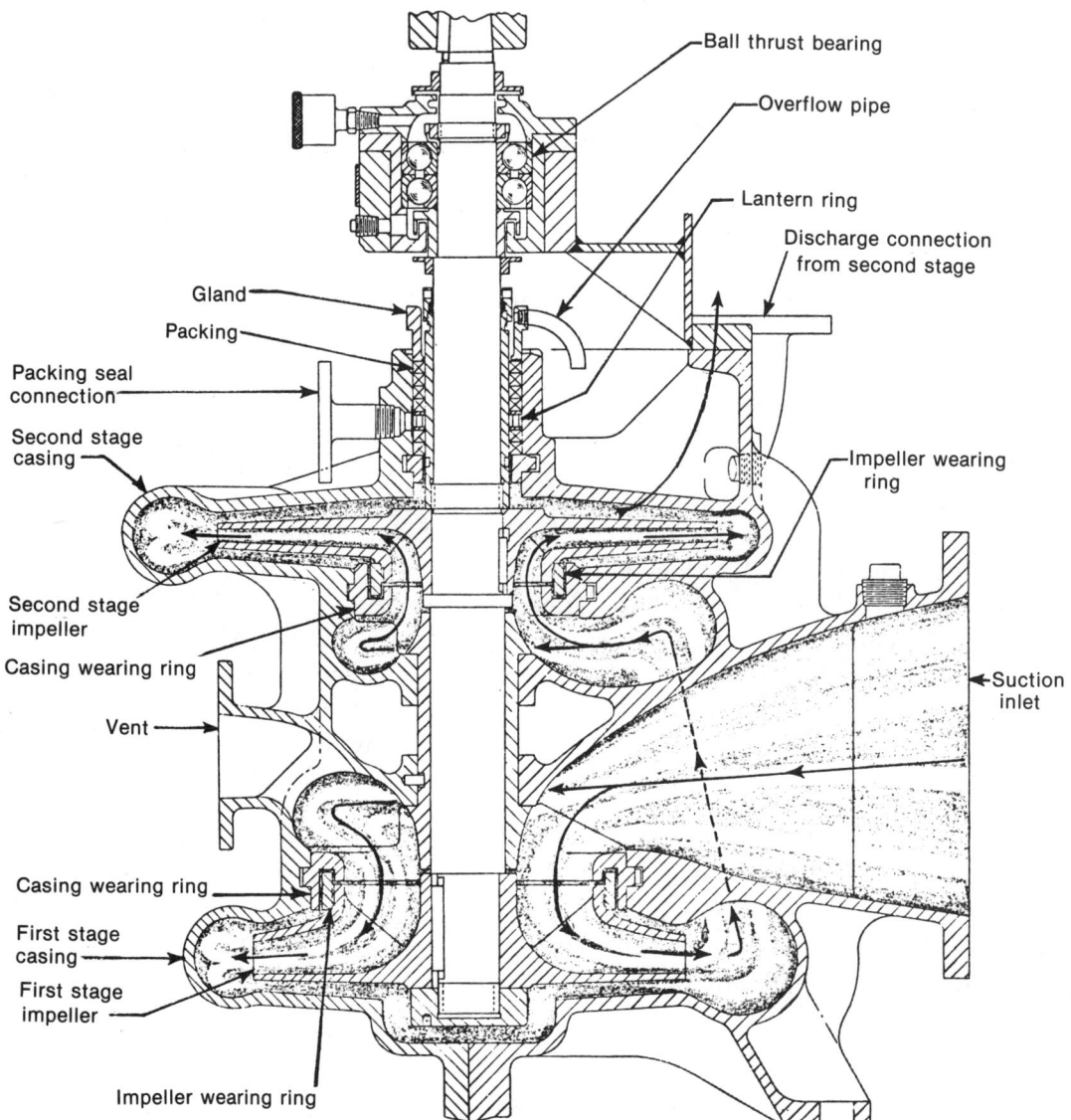

Figure 7-4. Main condensate pump cross section.

An air ejector is a type of jet pump having no moving parts and based on Bernoulli's principle. The flow through the air ejector is maintained by a jet of high-velocity steam passing through a nozzle, which further increases the velocity and decreases the pressure; the steam is taken from the 150 psi auxiliary steam systems in most ships.

The air ejector assembly (figure 7–5), used to remove air from the

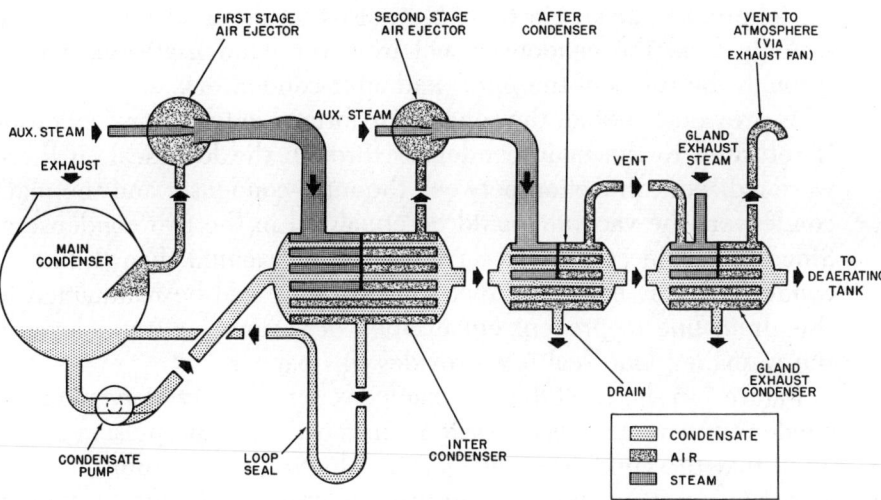

Figure 7–5. Steam and condensate flow in a representation of a main air ejector assembly.

main condenser, usually consists of a first-stage air ejector, an inter-condenser, a second-stage ejector, and an after-condenser.

In most installations, the gland exhaust condenser appears to be part of the air ejector assembly, since it is attached to the after-condenser. However, the gland exhaust condenser is functionally a separate unit even though it is physically attached to the air ejector after-condenser.

The first-stage air ejector takes a suction on the main condenser and discharges the steam/air mixture to the inter-condenser, where the steam content of the mixture is condensed. The resulting condensate drops to the bottom of the inter-condenser shell, and from there drains to the main condenser through a U-shaped loop seal line. The air passes to the suction of the second-stage air ejector, where another jet of steam entrains the air and carries it to the after-condenser. In the after-condenser, the steam is condensed and returned to the condensate system by way of the fresh water drain collecting tank. The air is then vented either to the atmosphere or to the gland exhaust condenser. The condensate from the main condenser is used as the cooling medium in the tubes of both the inter- and after-condensers. This serves two purposes: the steam/air mixture is condensed, and the condensate, now on its way to the boilers, begins its process of heat addition, absorbing the latent heat of condensation from the condensing steam mixture.

Note that the steam condensed in the air ejector condensers comes primarily from the auxiliary steam line—the steam used to activate the jet pumps (the air ejectors). Note also that the air ejectors remove air only from the condenser, not from the condensate that passes through the tubes of the inter- and after-condensers.

As previously noted, the condensate formed in the inter-condenser is returned to the main condenser through the loop seal. If there were a direct connection between the inter-condenser and the main condenser, the vacuum would be equalized in the two condensers. Since the main condenser carries a higher vacuum than the inter-condenser, it is necessary that some form of seal be maintained in this drain line to prevent equalization of vacuum. A water level in the U-shaped loop seal line provides this barrier.

Figure 7–6 shows a typical expanded schematic arrangement of the condensate and air ejector systems in most steam propulsion ships. Note that the condensed air ejector actuating steam from the first-stage air ejectors is returned to the condenser in the loop seal. The drains from the after-condenser and gland exhaust condenser enter the fresh water drain collecting tank and join the condensate from the condenser prior to entering the DFT, or, as an alternative, flow back into the main condenser. The latter method is not as efficient because the thermal energy that is given up by the drains to the condenser cooling water must be replaced in the boiler by burning more fuel.

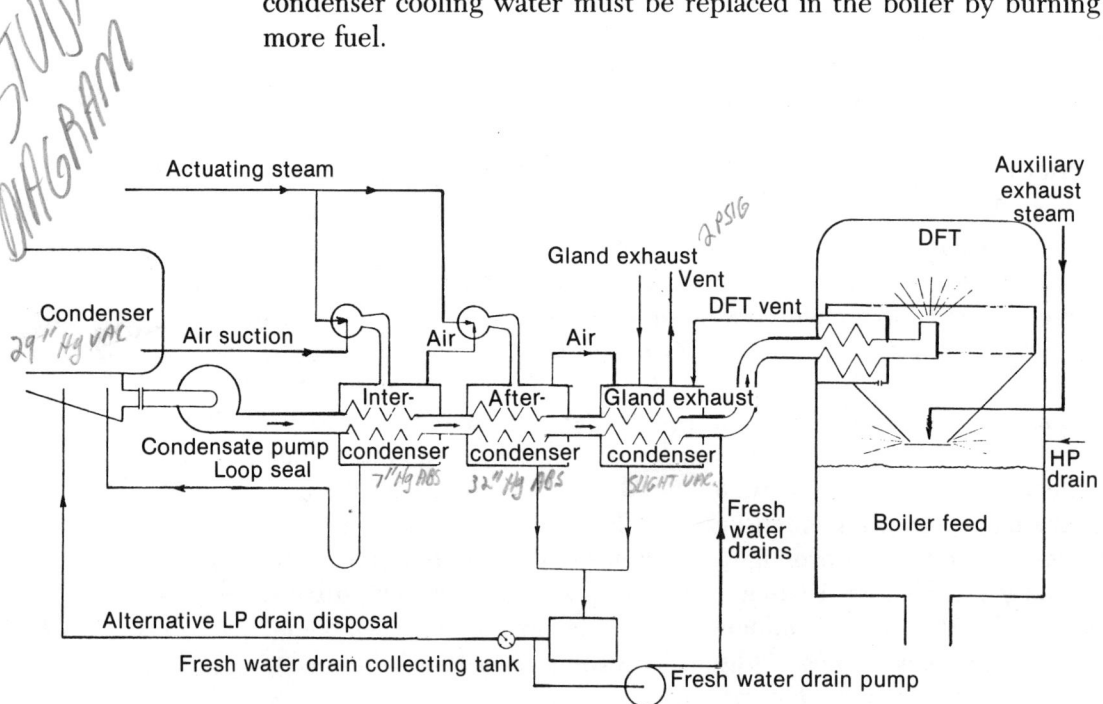

Figure 7–6. A typical schematic arrangement of the condensate and air ejector systems.

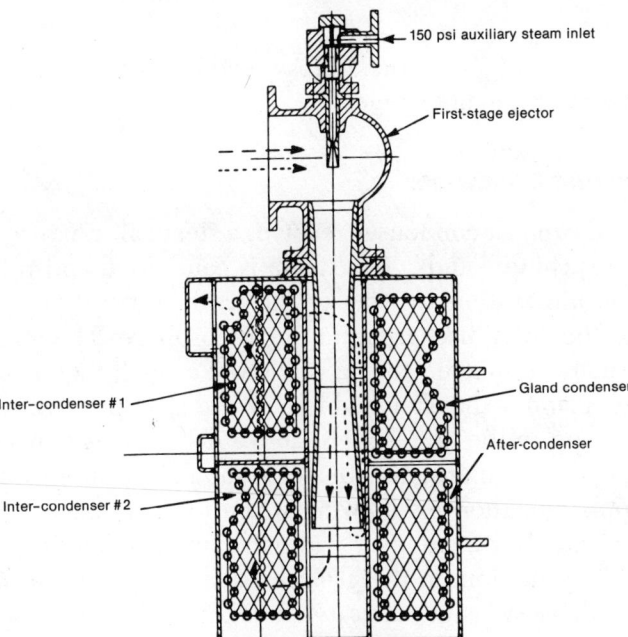

Figure 7–7. Cross-sectional view of the first stage main air ejector and condenser.

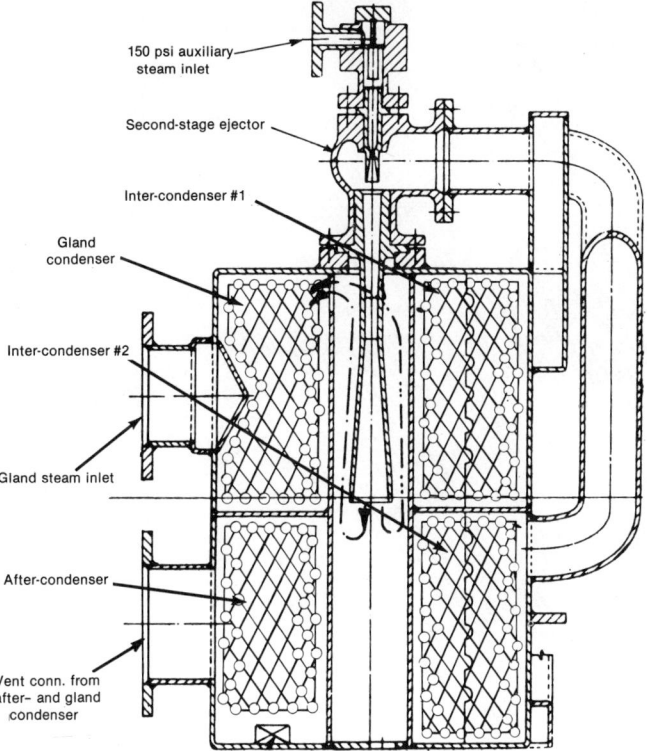

Figure 7–8. Cross-sectional view of the second stage main air ejector and condenser.

Main Condensate System

Figures 7–7 and 7–8 show a cross-sectional view of the first and second stages of a main air ejector.

Gland Exhaust Condenser

The gland exhaust condenser receives a steam/air mixture from the propulsion turbine glands. The steam is condensed and returned to the condensate system by way of the fresh water drain collecting tanks, and the air is discharged to the atmosphere. The atmospheric vent is usually connected to the suction of a small motor-driven fan called the gland exhauster, which provides a positive discharge through piping to the atmosphere above decks. This is necessary to avoid filling the engineroom with steam should the air ejector cooling water supply fail, thereby allowing steam to pass through the inter-condenser and after-condenser without being condensed.

The cooling medium in the gland exhaust condenser, as in the air ejector condensers, is condensate from the main condenser, on its way to the DFT.

8

Main Feed System

The DFT (deaerating feed tank, figure 8–1) is essentially a cylindrical tank having three sections. The lowest section is the storage space for the heated and deaerated feedwater. (Note: The storage section of the DFT is the beginning of the feed phase of the steam cycle.)

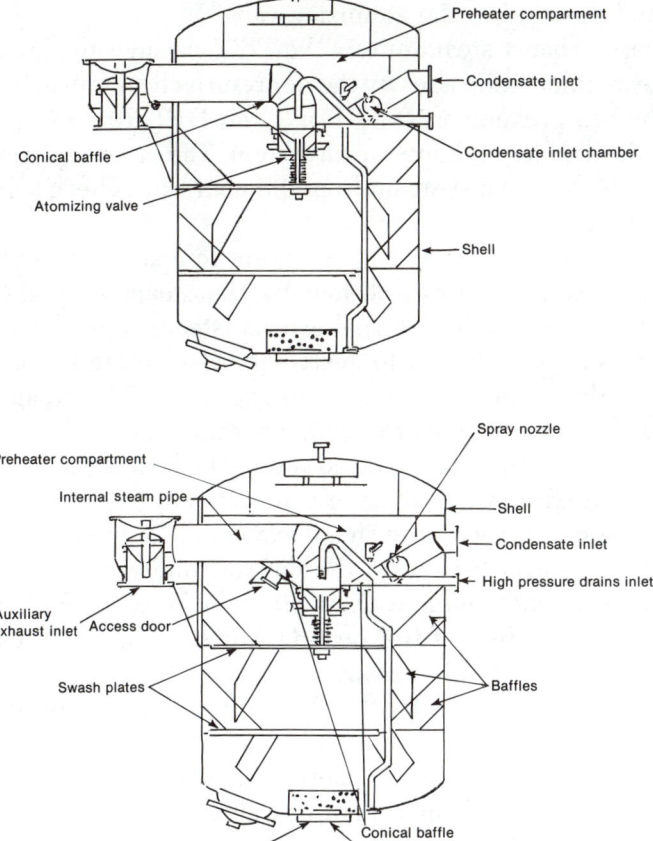

Figure 8–1. Internal components of a typical deaerating feed tank.

The middle section of the DFT contains the atomizing valve assembly, which serves to release entrapped air from the water in the final deaeration process.

The upper section of the DFT is essentially a preheater compartment, which contains the direct contact vent condenser, the condensate inlet chamber, spray nozzles, and various plates and baffles.

Condensate enters the DFT through the condensate inlet connection and then is directed to the condensate inlet chamber, which is a tube fabricated to form a manifold. Connections are evenly spaced around the circumference to accommodate eight spray nozzles.

The eight spray nozzles are threaded into the condensate inlet chamber. Condensate passing through the nozzles is broken up into a foglike spray and directed to the preheating compartment, which is integral with the upper section of the DFT shell and is accessible through the internal access door located in the conical water baffle.

The conical baffle is a collection point where condensate and the condensed steam vapors from the preheater compartment are collected and directed to the atomizing valve.

Auxiliary exhaust steam enters the DFT through the auxiliary exhaust steam inlet and passes by the nonreturn check valve. This valve is installed to prevent pressure inside the DFT from escaping back out into the auxiliary exhaust steam system. The internal steam supply pipe directs the exhaust steam from the nonreturn check valve to the atomizing valve.

After condensate leaves the preheating compartment, final deaeration and heating is accomplished by the steam atomizing valve, which is attached to the internal exhaust steam supply pipe. When steam pressure is sufficient to overcome valve spring tension, a disc will move down, opening the atomizing valve. The steam will be discharged at high velocity, carrying the condensate with it and spraying the mixture against an inverted cone, which changes its direction.

Condensate from the preheater compartment flows down the conical baffle and mixes with the steam leaving the atomizing valve. The high-velocity steam breaks up the flow of condensate into a foglike spray of steam and water, causing air to be released from the condensate by the action of the heating and scrubbing process of the steam leaving the atomizing valve.

High pressure drains enter the DFT and are directed by an internal pipe into the internal steam supply pipe just above the atomizing valve. The end of the pipe is closed off, and the side is slotted to prevent the flashing steam from discharging directly on the atomizing valve.

The heated and deaerated water, now called feedwater, is stored in the storage section of the DFT ready for use in the feed system. Feedwater is drawn into the main feed booster pump through the pump supply connection.

A protective screen has been fabricated to cover the pump supply connection. This screen prevents any foreign matter from entering the main feed booster pump. Swash plates and baffles are provided to keep to a minimum the change in water level caused by the roll of the ship.

Several openings are provided in the DFT shell (figure 8-2) for steam and water connections, access openings, vents, and safety and control devices. This shell is essentially a cylindrical tank constructed to accommodate both internal and external connections and components. The tank supports are integral with the shell and support the tank on its foundation.

The DFT is fitted with various external safety and control devices. A relief valve is installed to relieve excessive pressure in the shell of the DFT, caused by the introduction of high pressure drains. The spring loaded relief valve is attached directly to the shell. When the pressure inside the shell becomes greater than 30 psig, the valve lifts, relieving the excess pressure.

A vacuum breaker, located on the upper section of the DFT, serves to prevent the occurrence of subatmospheric pressure inside the shell of the DFT. A vacuum could occur in the DFT, if during normal

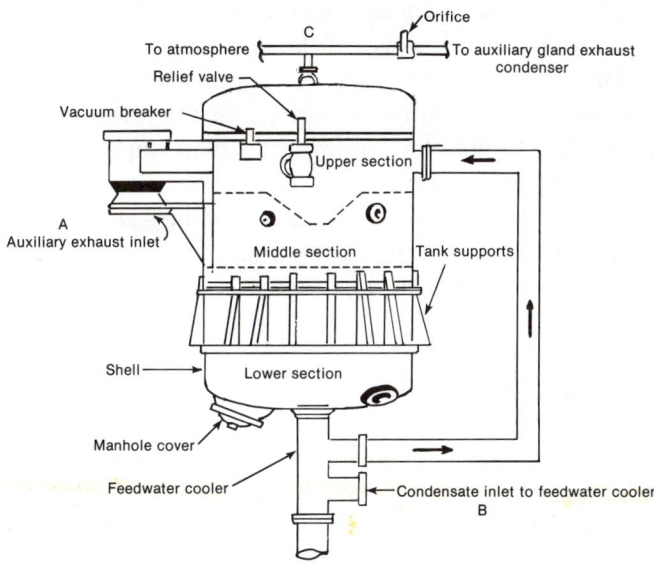

Figure 8-2. *External components of a typical deaerating feed tank.*

Main Feed System 115

operation under positive pressure, the steam supply were suddenly cut off or reduced to a point where the heating of the incoming condensate was insufficient to maintain at least atmospheric pressure (14.7 psia). If the pressure drops to approximately 13 psia, the vacuum breaker will open and admit air into the shell of the DFT.

The vacuum breaker consists of a body and an adapter piece assembly. The valve will open when the pressure on the underside of the disc is approximately 2 psi less than the pressure on its upper side.

Piping connections to steam inlet, condensate inlet, pump supply, high pressure drains, vents, and recirculating and level controls are made up as shown in figure 8–2. Refer to the figure as each pipe connection is discussed.

Auxiliary exhaust steam enters the DFT through connection A and passes through the nonreturn check valve.

Feedwater leaving the DFT passes through the feedwater cooler. The feedwater cooler lowers the temperature of the feedwater to prevent it from flashing into steam in the eye of the main feed booster pump. Condensate is the cooling agent that lowers the temperature of the feedwater. The condensate leaves the feedwater cooler and goes directly to the condensate inlet of the DFT.

Condensate enters the DFT through connection B, the condensate inlet.

Connection C is the vent connection. The DFT may be vented to either the atmosphere or the auxiliary gland exhaust condenser. When the DFT is being warmed, all air is released to the atmosphere. Normal operational conditions call for all air to be expelled to the auxiliary gland exhaust condenser.

The vent line to the auxiliary gland exhaust condenser contains an orifice plate that limits the volume of steam vapors permitted to escape from the DFT.

High pressure drains are introduced into the DFT to ensure that a sufficient supply of steam pressure is available to open the atomizing valve, and thus, continue the heating and deaerating process of the condensate.

Main Feed System

The boiler feed system includes the feed booster pump, the main feed pump, and the piping required to carry water from the deaerating feed tank to the boilers. The deaerated feedwater from the deaerating feed tank is pumped to the boiler by the feed booster pump and the main feed pump. The feed booster pump provides a positive suction pressure for the main feed pump and thus prevents the hot water from flashing into steam at the main feed pump suction.

It is debatable whether the deaerating feed tank is part of the condensate system or part of the boiler feed system, since the tank is generally taken as the dividing line between the two systems. The water is called condensate between the condenser and the deaerating feed tank. It is called feedwater or boiler feed between the deaerating feed tank and the economizer of the boiler.

Four main types of feed systems have been used in naval ships. As boilers have been designed for higher operating pressures and temperatures, the removal of dissolved oxygen from the feedwater has become increasingly important, because the higher pressures and temperatures accelerate the corrosive effects of dissolved oxygen. Each new type of feed system has represented an improvement over the one before it in reducing the amount of oxygen dissolved or suspended in the feedwater.

Practically all modern naval ships have pressure-closed feed systems, and this is the only type discussed here. Pressure-closed systems are used on board all naval ships having boilers operating at 600 psi and above. In a pressure-closed system, all feed lines throughout the system are under positive pressure, and the system is closed to prevent the entrance of air.

Our primary interest in this system is the pumps and their means of operation. Remember that the feedwater in the DFT has been heated very close to its saturation temperature corresponding to the DFT pressure. If a slight reduction in pressure should occur, the feedwater would flash to steam. If this should occur in the feed pumps, they would become "vapor-locked." To prevent such an occurrence, the first pump in the line is located physically below the DFT in order to ensure that a positive pressure head is always provided to the pump. This first pump is called the main feed booster pump (MFBP). This pump is a relatively small pump designed to operate at a low speed in order to reduce the possibility of developing a pressure less than saturation pressure at its suction, and the resultant "vapor lock."

Figure 8–3 depicts a recirculating line going from the discharge side of the MFBP back to the DFT. This line ensures that there will always be fluid flowing through the pump when it is in operation, regardless of the feedwater demand by the boiler. Continuous flow through the pump is necessary to prevent pump overheating and subsequent damage. The recirculating line is designed to permit about 5 percent of rated pump capacity to flow while the pump is in operation.

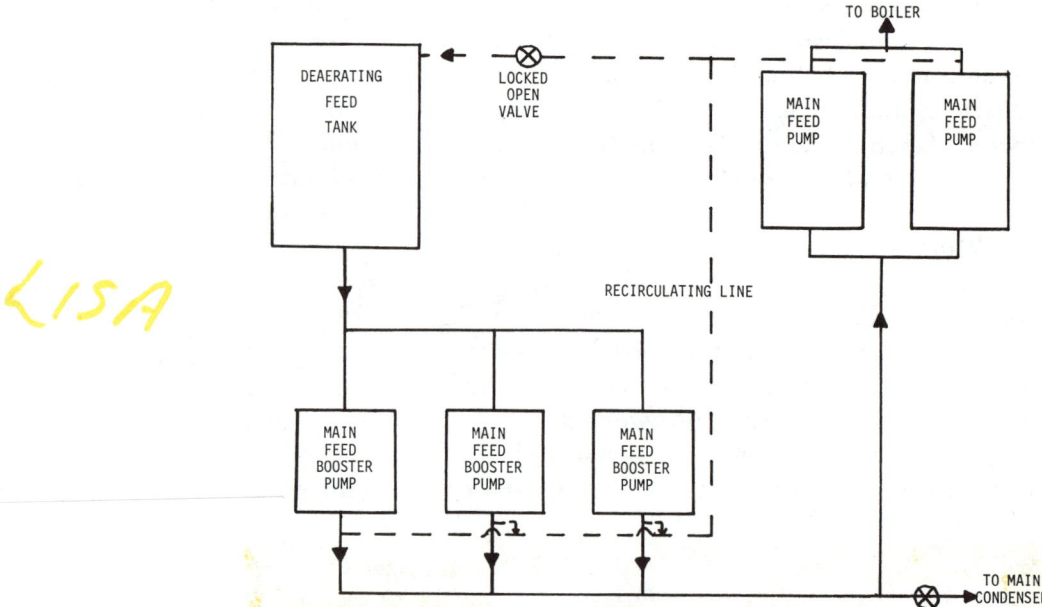

Figure 8–3. *Schematic diagram of a main feed system.*

Referring to figure 8–3 again, notice that the MFBP discharges to the suction side of the main feed pump (MFP), which is a two-stage steam-turbine-driven pump. The main feed pump discharges to the boiler at a pressure 20–25 percent above the boiler pressure. Recirculation for the MFPs is also required, for the reasons stated above.

A typical shipboard installation consists of three MFBPs and three MFPs for each propulsion plant (two boilers, one turbine, and its auxiliaries). All of the MFBPs are electric-motor-driven. The total capacity of the main feed pumps must be 150 percent of the boiler requirement at full power plus the capacity required for recirculation. Each MFBP should be sized for the maximum capacity of the MFP it serves and should include sufficient margin for internal losses and recirculation.

Practically all naval ships have more than one feedwater system, with cross-connecting lines and valves arranged so that the systems may be operated either split-plant or cross-connected. In warming up or securing a plant, it is often necessary to transfer feedwater from one plant to another. For example, it might be necessary to transfer the feedwater so that one plant will not have to take on cold makeup feed while another plant is discharging hot excess feed.

In recent years, there has been an ever increasing need to cut costs. On board ship, fresh water for use by the crew and feedwater to be generated into steam to drive the ship's turbines and other machinery and equipment are costly items. So the development of systems to recover steam used by that machinery and equipment has become necessary, involving converting the steam back into water and returning it to the steam cycle through various systems. The fresh water drain collection system is such a system.

The fresh water drain collecting system (figure 8–4) collects condensate from low-pressure steam systems and steam-driven machinery in the fireroom and engineroom, which requires continuous or intermittent drainage.

The steam drains are collected from various systems throughout the engineering plant by means of funnel drains with swing-check valves installed immediately downstream, which prevent drains from backing up through the funnel and being lost to the bilges. At this point, the fresh water drains enter the fresh water drain main.

Once the fresh water drains enter the funnel, the drains travel through a network of piping by gravity flow to the fresh water drain collecting (FWDC) tank.

The purpose of the fresh water drain collecting tank is to provide a common collection point for fresh water drains.

The FWDC tank is constructed of steel, has a capacity of 170 gallons, and can be vented to two different places. Normally, the

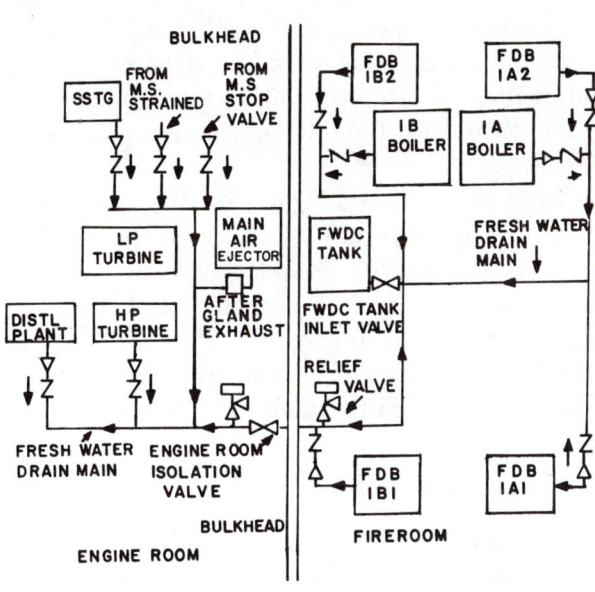

Figure 8–4. Schematic diagram of a fresh water drain collecting system.

Main Feed System

FWDC tank is vented to the auxiliary gland leak-off condenser; or it can be vented to the atmosphere. The FWDC tank is also equipped with an overflow to the contaminated drain system.

Fresh water is normally removed from the FWDC tank by use of the FWDC pumps (1A and 1B).

The FWDC pump discharge regulating valve is an air pilot actuated regulator. This valve controls the fresh water drain discharge going to the main condensate system.

The air pilot is operated by control air and is actuated through the pressure differential of the constant-pressure leg and the variable-pressure leg. The regulated control air discharged from the air pilot actuates the FWDC pump discharge regulating valve.

The fresh water drains leaving the FWDC tank through the vacuum drag line are controlled by the vacuum drag line float control valve. This valve will close when the water in the FWDC tank decreases to a predetermined level. If the FWDC tank were allowed to go dry, it would cause a loss of vacuum in the main condenser.

The vacuum drag line cutout valve is located in the engineroom where the vacuum drag line enters the main condenser. This valve will be opened *only* when the fresh water drain tank pumps are incapable of maintaining the DFT water level.

The water level inside the FWDC tank is maintained by the FWDCT pump and automatic discharge regulating valve.

There are two electric FWDCT pumps (1A and 1B) (figure 8–5). During normal operation, one pump will be in the constant-run position on the motor controller, and the other will be in the standby position. When the pump that is on constant run is operating by itself, the automatic discharge regulating valve opens and closes automatically, regulating the FWDC tank between its minimum and maximum levels.

When the level in the tank is 6 inches, the automatic discharge regulating valve is fully closed. The reason for this is to prevent the pump, which is on constant run, from losing suction and overheating. When the level inside the tank is 12 inches or above, the automatic discharge regulating valve is fully open. When the water level inside the FWDC tank rises to 21 inches, the second pump automatically cuts on to lower the level in the tank to 12 inches.

Makeup and Excess Feed System

During normal operation of a steam propulsion plant, the need to add water to or remove it from the steam cycle is always present. For this purpose, there is a makeup and excess feed system (figure

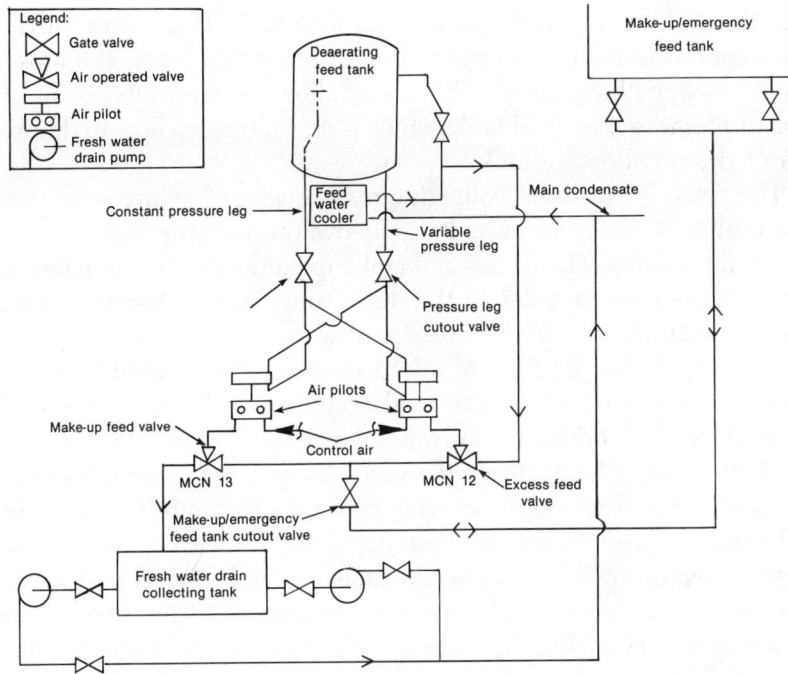

Figure 8–5. Schematic diagram of a makeup/excess feed system.

8–5), which automatically adds or removes water. This is accomplished by maintaining a specific level in the deaerating feed tank.

The makeup and excess feed system starts at the makeup and emergency feed tanks. These tanks serve as a storage area for feedwater until it is needed to replace feedwater losses in the steam cycle. During normal operation, one of the tanks is lined up to supply makeup feedwater when needed. This tank also receives the excess feedwater from the steam cycle. At the same time, the other tank is in "ready standby" for immediate use.

Feedwater for use in the makeup feed system leaves the makeup/emergency feed tanks by gravity flow to the makeup feed valve. At each tank, there is an isolation valve which allows for only one tank at a time to be on suction. Located in the piping system between the tank suction valves and the makeup feed valve is the makeup/emergency feed tank cutout valve. This valve must be open to allow feedwater to flow from the makeup/emergency feed tank, which is on suction, to the makeup feed valve.

The deaerating feed tank water level control system automatically positions the makeup and excess feed valves in the system. The makeup and excess feed valves must work together to properly maintain the water level in the deaerating feed tank. If the water level in the deaerating feed tank approaches its minimum level, the makeup feed valve will automatically open, allowing feedwater from the makeup/emergency feed tank, which is on suction, to flow to the fresh water drain collecting tank.

The fresh water drain collecting tank receives low pressure drains from other steam systems and also feedwater from the makeup/emergency feed tanks. The drains and makeup feedwater are then pumped into the condensate inlet to the deaerating feed tank where the increased water level can be visually observed.

Coming off the condensate inlet to the deaerating feed tank is the excess feed valve. This valve is also controlled by the level in the deaerating feed tank. As the water level approaches the maximum limit, the excess feed valve will open, allowing the excess condensate to be diverted from the deaerating feed tank through the excess feed valve back to the makeup/emergency feed tank that is on suction.

In order to open and close the makeup and excess feed valve, a control system is used. This system has a constant-pressure leg and a variable-pressure leg. In order to isolate this system from the deaerating feed tank for maintenance, a cutout valve is installed in each leg, and each leg is connected to an air pilot. The air pilots measure the difference in weight of the constant- and variable-pressure legs (two columns of water) and send out an air signal proportional to the water level in the deaerating feed tank to the makeup and excess feed valves.

When the water level in the deaerating feed tank approaches minimum level, the drop in water level will be sensed at the air pilots. At this time, the air pilots will send out a signal to open the makeup feed valve, allowing more feedwater to enter the steam cycle. At the same time, the excess feed valve is kept closed by an air signal. When the water level approaches its maximum level, the air pilots will send out a signal to open the excess feed valve and allow water to be taken away from the steam cycle and also keep the makeup feed valve closed. The makeup and excess feed valves must work together in order to maintain the proper level in the deaerating feed tank and have an efficient steam cycle.

The Automatic Boiler Control System

Man is constantly striving to find safer and easier ways of accomplishing tasks. For the boiler technician, the Navy has made vast improvements in making his job safer, easier, and more comfortable since the days of coal-fired boilers. One such significant improvement has been the introduction of automatic boiler control systems (ABCs) for most of the Navy's combat fleet.

The term automatic boiler control system is a general term used to describe all fireroom pneumatic (air) control systems. There are three ABC subsystems: automatic combustion control (ACC), feedwater regulator control (FWC), and the main feed pump differential control subsystems.

In order for any control system to be completely automatic, whether it be pneumatically operated for controlling the boilers of a ship, or a simple electrical system for controlling home heating, it must be able to perform the four functions of a closed loop feedback system: *measure, compare, compute,* and *correct*.

In a home heating system, these four functions are accomplished by the thermostat. The thermostat is constantly measuring the surrounding air temperature; it is comparing that temperature to the desired temperature setting; it is computing whether or not a change in operating condition of the furnace is necessary; and, finally, the system will make the required correction by starting or stopping the furnace when needed.

The home-heating-system example is an automatic control system in the simplest form; however, both the ABC system and the home heating system must measure, compare, compute, and correct operating parameters for the system that is being controlled.

The purpose of the boiler automatic combustion control (ACC) system, as shown in figure 8–6, is to maintain the boiler steam drum pressure at a constant 1275 psig during steady and changing steam conditions. The ACC system accomplishes this task by:

1. Constantly measuring the steam drum pressure and combustion air flow. (See air flow and steam drum pressure sensing lines in the drawing.)
2. Comparing steam drum pressure to the desired pressure (1275 psi).
3. Computing the amount of change, if any, in furnace combustion.
4. Correcting furnace combustion as needed.

In the home-heating-system example, the control system turned the furnace on or off as needed. This on/off type of control is impractical for propulsion boilers. Rather than turning the boiler on and off, the automatic combustion control system regulates furnace combustion by proportionally increasing or decreasing energy input (com-

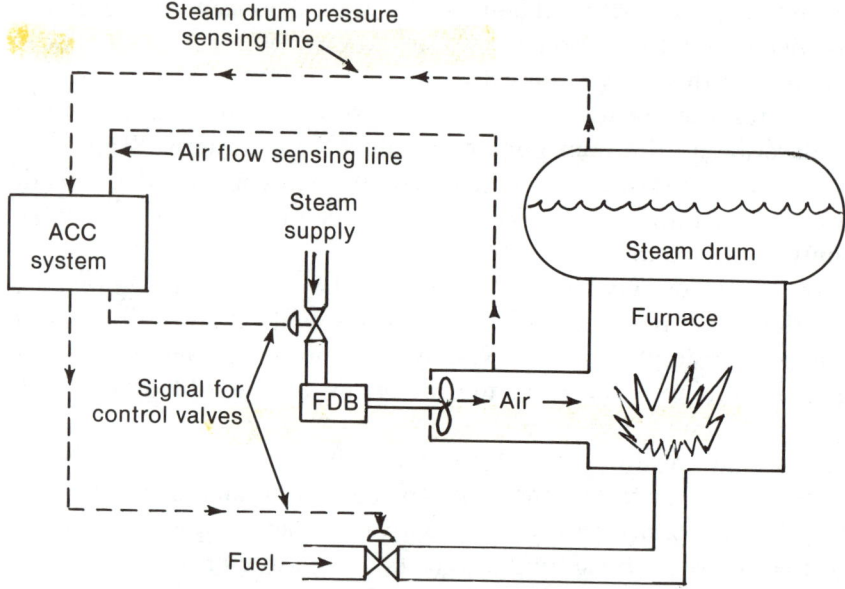

Figure 8–6. *Schematic diagram of a boiler automatic combustion control system.*

bustion), depending upon the direction of change in energy output (steam), where the input/output relationship is:

$$\text{Btu's input (fuel)} = \text{Btu's output (steam)}$$

Note: The above formula is true only in theory and not in actual practice because boilers are not 100 percent efficient. In actual practice, the formula would read:

$$\text{Btu's input} - \text{Btu's loss} = \text{Btu's output}$$

When the steam demand on a boiler is increased, the steam drum pressure in the boiler tends to decrease. In measuring the steam drum pressure, the ACC system senses this decreasing pressure and responds by ordering an increase in furnace combustion to return the boiler to normal operating pressure.

The mixture of fuel and air in a boiler furnace must be in the proper proportions to assure complete combustion. (Too much air—the fire goes out; too much fuel—black smoke.) The function of the ACC system is to regulate the amount of fuel and air being supplied to the furnace, for all normal conditions of boiler load, or steaming rate. In addition to steam drum pressure, the ACC system needs information about boiler operating conditions; therefore, the system is designed to measure combustion air flow as well.

Just how does the ACC system control fuel flow and combustion air flow? The ACC system accomplishes this in the same way you would if you were manually controlling the boiler; that is, by positioning valves to produce the desired results.

Combustion air flow is controlled by positioning a steam supply valve on the forced draft blower turbine, which, in turn, regulates the speed of the blower fan. The faster the turbine is operated, the greater the air flow will be to the boiler. Fuel flow is controlled by positioning a fuel supply valve to the burners. Although these pneumatically operated control valves differ in design from the manually operated handwheel-type valve, they serve the same purpose.

Although one of the functions of the automatic combustion control system is to maintain fuel flow and air flow in the correct proportions at all times, the control of fuel and air flow by the ACC system isn't quite as simple as it may first appear because, during a change in boiler steam demand, the initial inertia of the forced draft blower turbine and fan must be overcome before a change in air flow is achieved. On the other hand, the reaction of the fuel control valve is instantaneous. During an increase in boiler steam demand, because of the delayed response of the force draft blower, the ACC system must provide a means to hold back or delay the opening of the fuel control valve until air flow has had a chance to catch up. The measurement of air flow provides the automatic combustion control system with the needed feedback to perform this function. This feedback is used to assure that enough combustion air is available during all *changing* boiler conditions, and the automatic combustion control system provides this function by:

1. Ensuring that air flow precedes fuel flow during any increase in boiler steam demand (figure 8–7).
2. Ensuring that fuel flow precedes air flow during any decrease in boiler steam demand (figure 8–8).

Earlier, you learned that the boiler water level must be maintained near the center of the steam drum. Boiler water level is another

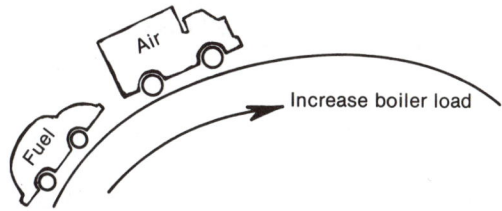

Figure 8–7. Air flow precedes fuel flow.

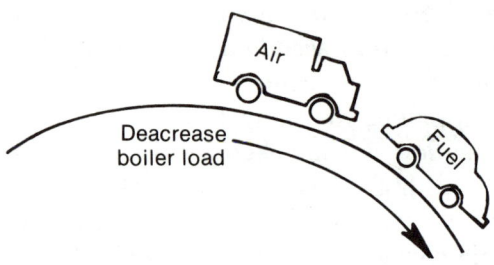

Figure 8–8. Fuel flow precedes air flow.

condition that is automatically controlled on a 1200 psi boiler. The feedwater regulator control system (FWC) automatically does this by:

1. Measuring the feedwater flow to the boiler, the steam flow from the boiler, and the steam drum water level.
2. Comparing the measured steam drum water level to the desired water level (set point), and comparing the feedwater flow into the boiler with the steam flow out of the boiler.
3. Computing the required change, if any, to the rate of feedwater flow.
4. Correcting the feedwater flow rate as required.

In the feedwater regulator control system, the meaurement of steam flow out of the boiler and feedwater flow into the boiler is just as important as the measurement of the steam drum water level, if not more so (figure 8–9). Assume that a boiler is steaming at a steady

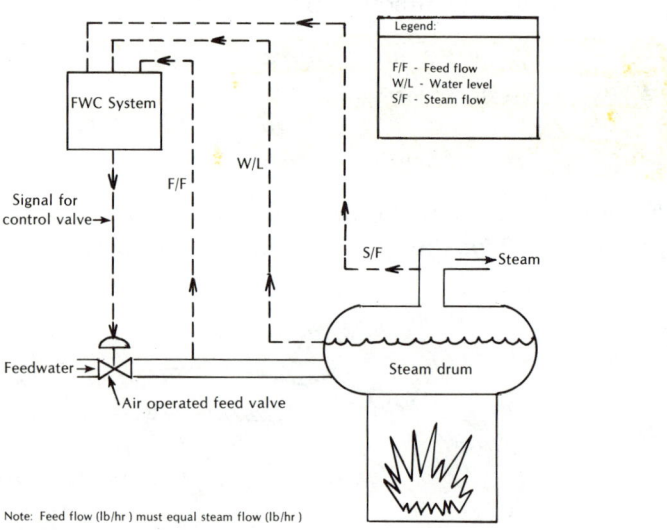

Figure 8–9. Feedwater control system.

rate and the water level in the steam drum is normal. Further assume that the FWC system is adding feedwater to the boiler at a rate of 10,000 lb/hr. From these assumptions, you can conclude that the steam must be leaving the boiler at a rate of 10,000 lb/hr.

Any time there is a change in the rate of steam leaving a boiler, the FWC system immediately senses the change and increases or decreases the rate of feedwater flow to equal the steam flow. The FWC system does this without waiting for an actual change in the steam drum water level. Because of this immediate reaction by the feedwater regulator control system, response time is greatly reduced. At the same time, the part of the system that measures steam drum water level makes minor corrections to the feed flow rate to assure a normal water level.

The FWC system must also compensate for boiler phenomena commonly referred to as shrink and swell. Shrink, or swell, is the descending, or rising, of the steam drum water level when a change in boiler firing rate occurs. For example, when the steam demand on a boiler is increased, say from 10 to 60 percent the increased firing rate causes an increase in the activity of the boiling water. This increased activity results in many more steam bubbles in the boiler water so that the water level in the steam drum rises even though feed flow has not changed! This phenomenon is called swell. (See figure 8–10.) (Note: The bubbles also get somewhat larger because of the slight decrease in steam drum pressure.) Shrink is the reverse of swell. (See figure 8–11.)

Earlier in this chapter, you learned that the feedwater regulator control system responds to an increase in boiler steam demand by immediately increasing the feedwater flow. However, as the boiler water level begins to rise above normal (because of swell), the "supervisory signal" from the measured steam drum water level responds by slightly decreasing feedwater flow, thus compensating for the

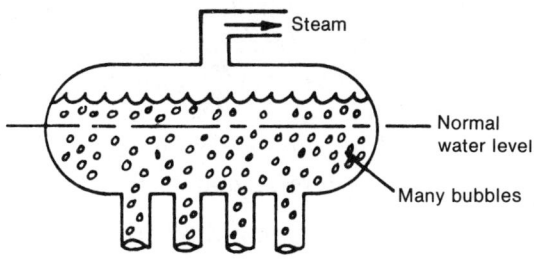

Figure 8–10. The swell phenomenon.

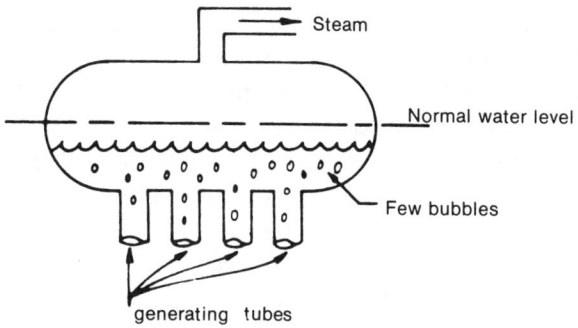

Decreasing to a low firing rate

Note: Both figures 8-10 and 8-11 have the same volume of water.

Figure 8–11. The shrink phenomenon.

swell. As the water level approaches normal, the feedwater flow is gradually increased, so that, when the water level is normal, both the steam flow from the boiler and the feedwater flow to the boiler are again equal. To compensate for shrink, this action by the FWC system is reversed.

The actual control of feedwater flow to the boiler is accomplished by an air-operated diaphragm control valve located in the main feed piping system, between the main feed pump and the boiler. This valve is normally referred to as the automatic feedwater check valve.

The feedwater regulator control system used in all 1200 psi ships is often referred to as a three-element feed system because it measures three different parameters to control the steam drum water level in the boiler.

To ensure satisfactory operation of the feedwater regulator control system, a constant boiler feedwater supply pressure is necessary. The main feed pump differential control system automatically provides this service by maintaining the discharge pressure of the main feed pump at a constant 75 psi above boiler steam drum pressure. This means that any time the boiler steam drum pressure varies above or below its normal 1275 psig operating pressure, the difference (differential) between main feed pump discharge pressure and boiler steam drum pressure will remain unchanged (figure 8–12).

The differential control system automatically does this by:

1. Measuring the main feed pump discharge pressure and the steam drum pressure;
2. Computing the difference between the two pressures and comparing it to the desired differential of 75 psi;

3. Computing the required air signal necessary to change the speed of the main feed pump;

4. Correcting the main feed pump speed by changing the steam flow to the pump turbine, causing a proportional change in the pump's discharge pressure.

Any one of the ABC subsystems may be controlled using one of three different modes of control: local manual, remote manual, or automatic.

Previously in this chapter, you learned that the ABC subsystems "effect control" by positioning air-operated control valves to produce the desired direction of change of air, fuel, and water when the load

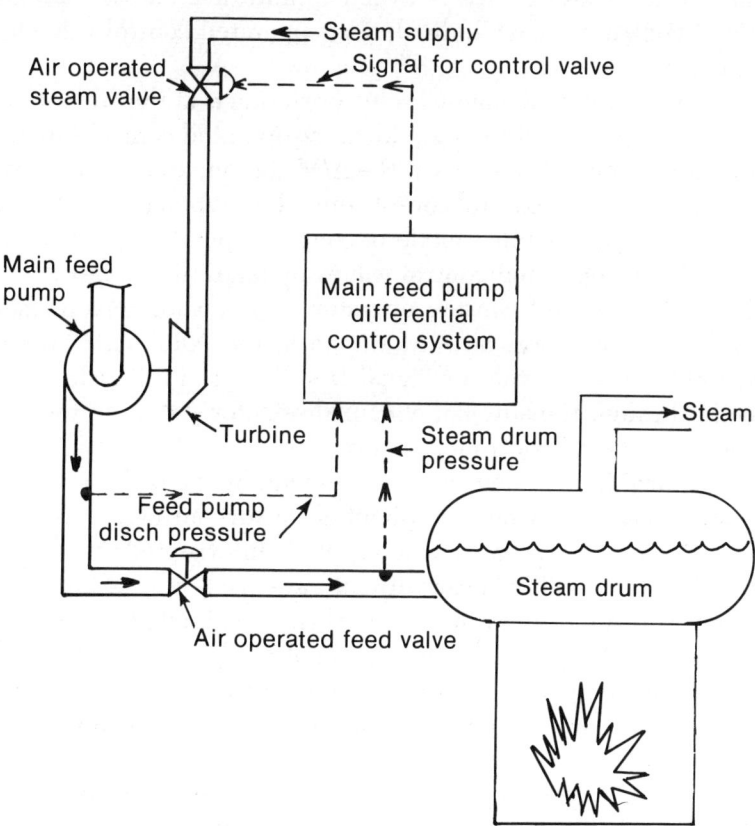

Figure 8–12. Main feed pump differential control system.

on a boiler changes. However, if for any reason a subsystem cannot be controlled using air, it must be locally controlled by hand. When this type of manual control is used, the system is said to be in "local manual mode of control."

An example of local manual control would occur when the boiler operators are directly controlling the feed flow rate to maintain the steam drum water level at normal using the manual feed check valve.

Local manual control of any of the ABC subsystems is normally considered to be an emergency condition resulting from control system breakdown or failure of the control air system and related backup air systems. The only exception occurs during initial machinery lineup when individual machinery components may be briefly placed in local manual mode of control.

Of all the machinery components in the fireroom of a 1200 psi ship, you will probably be most impressed by the boiler control console. This console, located in the boiler control booth, consists of numerous automatic/manual selector stations (A/M stations) that allow the fireroom operators the choice between remote manual and automatic mode of control for each individual air-operated control valve in the ABC system.

When the automatic/manual selector station is in the "remote manual" position for any of the individual air-operated control valves, the "automatic" signal dead-ends at the A/M station, and a new air signal is introduced by the control console operator. When the A/M stations are in the remote manual mode of control, operating personnel can control the air-operated control valves by hand (figure 8–13).

The boiler control console operator relinquishes remote manual control when he places an A/M station in the "automatic" position. When this is done, the air signal from the control system passes directly through the automatic/manual selector station to the air-operated valve being controlled (figure 8–14).

To perform its function properly, the automatic boiler control system must have a constant supply of clean, dry air of at least 60 psi. But what happens when this air supply is interrupted? Without air, the ABC system is like a car without gas—useless.

When control air pressure is less than 60 psi, all of the ABC subsystems must be placed in local manual mode of control. The procedure of actually shifting from automatic to local manual mode of control can take even experienced fireroom operators several minutes to accomplish.

There is an emergency feature that helps to prevent the automatic boiler control system from "running amuck" during a loss of control air casualty. This emergency feature is the air lock system. Its function

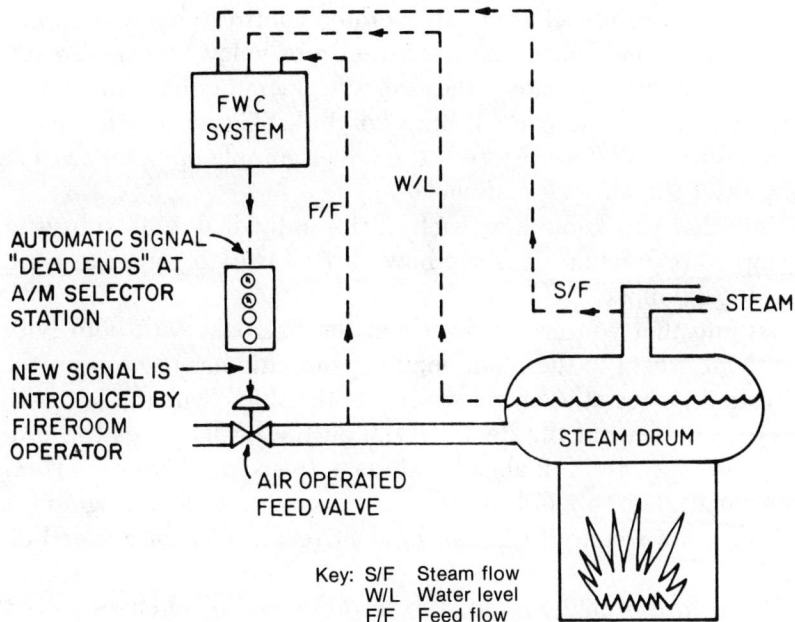

Figure 8–13. FWC system in "remote manual."

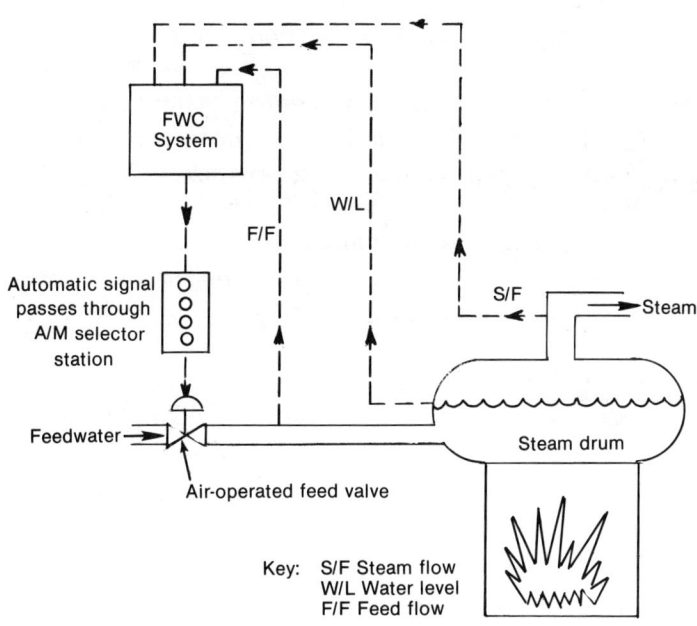

Figure 8–14. FWC system in "automatic."

Main Feed System 131

is to automatically lock each air-operated control valve in its present position any time control air pressure drops below 60 psi. When the air lock system is actuated, the fireroom operators have just enough time to achieve local manual control of the ABC system. After control air pressure has been restored, the boiler console operator can manually reset the air lock system.

Now that you know how each of the individual ABC subsystems performs its function, let's see how they all work together to control the ship's boilers.

Assume that you are on watch in the fireroom with both boilers supplying steam to the main engine. Your ship has just left port and is headed for sea at ⅔ ahead speed. As the ship clears the last harbor buoy, the officer of the deck (OOD) on the bridge gives the order, "Increase speed to full ahead and make turns for 25 knots." The lee helmsman rings up full ahead on the engine order telegraph and indicates the required number of shaft revolutions for a speed of 25 knots.

Down in the engineroom, the throttleman on watch answers the order for full ahead and opens his ahead throttle valve to allow more steam to flow into the turbines of the main engines.

As the demand for steam on the boilers is increased, the pressure in the boilers will tend to decrease. To this point, the rate of combustion hasn't changed. However, as the steam pressure in the boilers starts to decrease, the ACC system begins to measure the change. Now let's watch the automatic combustion control system swing into action.

When the automatic combustion control system senses the drop in boiler pressure, it orders an increase in furnace combustion to return the boilers to normal operating pressure.

To return the boilers to normal operating pressure, the ACC system must order an increase in the amount of fuel flow and combustion air flow. But wait—in order to maintain the fuel and air in the proper proportion, the ACC system delays the opening of the fuel control valve.

Any time an increase in furnace combustion is required, air flow must precede fuel flow to assure that enough air is available for complete combustion of the fuel. When the ACC system has corrected furnace combustion, the boiler steam pressure again returns to normal operating pressure.

When the throttleman opened his ahead throttle valve, the increase in steam flow was also sensed by the feedwater regulator control system.

As the FWC system measures the increase in steam flow from the boilers, it transmits a pneumatic signal that orders an increase in the amount of feedwater flow. Now that the feedwater flow to the boilers is equal to the steam flow leaving the boilers, the steam drum water level should be normal—right? Well, not quite—because there is one additional factor that hasn't been considered, water level swell.

When swell occurs, the measurement of the steam drum water level by the FWC system causes the control system to correct for the swell. This is achieved by slightly decreasing the feedwater flow to the boiler until the water level returns to normal. When the swell has been corrected, by returning the steam drum water level to normal, the FWC system will assume a balanced state, with feedwater flow and steam flow both being equal.

The main feed pump differential control system is also affected when the throttlemen opens his ahead throttle valve. Recall that the purpose of this system is to maintain the main feed pump discharge pressure at a constant 75 psi above the boiler steam drum pressure.

As the FWC system causes an increase in feedwater flow to the boiler, the main feed pump discharge pressure will have a tendency to decrease. The differential control system senses this drop in pump discharge pressure and responds by increasing the steam flow to the main feed pump turbine, thus maintaining the desired differential of 75 psi.

All of the ABC subsystems have now responded and corrected for the increase in boiler steam demand. The ABC system at this point is said to be "in a balanced condition," and it will remain in balance, except for minor corrections, until another change in steam demand occurs.

9

Machinery Arrangement and Plant Layout

To understand a shipboard propulsion plant, it is necessary to visualize the general configuration of the plant as a whole and to understand the physical relationships among the various units. This chapter provides general information on the distribution and arrangement of propulsion plants and on the arrangement of the major engineering piping systems that connect and serve the various units of machinery.

It is important to note that the information given in this chapter is general rather than specific. No two ships—not even sister ships—are exactly alike in their arrangement of machinery and piping. The examples given in this chapter are based on the arrangements used in various kinds of ships, large and small. These examples give some idea of the variety of arrangements that may be found on board steam-driven and gas-turbine surface ships, and they indicate the basic functions of the machinery and piping; but the examples cannot provide an exact picture of the machinery and piping on board any one ship. For detailed information concerning the arrangements in any particular ship, it is necessary to consult the ship's blueprints, various ship's manuals, and the manufacturers' technical manuals that cover the engineering equipment and piping systems installed in the ship.

Arrangement of Propulsion Machinery

The propulsion machinery on board conventional steam-driven surface ships includes (1) the propulsion boilers, (2) the propulsion turbines, (3) the condensers, (4) the reduction gears, and (5) the pumps, forced draft blowers, deaerating feed tanks, and other auxiliary machinery units that directly serve the major propulsion units. In most steam-driven surface ships other than oilers, tankers, and certain auxiliaries, the propulsion machinery is located amidships. Turbogenerators and their auxiliary condensers are usually located in the propulsion machinery spaces; other engineering equipment that is not directly associated with the operation of the major propulsion units may be located in or near the propulsion machinery spaces or in other parts of the ship, as space permits.

A word about terminology may be helpful at this point. The boilers in a propulsion plant may be identified as *propulsion* boilers (or occasionally as *main* boilers) when it is necessary to distinguish between propulsion boilers and the auxiliary boilers that are installed in some ships. The turbines are identified as *propulsion* turbines when it is necessary to distinguish between them and the many auxiliary turbines that are used in all steam-driven ships to drive pumps, forced draft blowers, and other auxiliary units. The propulsion turbines are also sometimes referred to as the *main engines,* although this usage is not considered particularly desirable. The term *propulsion unit* is correctly used to identify the combination of propulsion turbines, main reduction gears, and main condenser in any one propulsion plant; however, the term propulsion unit may also be used in a more general sense to indicate any major unit in the propulsion plant.

Each propulsion shaft has an identifying number which is based on the location of the shaft, working from starboard to port. The shaft nearest starboard side is the No. 1 shaft, the one next inboard is the No. 2 shaft, and so forth. In recent ships, the propulsion machinery that serves each shaft is given the same number as that shaft. For example, the No. 2 shaft is served by the No. 2 propulsion unit and the No. 2 boiler. Where two similar units serve one shaft, the identifying number is followed by a letter. If two boilers serve the No. 3 propulsion unit and the No. 3 shaft, for example, the boilers would be identified as No. 3A and No. 3B. Where letters are used, they are used in sequence going from starboard to port and then from forward to aft.

In older ships, the practice of identifying propulsion units by the number of the shaft they serve is slightly different. In general, each propulsion unit is numbered to correspond with the number of the shaft it serves; but the numbering of the boilers is generaly not the same as the numbering of the propulsion units and the shafts. In an older ship, for example, the No. 1 boiler and the No. 2 boiler might serve the No. 1 propulsion unit and the No. 1 shaft, while the No. 3 boiler and the No. 4 boiler would serve the No. 2 propulsion unit and the No. 2 shaft.

The functional relationships of the major propulsion units and of many auxiliaries are shown in figures 9–1 through 9–10. These illustrations do not always indicate the actual location of the machinery units; indeed, the physical location is often surprisingly different from the location that might be assumed from a figure of this type. In considering the physical arrangement of machinery, however, we must keep the functional relationships clearly in mind. The three major piping systems shown in figures 9–12 through 9–15 are the main steam system, the auxiliary steam system, and the auxiliary exhaust system; again, a functional rather than a physical relationship is indicated. The three systems are discussed in more detail later in this chapter; at this point it is only necessary to note the relationships of these vital systems to the propulsion units and auxiliaries.

The propulsion machinery spaces may be physically arranged in several ways. Some ships have firerooms, containing boilers and the

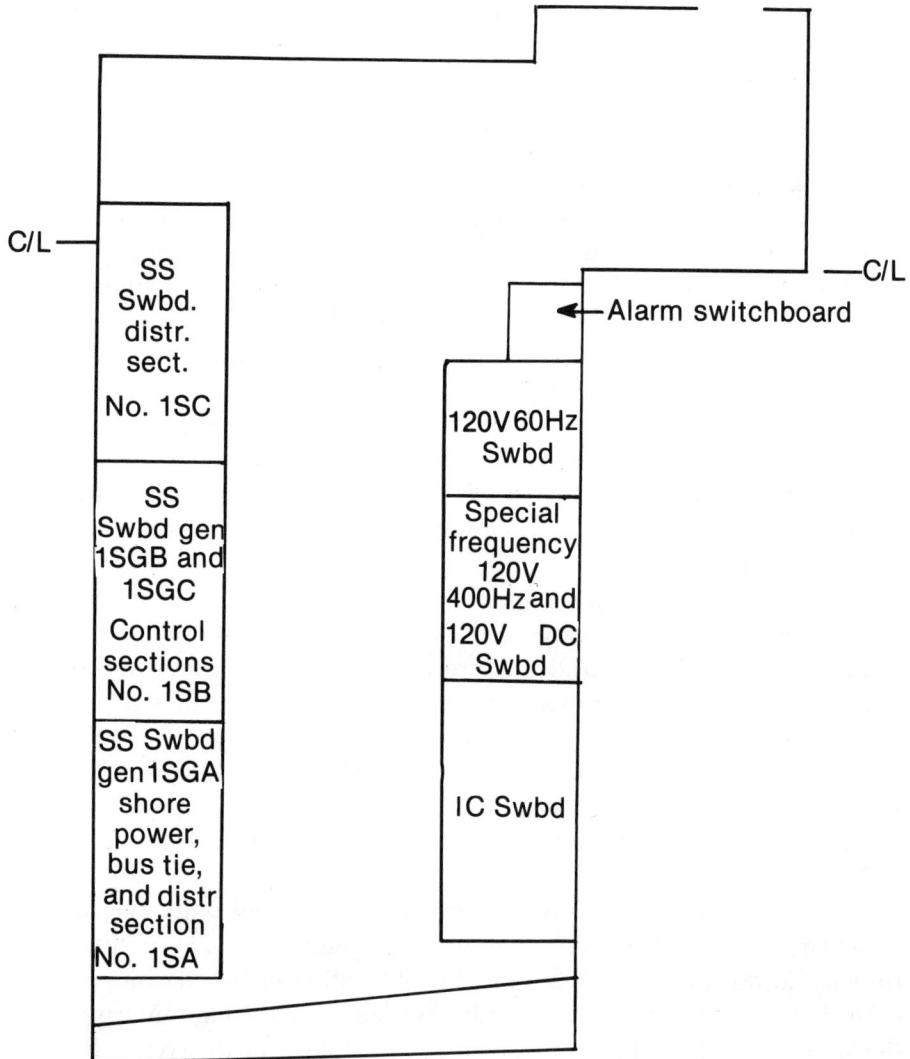

Figure 9–1. *FF 1078 machinery arrangement first platform level (electrical central).*

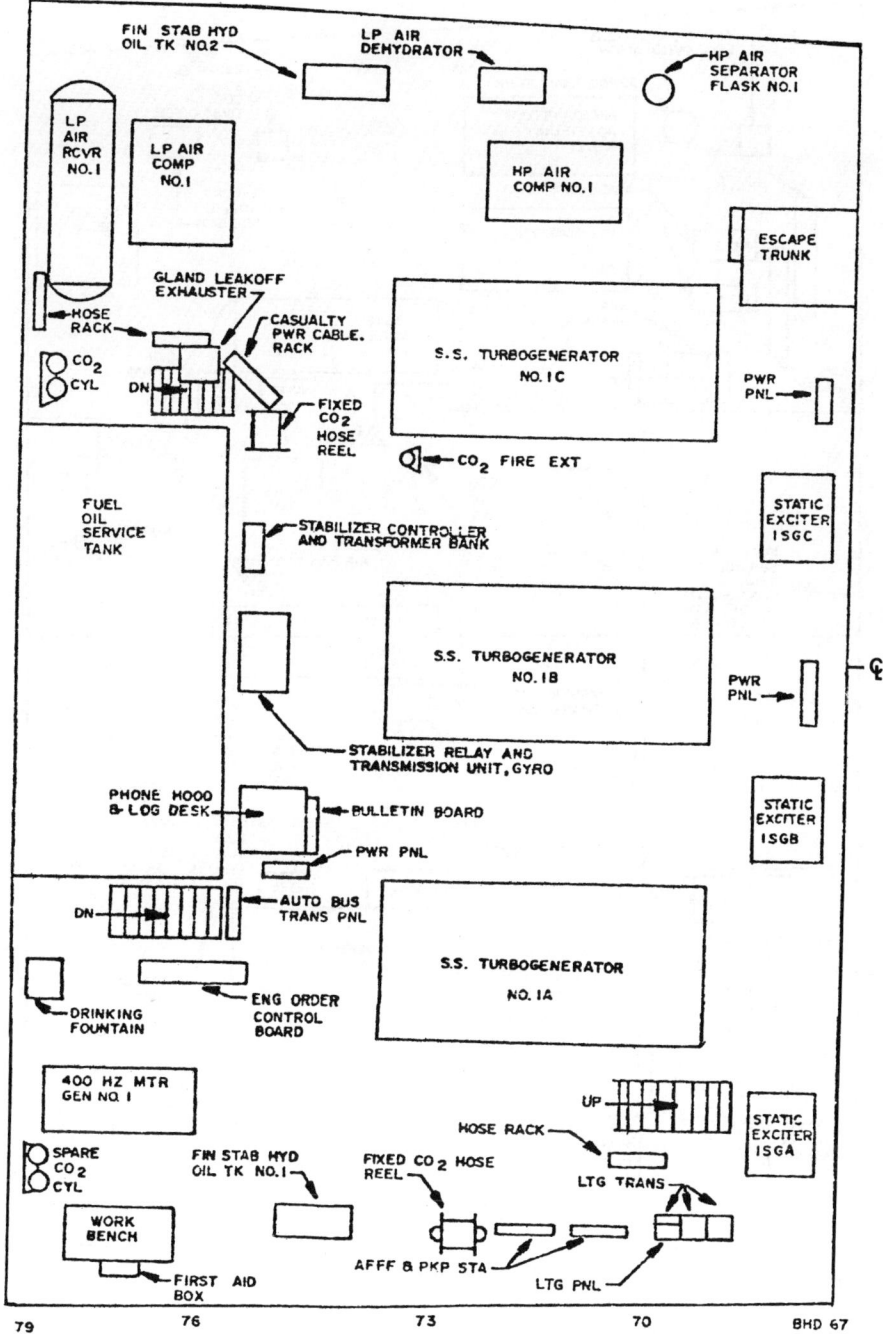

Figure 9–2. FF 1078 machinery arrangement upper level (auxiliary machinery room #1).

Machinery Arrangement and Plant Layout

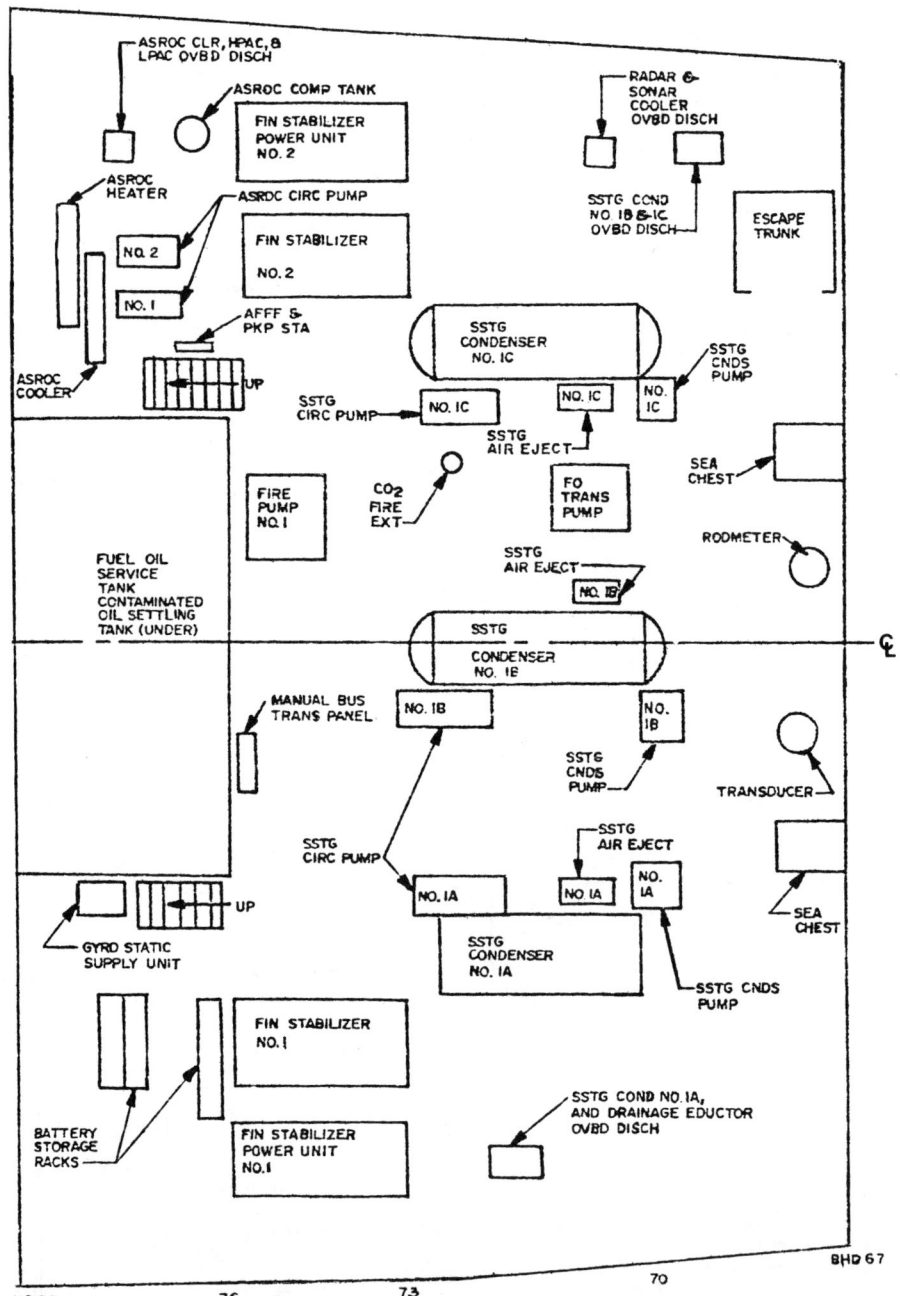

Figure 9-3. FF 1078 machinery arrangement lower level (auxiliary machinery room #1).

138   Introduction to Naval Engineering

stations for operating them, and enginerooms, containing propulsion turbines and the stations for operating them. In some ships, one fireroom serves one engineroom; in others, two firerooms serve one engineroom. Instead of firerooms and enginerooms, many large ships of recent design have spaces called machinery rooms. Each machinery room contains both the boilers and the propulsion turbines that serve a particular shaft. In some recent ships that have certain automatic controls, the propulsion machinery is very largely operated from separate enclosed operating stations located within the machinery room.

No matter what arrangement of machinery spaces is used, the propulsion machinery is usually on two levels. The condensers and

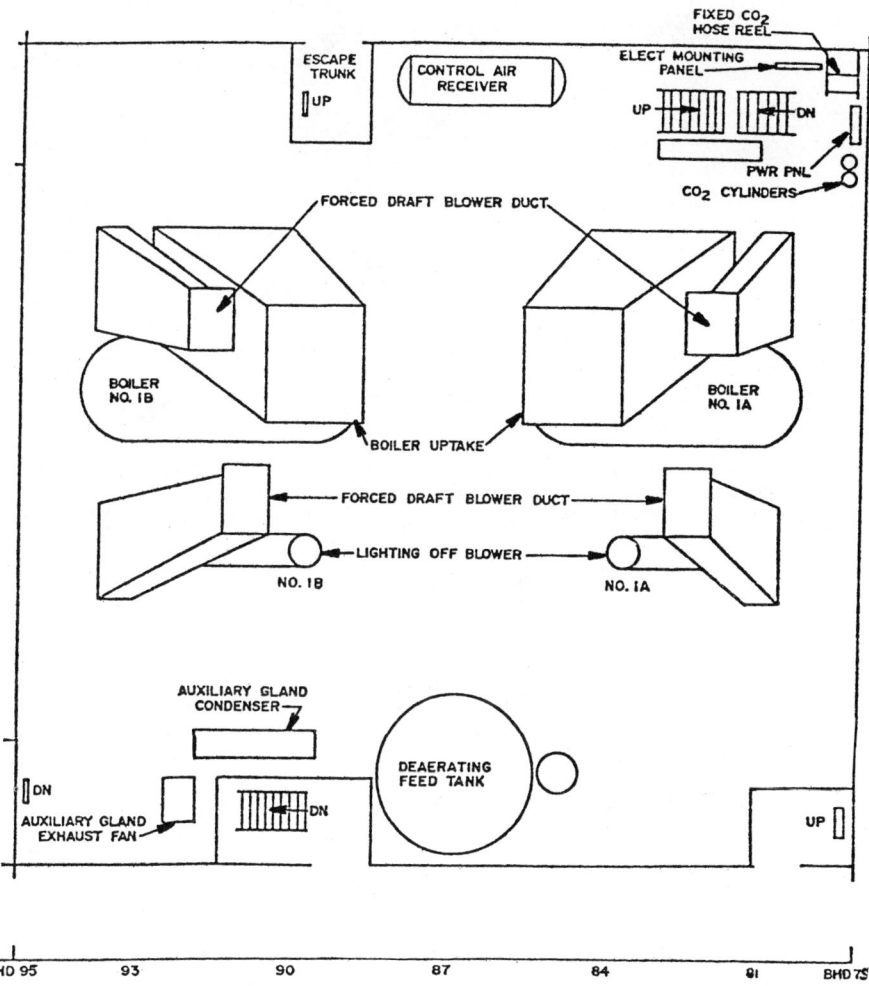

Figure 9–4. FF 1078 machinery arrangement second deck level (fireroom).

Machinery Arrangement and Plant Layout **139**

the main reduction gears are on the lower level. The propulsion turbines and the high-speed pinion gears to which they are connected are on the upper level with the lower pressure turbine exhaust directly over the condenser. The boilers occupy both the lower level and the upper level; the stations for firing the boilers (sometimes

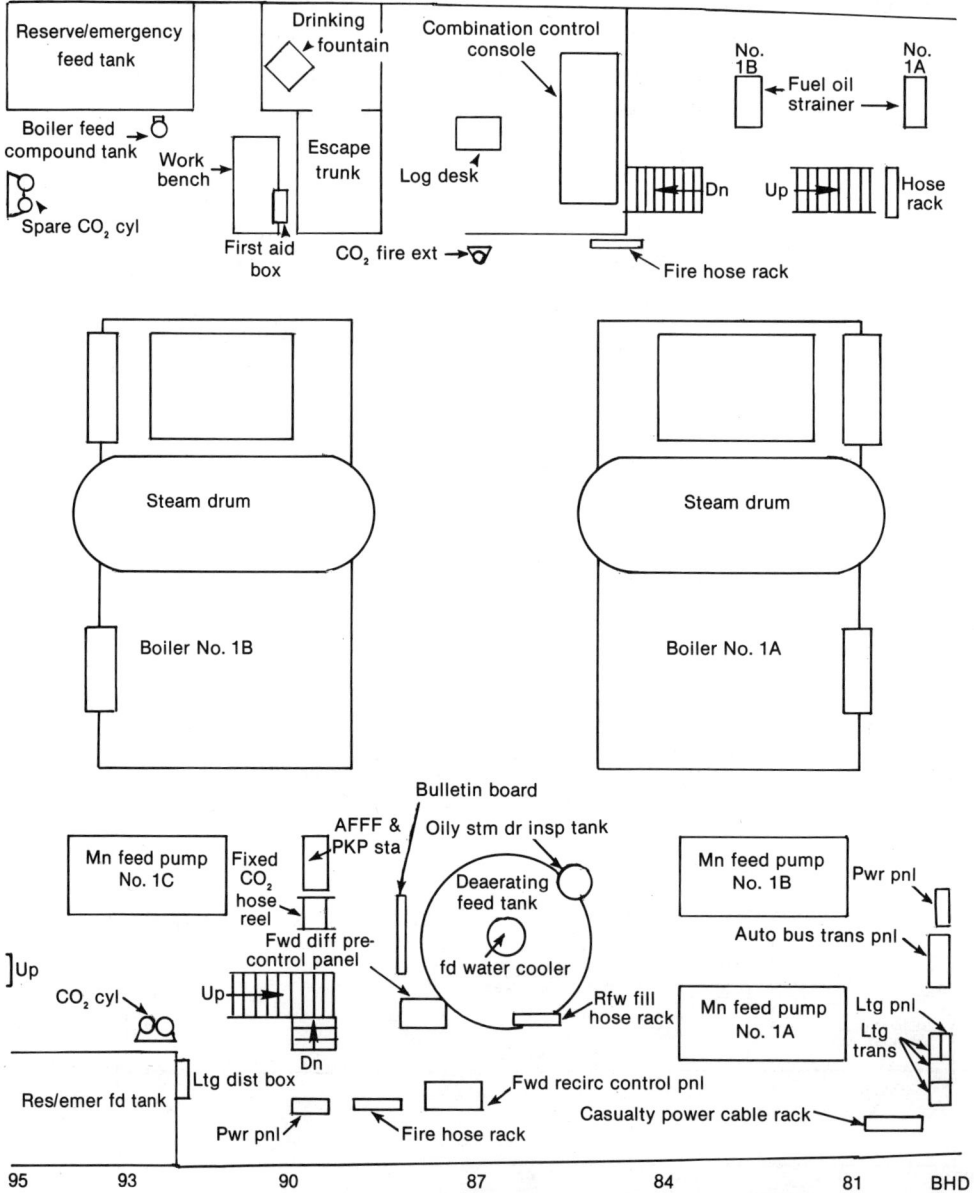

Figure 9–5. *FF 1078 machinery arrangement upper level (fireroom).*

140   Introduction to Naval Engineering

referred to as the firing aisle) are on the lower level, while the stations for operating the valves that admit feedwater to the boilers are on the upper level. The boilers are usually located on the centerline of the ship, or else they are distributed symmetrically about the cen-

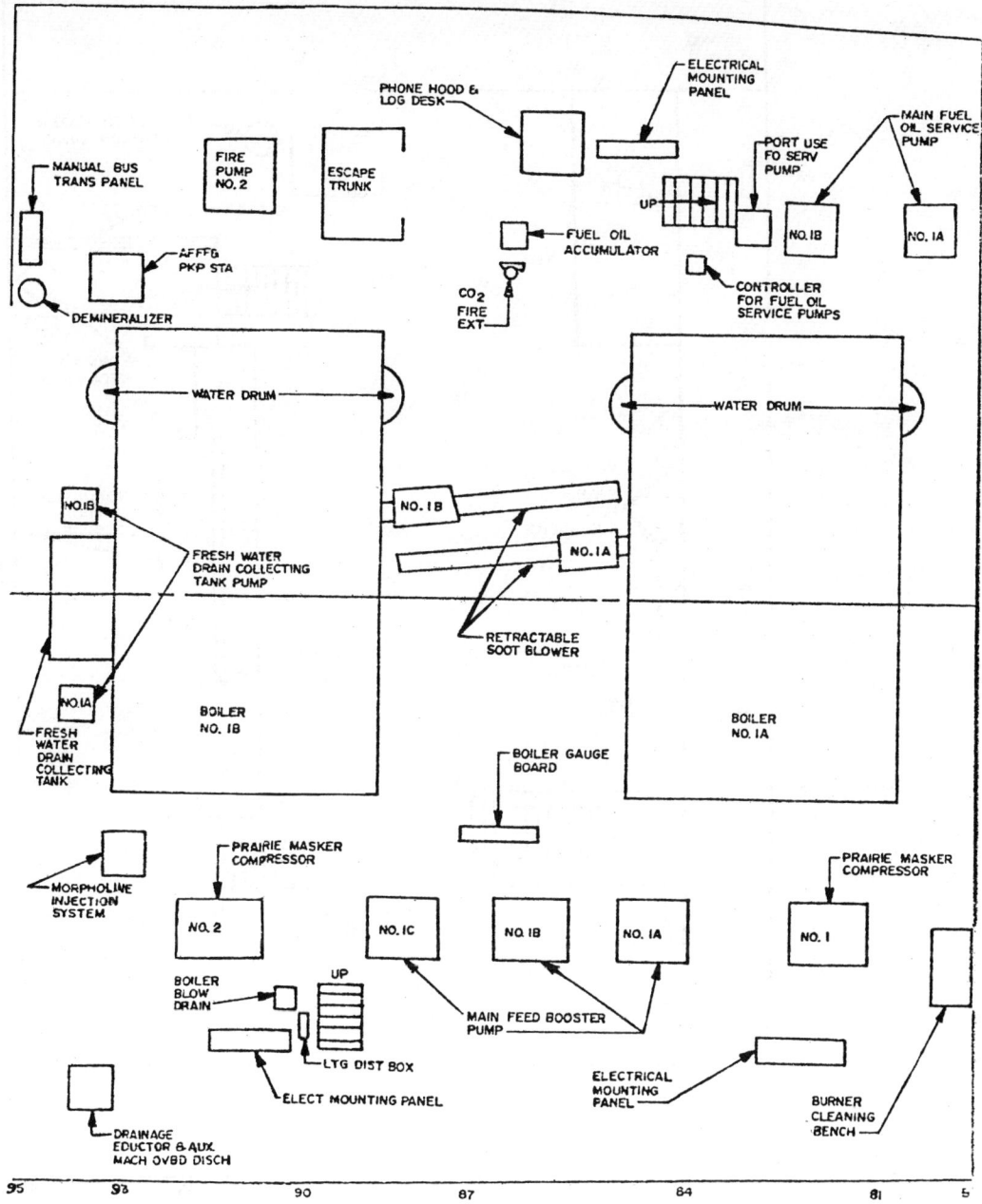

Figure 9–6. FF 1078 machinery arrangement lower level (fireroom).

terline. The long axis of the boiler drums runs fore and aft rather than athwartship. Other machinery, including the propulsion auxiliaries, is arranged in various ways as space and weight considerations permit.

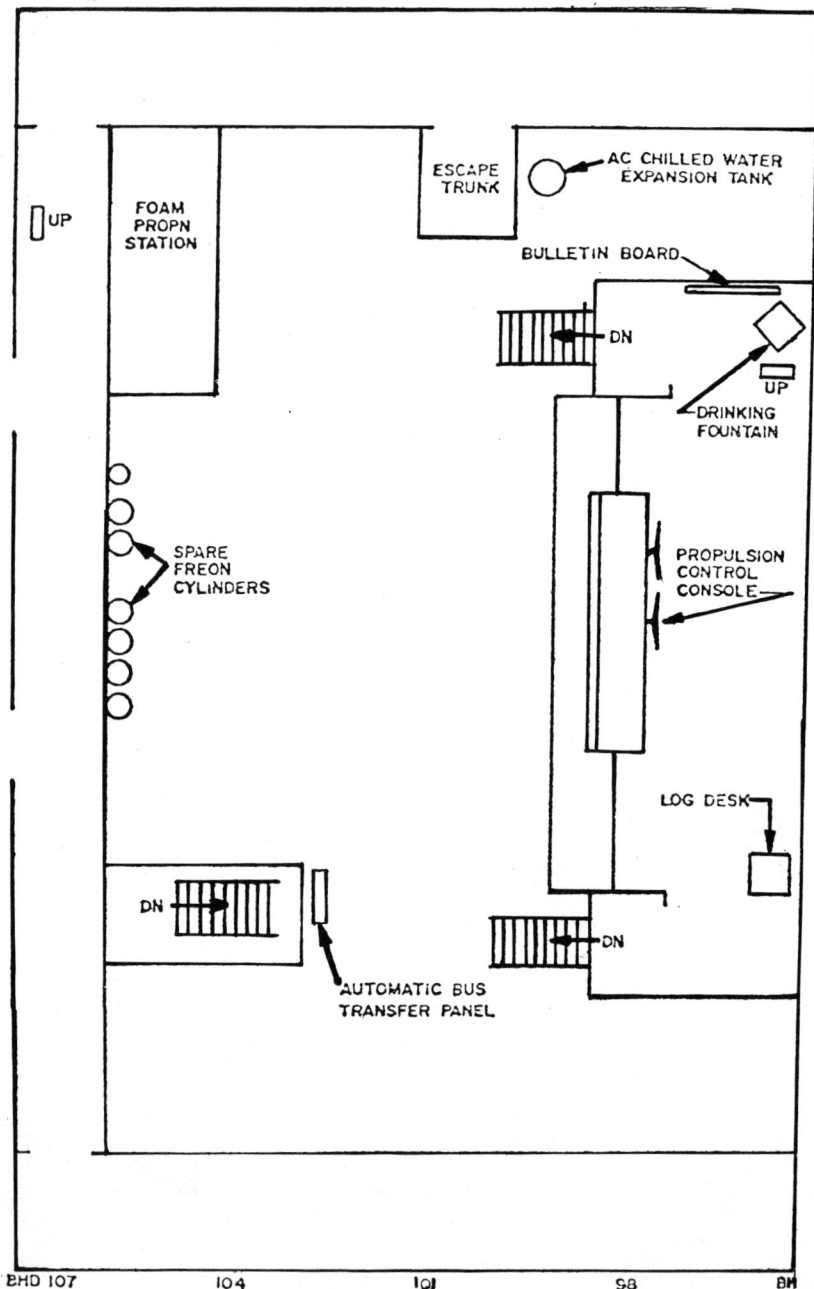

Figure 9-7. *FF 1078 machinery arrangement second deck (engineroom).*

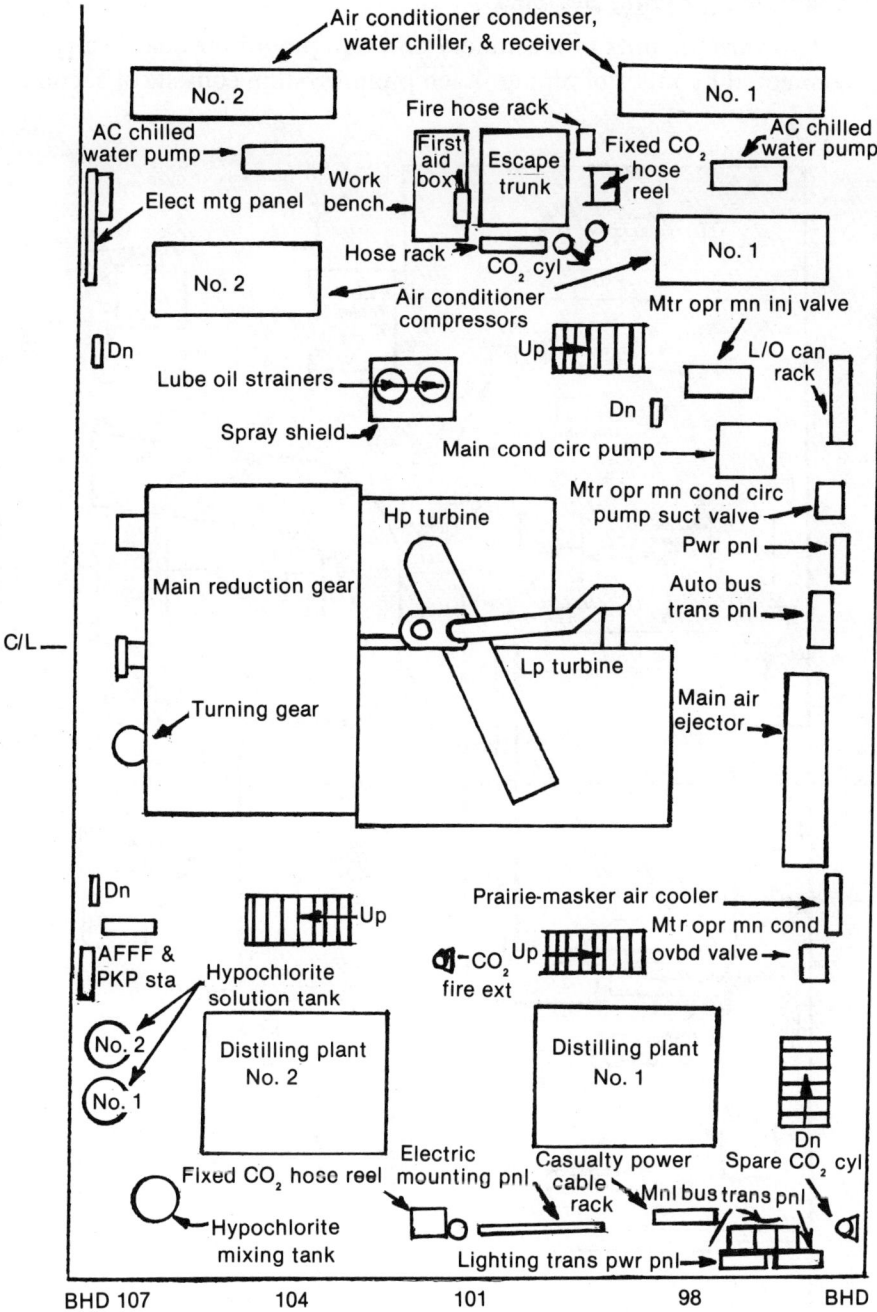

Figure 9-8. FF 1078 machinery arrangement upper level (engineroom).

Machinery Arrangement and Plant Layout **143**

Engineering Piping Systems

The various units of machinery and equipment on board ship are connected by miles of piping. Each piping system consists of sections

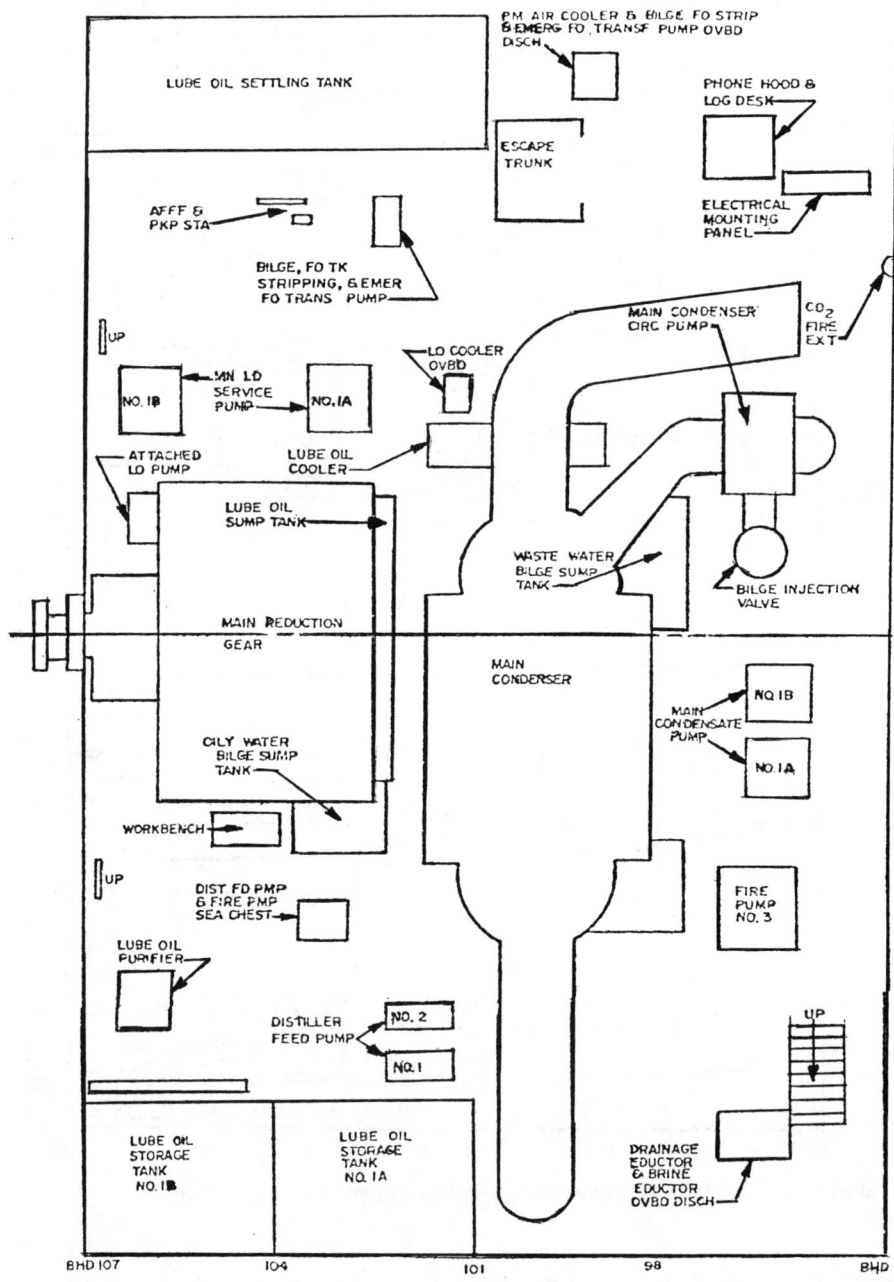

Figure 9–9. FF 1078 machinery arrangement lower level (engineroom).

144 Introduction to Naval Engineering

of pipe or tubing, fittings for joining the sections, and valves for controlling the flow of fluid. Most piping systems also include a number of other fittings and accessories such as vents, drains, traps, strainers, relief valves, gages, and instruments. We are concerned with piping system standard symbols, piping system markings, and general arrangement and layout of the major engineering piping systems on board ship.

Piping system standard symbols are used to indicate machinery

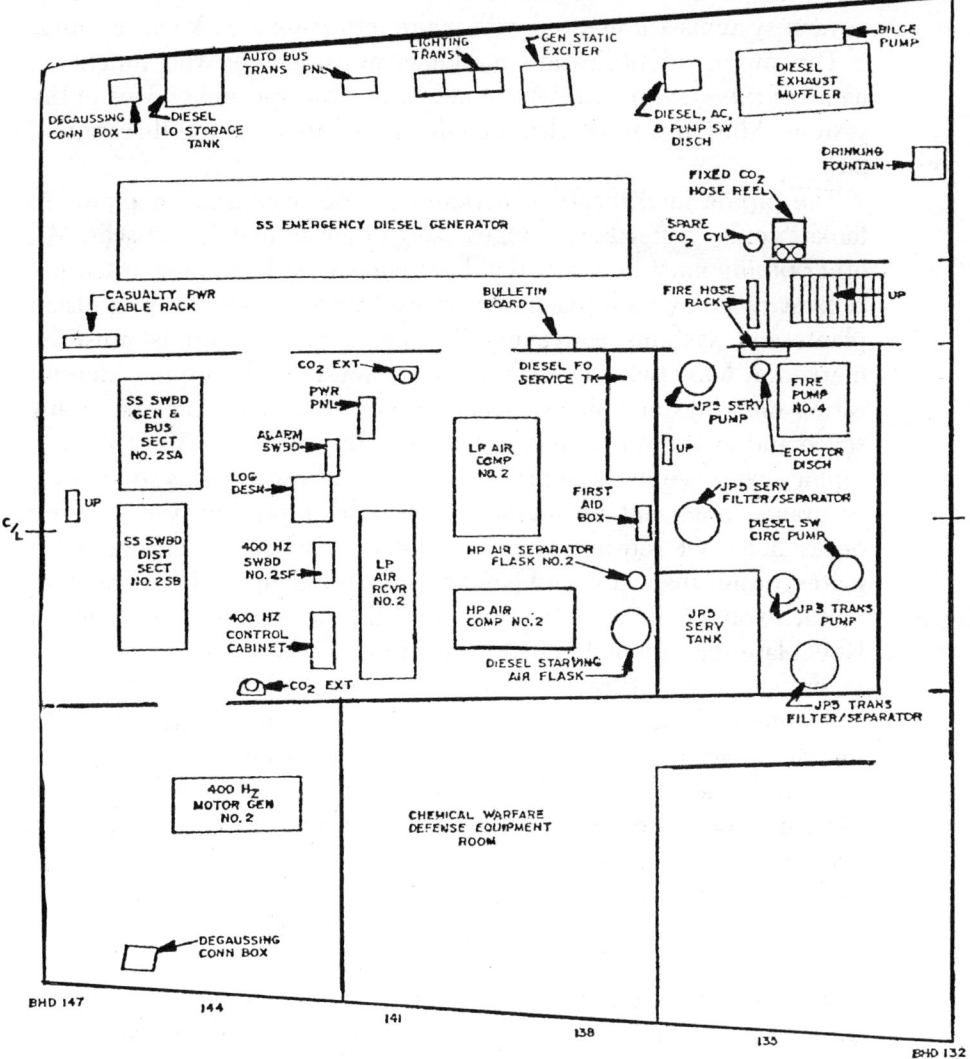

Figure 9–10. FF 1078 machinery arrangement (auxiliary machinery room #2).

units, piping connections, valves, gages, strainers, steam traps, and other items on engineering blueprints and drawings. In some cases, there is deviation from these symbols on blueprints and drawings; but the basic principles of representation are usually followed. Most plans or drawings that utilize special symbols include a legend or list of symbols.

Standard piping system markings are used to mark each shipboard piping system at suitable intervals along the entire length of the system. The markings may be applied with paint and stencils, or prepainted vinyl cloth markers may be used. The markings are in black letters on a white background for all systems except oxygen; oxygen systems are marked with white letters on a dark background.

The piping identification markings must include the functional name of the system and, where necessary, the specific service of the system. Markings must also include arrows to show the direction of flow.

The piping identification markings are not required for piping in tanks, voids, cofferdams, bilges, and other unmanned spaces. All other piping must be marked at least once in each manned space and at least twice in each machinery space. Systems serving propulsion plants and systems conveying flammable or toxic fluids must be marked at least twice in each space. When feasible, piping identification markings are placed near the entry and near the exit to any space and at the junction of interconnecting systems. Short runs of piping that serve an immediately obvious purpose, such as short vents or drains, need not be marked. As a rule, piping on the weather decks does not require marking; if it does require marking, label plates (rather than stenciled paint or prepainted vinyl labels) are used.

Each valve is marked on the rim of the handwheel, on a circular label plate secured by the handwheel nut, or on a label plate attached to the ship's structure or to adjacent piping. The valve label gives the name and purpose of the valve, if this information is not immediately apparent from the piping system marking, and it gives the location of the valve. The location is indicated by three numbers that give, in order, the vertical level, the longitudinal position, and the transverse position. Consider, for example, a drain bulkhead stop valve that is labeled:

2–85–1

The location of this valve is indicated by these numbers. The first number indicates the vertical position—in this case, the second deck. The second number indicates the longitudinal position by giving the frame number—in this case, frame 85. The third number indicates the transverse position—starboard side if the number is odd, port side of the number is even. The numbers indicating transverse position begin at the centerline of the ship and progress out toward the sides. For example, a second drain bulkhead stop installed on the same level and at the same frame but farther to starboard, would be identified as:

2–85–3

In either case, of course, the valve would also be identified as to system (drain bulkhead stop, in these examples) if the piping system identification did not make the system obvious.

A slightly different system of marking is used for identifying main line valves, cross-connection or split-plant valves, and remote-operated valves in vital engineering piping systems. Instead of being identified by location, these valves are assigned casualty control identification numbers, by systems, as:

Main steam	MS1, MS2, MS3, etc.
Auxiliary steam	AS1, AS2, AS3, AS4, etc.
Auxiliary condensate	ACN1, ACN2, etc.
Auxiliary exhaust	AE1, AE2, AE3, etc.
Fuel oil service	FOS1, FOS2, etc.

In newer ships, the system for marking valves in the vital engineering systems is slightly different, consisting of a three-part designation in the following sequence: (1) a number designating the shaft or plant number; (2) letters designating the system; and (3) a number, or a combination of a number and a letter, indicating the individual valve. Individual valve numbers are assigned in sequence, beginning at the origin of a system and going in order to the end of the system, excluding branch lines. In other words, the first valve in the main line is No. 1, the second is No. 2, and so forth. Since parallel flow paths frequently exist, it is often necessary to assign a shaft number and a system designation to the parallel flow paths as well as to the basic main line of the system. The valves in the parallel flow paths are then numbered in sequence; identical numbers are used for valves that perform like functions in each of the parallel flow paths, but a letter suffix is added to distinguish between the similar valves. This system of identification is illustrated for part of a main steam system in figure 9–11.

It is of utmost importance that all engineering personnel (officer and enlisted) become familiar with the valve markings used in the

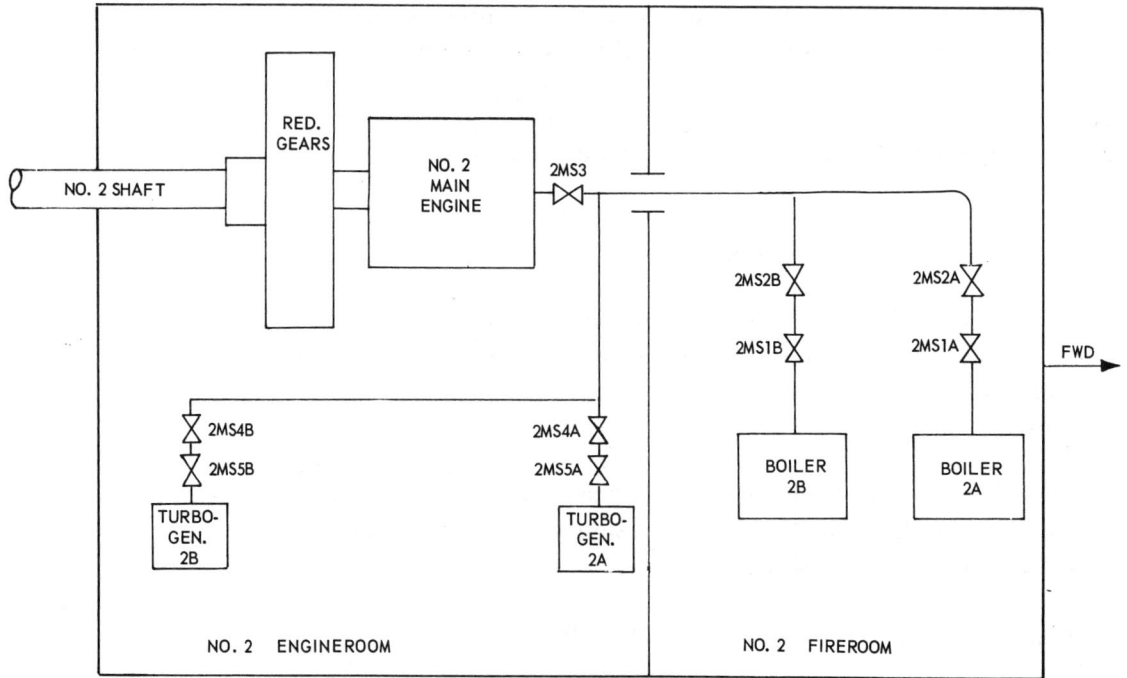

Figure 9–11. Principle of valve identification in engineering piping systems.

vital engineering sytems. Use of the identification numbers tends to prevent confusion and error when the plant is being split or cross-connected and when damaged sections are being isolated, since it provides a means of ordering any particular valve to be opened or closed without taking time to describe the actual physical location of the valve. However, the identification markings cannot serve their intended purpose unless all engineering personnel are thoroughly familiar with the physical location and the identification number of each valve they may be required either to operate themselves or to order opened or closed.

Most shipboard piping is painted to match and blend in with its surrounding bulkheads, overheads, or other structures. In a very few systems, color is used in a specified manner to aid in the rapid identification of the systems. For example, JP–5 piping in interior spaces is painted purple. Gasoline valves in interior spaces are painted yellow, except for moving parts of the valves; in exterior locations, part of the valve handwheel or the operating lever is painted yellow. Green is similarly used to identify oxygen, and red is used for fireplugs and foam discharge valves.

Main Steam Systems

The main steam system (figure 9–12) is the shortest and simplest of all the major engineering piping systems on board ship. This statement is true regardless of the steam pressures involved. With the advent of the 1200 psi main steam system there is a tendency to regard high-pressure main steam systems as basically different from (and mysteriously more complex than) the lower-pressure systems. In reality, a 1200 psi main steam system serves the same basic purpose as a lower-pressure system, and differs only in minor details, as noted in subsequent discussion. The major difference between high-pressure main steam systems and lower-pressure systems is in the materials used for piping and fittings; in general, the metals for 1200 psi systems must be designed to withstand operating temperatures approximately 100 to 200°F higher than the operating temperatures of the lower-pressure systems.

In most ships, any piping that carries superheated steam is considered part of the main steam system. On board many ships, the main steam system includes only the piping that carries superheated

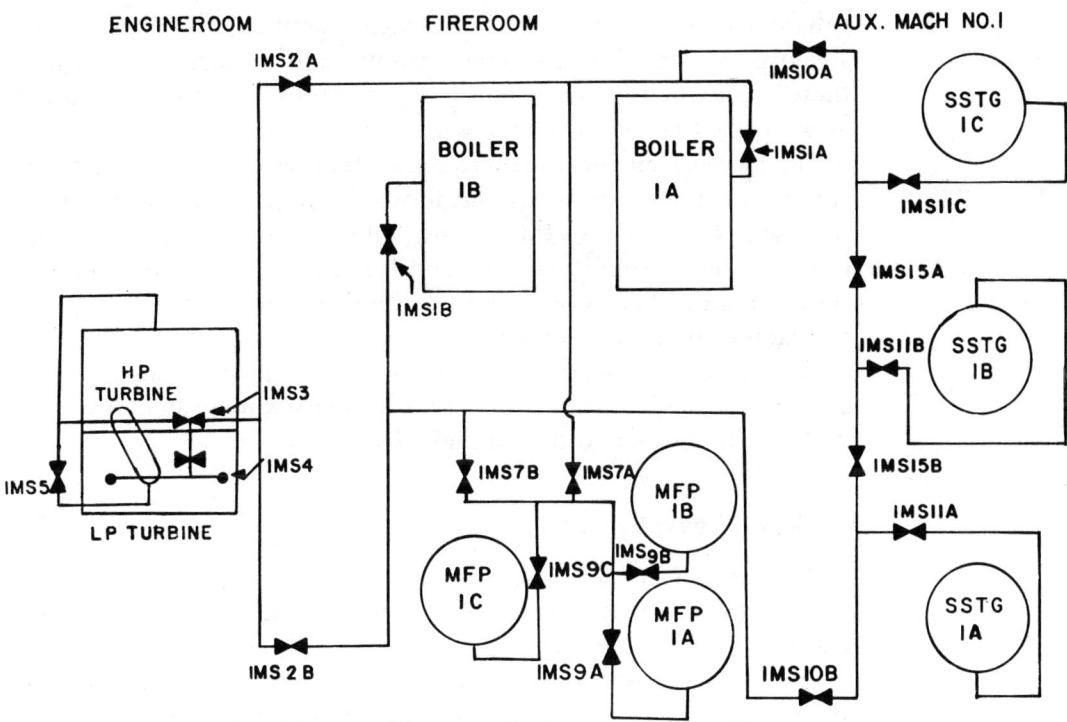

Figure 9–12. Schematic diagram of a 1200 psi main steam system (superheated).

steam from the boilers to the propulsion turbines, the turbogenerators, and the boiler soot blowers. On board some recent ships (both 600 psi and 1200 psi) the main steam system supplies superheated steam to several other units as well. For example, some carriers use superheated steam to supply steam catapult systems; also, some carriers and other ships use superheated steam to operate forced draft blowers, main feed pumps, main circulating pumps, and other auxiliaries. Soot blowers are not supplied from the main steam system in some ships that have 1200 psi main steam systems; instead, steam for the soot blowers is taken from the 1200 psi auxiliary steam system.

There are usually two boilers in each fireroom. Each boiler is provided with a boiler stop valve which can be operated either locally from the fireroom or remotely from the main deck. A second line stop valve in each fireroom provides two-valve protection for the boiler when it is not in use, and permits effective isolation in case of damage. This type of two-valve protection is standard for all boilers installed in U. S. naval ships.

For ahead operation, the superheated steam passes through a main steam strainer, a guarding valve, and a throttle valve before entering the high pressure turbine. From the high pressure turbine, the steam passes through a crossover pipe to the low pressure turbine; then it exhausts to the condenser. For astern operation, the superheated steam passes through the steam strainer and through a stop valve; then it goes to the steam chest of the astern elements, which are located at either end of the low pressure turbine.

The forward and after main steam systems are connected by cross-connection piping between the forward engineroom and the after fireroom. By means of this piping, either boiler can be used with either or both propulsion units and turbogenerators. Thus, the two propulsion plants can be operated either independently (split-plant) or together (cross-connected).

Note that superheated steam for the soot blowers goes from the superheater outlet piping into a soot blower steam header. Branches go from the header to the individual soot blowers.

Auxiliary Steam Systems

Auxiliary steam systems (figures 9–13 and 9–14) supply steam at the pressures and temperatures required for the operation of many systems and units of machinery, both inside and outside the engineering spaces. Although auxiliary steam is often called saturated steam, it has some degree of superheat in some auxiliary steam systems. Constant and intermittent service steam systems, steam smoth-

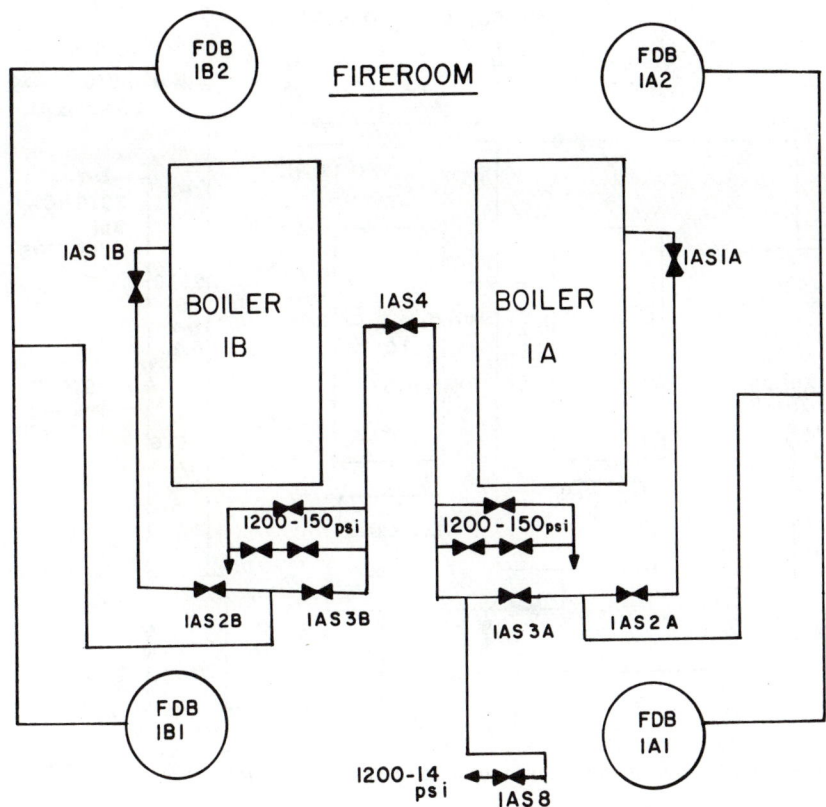

Figure 9–13. Schematic diagram of a 1200 psi desuperheated steam system.

ering systems, whistles and sirens, fuel oil heaters, fuel oil tank heating coils, air ejectors, forced draft blowers, and a wide variety of pumps are typical of the systems and machinery that receive their steam supply from auxiliary steam systems in most steam-driven ships. As previously noted, the units are not the same in all ships. Some recent ships use main steam instead of auxiliary steam for the forced draft blowers and for some pumps. In some ships, turbine gland sealing systems receive their steam supply from an auxiliary steam system; in other ships, the source of supply is the auxiliary exhaust system. In general, an increasing use of electrically driven (rather than turbine-driven) auxiliaries has led to the simplification of auxiliary steam systems in recent ships.

Auxiliary Exhaust Systems

The auxiliary exhaust system (figure 9–15) receives exhaust steam from pumps, forced draft blowers, and other auxiliaries which do not

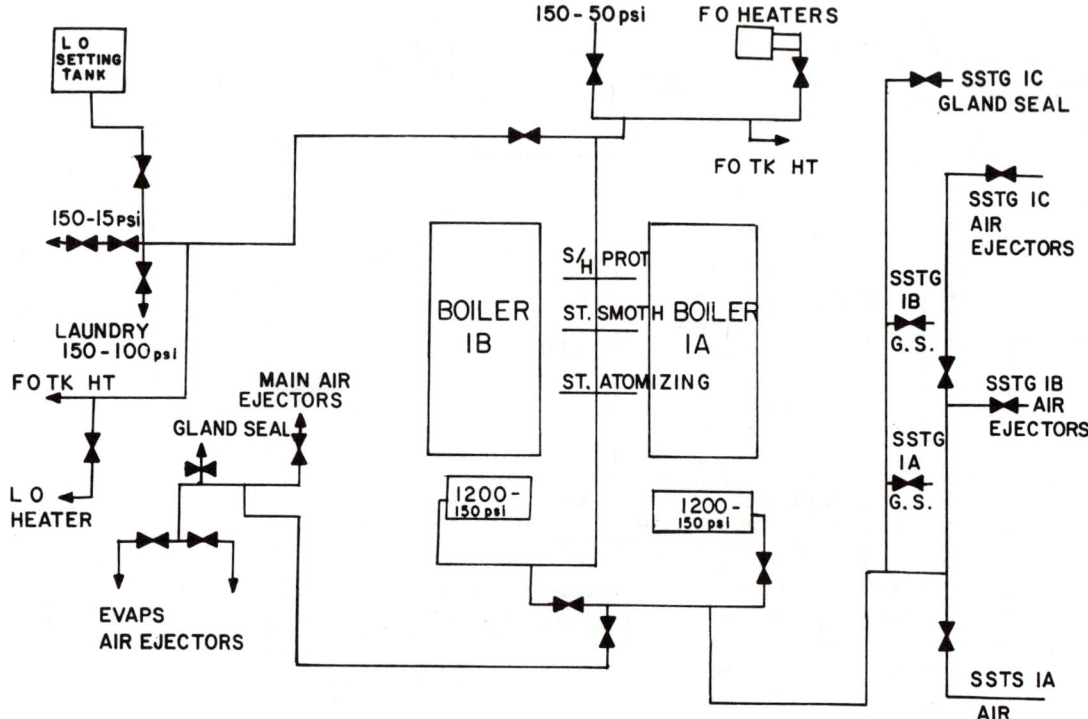

Figure 9–14. Schematic diagram of a 150 psi steam system.

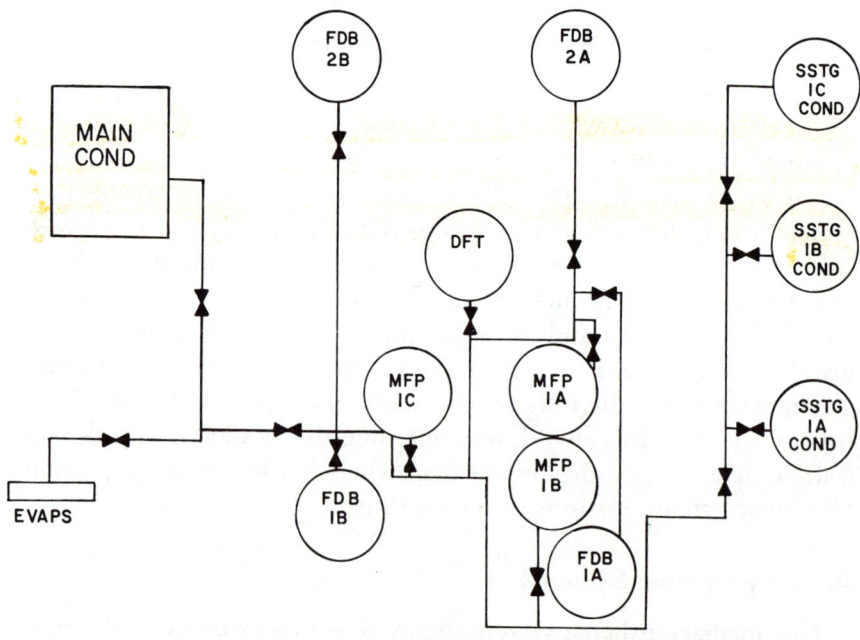

Figure 9–15. Schematic diagram of an auxiliary exhaust steam system.

exhaust directly to a condenser. Auxiliary exhaust steam is used in various units, including deaerating feed tanks, distilling plants, and (in many ships) turbine gland seal systems.

The pressure in the auxiliary exhaust system is maintained at about 15 psig. If the pressure becomes too high, automatic unloading valves (dumping valves) allow the excess steam to go to the main or auxiliary condensers; in the event of failure of these unloading valves, relief valves allow the steam to escape to the atmosphere. If the pressure in the auxiliary exhaust system drops too low, makeup steam is supplied from an auxiliary steam system (usually the 150 psi system) through augmenting valves.

The auxiliary exhaust system must be clearly distinguished from the various auxiliary steam systems. Even though the auxiliary exhaust system is a steam system, it is not considered an auxiliary steam system.

USS *Spruance* (DD 963)

Introduction and Plant Lay-Out (figures 9–16 through 9–20)

The USS *Spruance* (DD 963) is a twin engine/twin screw [controllable reverse pitch (CRP) propellers], gas turbine destroyer. The *Spruance* has two main enginerooms and two auxiliary machinery rooms plus a number three generator room, and a central control station (CCS). Each shaft is coupled to a main reduction gear assembly (MRGA), which is driven by two gas turbine modules (GTMs) through two clutch/brake assemblies, one for each GTM. (Note: The clutch brake assembly is an integral part of the MRGA, permitting operation of one or both GTMs.)

Gas Turbine Modules (GTMs)

The main propulsion gas turbine engines are housed in modules that provide:

1. Engine mounting
2. Engine cooling
3. Airborne noise reduction
4. Fire-extinguishing capability

Propulsion Gas Turbine Engines

The *Spruance* has four propulsion gas turbine engines, which are GE LM 2500 marine engines rated at 21,500 BHP each. The LM 2500 has been developed from the USAF/C5A and DC–10 turbo fan engine. It can achieve a no-load idle speed from cold iron in 60 seconds and full power from no-load idle speed in 30 seconds.

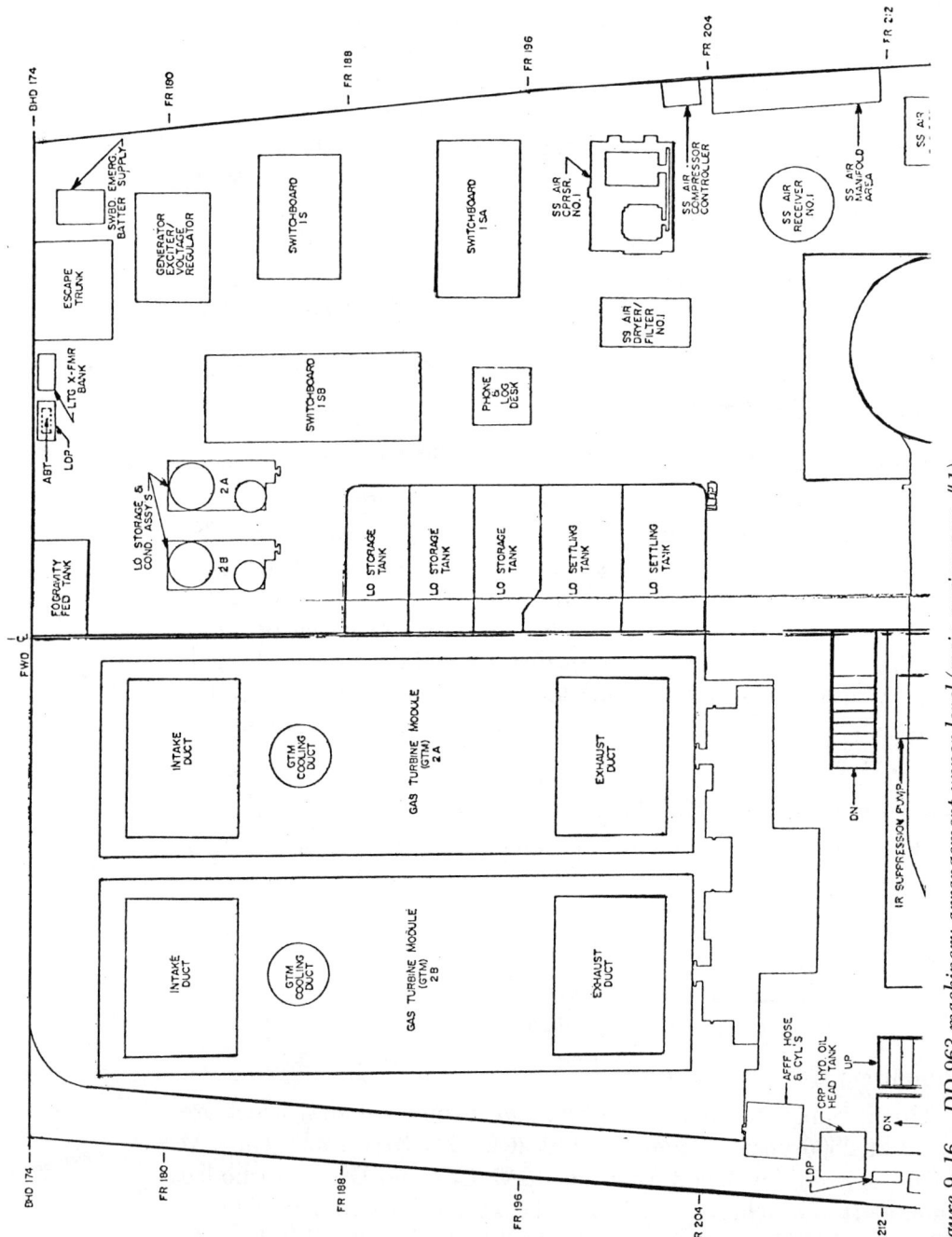

Figure 9-16. DD 963 machinery arrangement upper level (main engineroom #1).

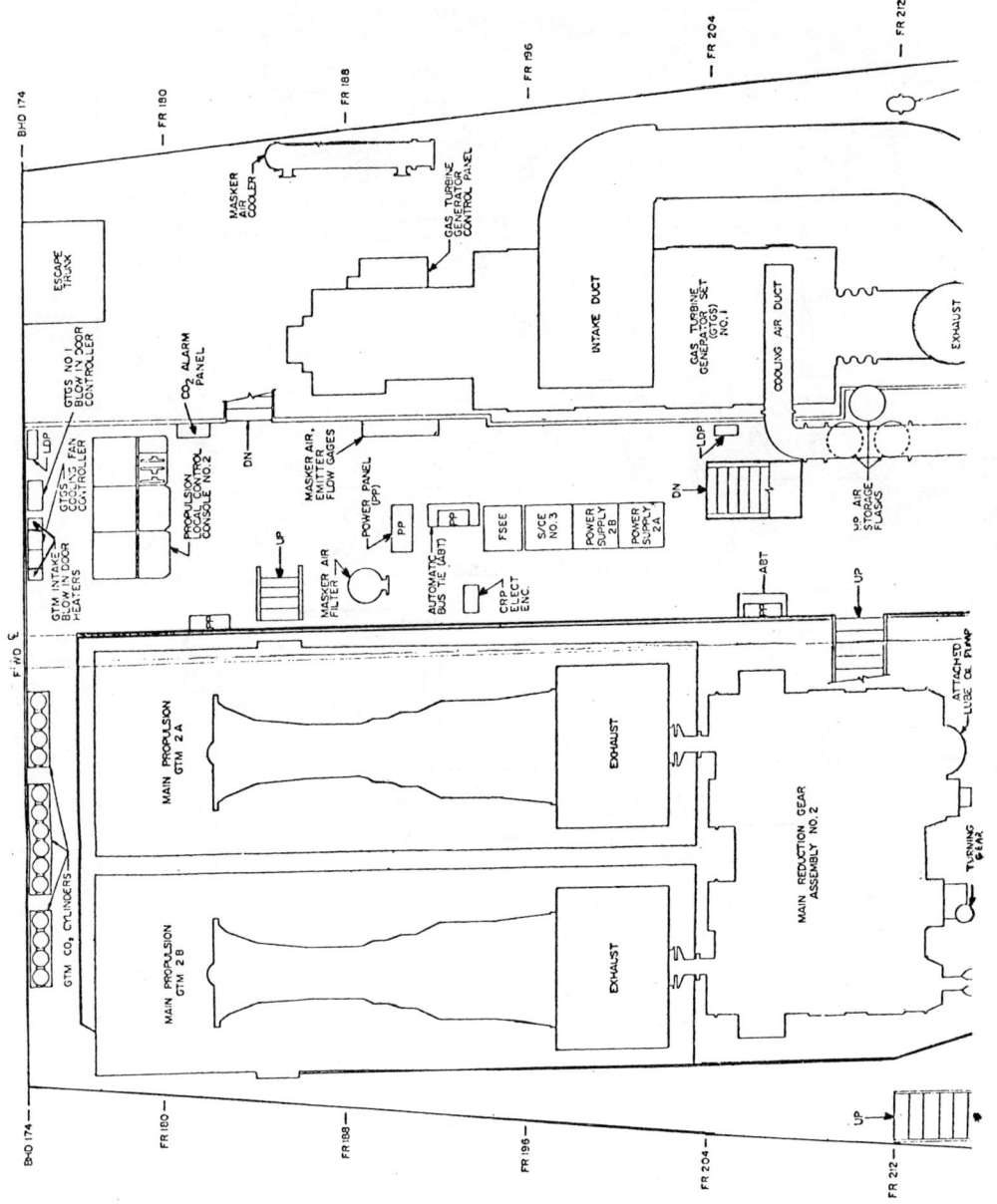

Figure 9-17. DD 963 machinery arrangement intermediate level (main engineroom #1).

155

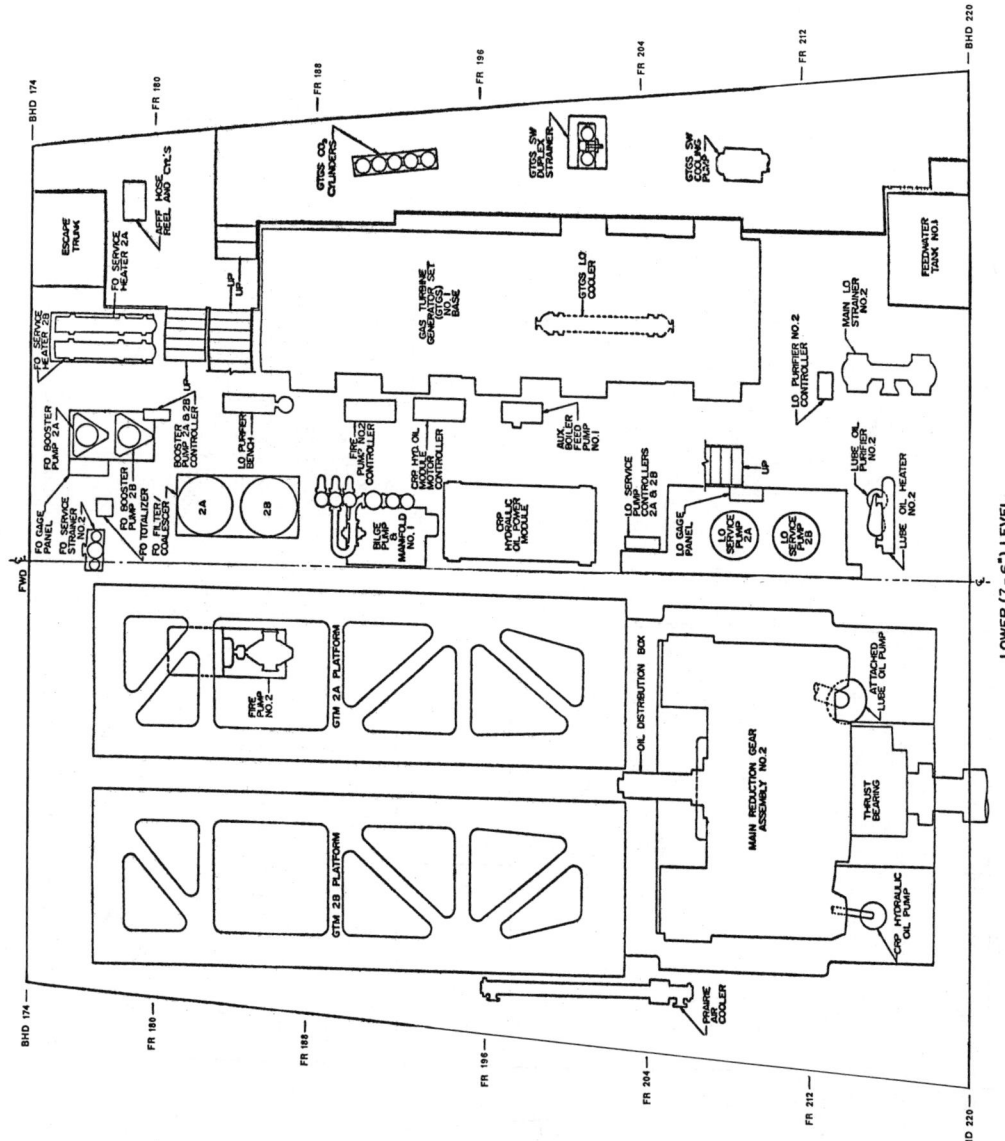

Figure 9-18. DD 963 machinery arrangement lower level (main engineroom #1).

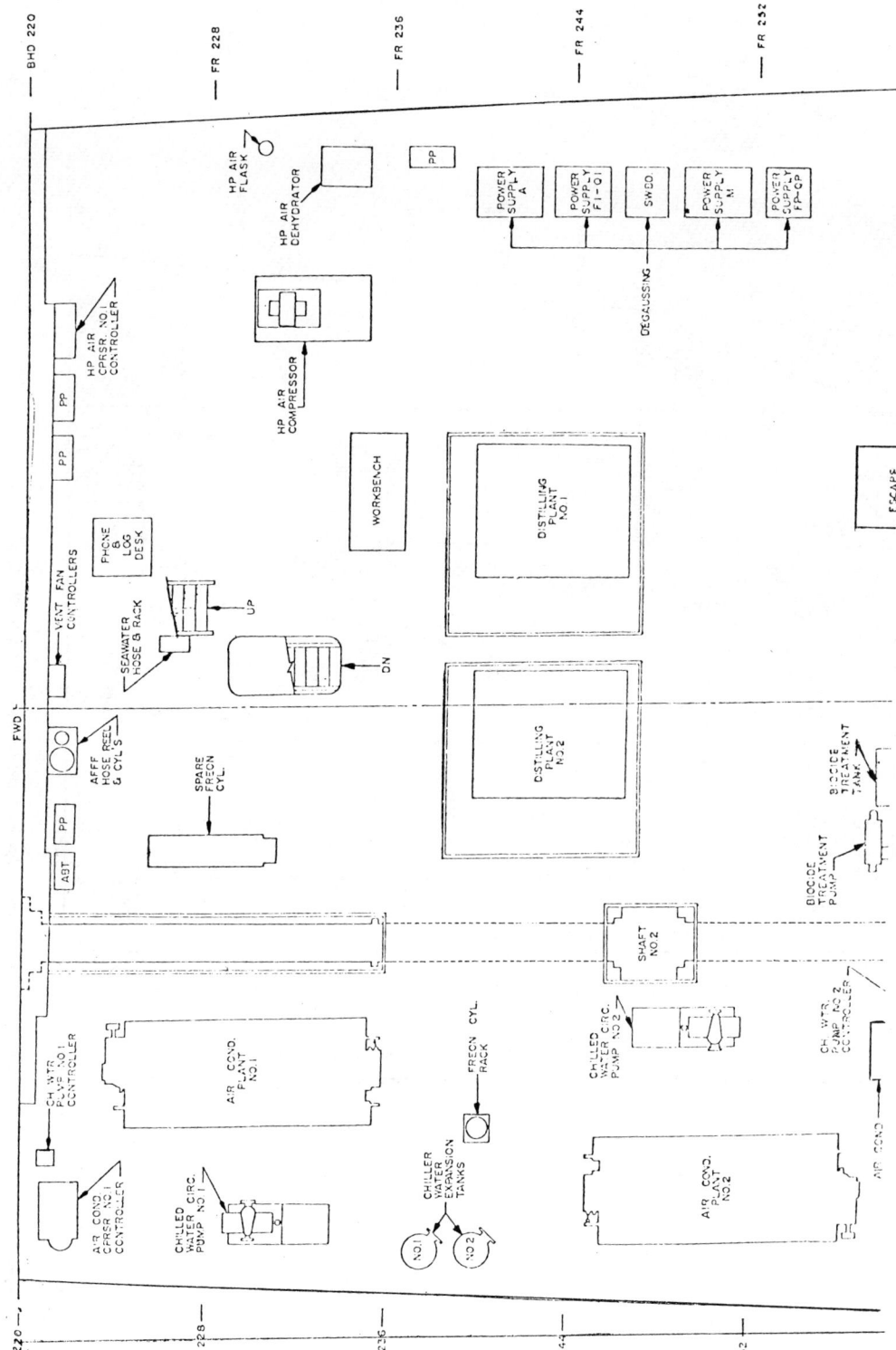

Figure 9-19. DD 963 machinery arrangement upper level (auxiliary room #1).

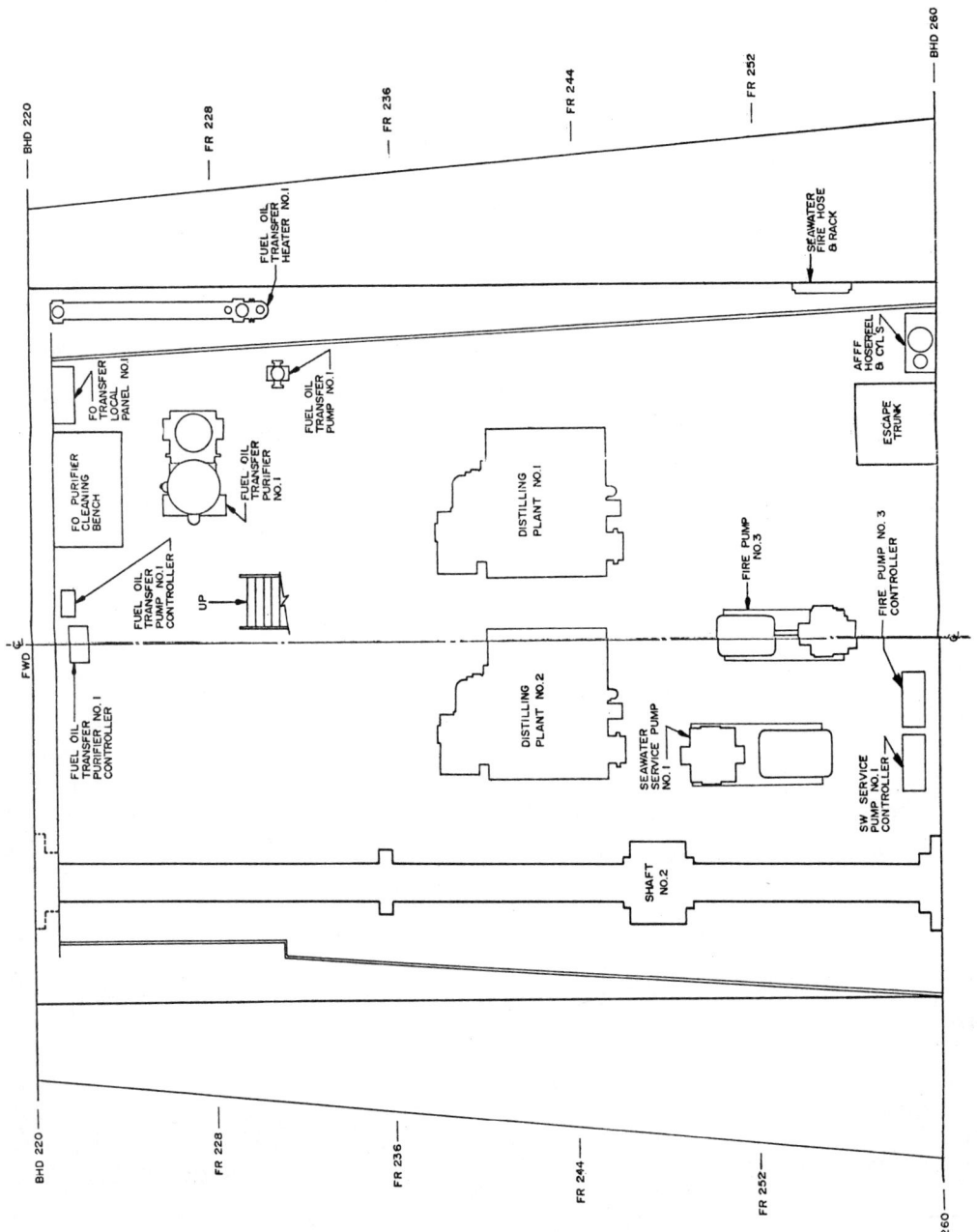

Figure 9–20. DD 963 machinery arrangement lower level (auxiliary room #1).

Main Propulsion Intake and Exhaust System

The intake system provides:

1. High volume of air
2. Moisture separation
3. Silencing
4. Anti-icing protection
5. Engine cooling air

Additionally, the intakes provide for the removal of the main propulsion engines. The exhaust system:

1. Routes exhaust gas to atmosphere.
2. Is designed to prevent injection into air intake ducts.
3. Has sidewall baffles for noise suppression.
4. Has an infrared radiation suppression system to cool exhaust to minimize detection by infrared detectors.

Main Reduction Gear Assemblies (MRGAs)

The two MRGAs are of a double reduction, locked-train design. Each MRGA can combine the power of one or two GTMs with a 21 to 1 ratio of reduction for transmission to the propeller shaft. The two clutch/brake assemblies connect to the two high-speed inputs from the GTMs with the first reduction pinion. This enables:

1. Secure engine
2. Split plant (A or B)
3. Full power (all four on)

(Note: The GTMs in relation to the MRGAs are reversed in order to allow opposite rotation of the port and starboard shafts.)

Controllable Reversible Pitch (CRP) Propellers

The two propellers are five-bladed and are the reversible pitch type; reversible pitch propellors are required because it is impossible to change the direction of the gas turbine. Changing pitch is accomplished by the hydraulic blade positioning mechanisms located in propeller hubs. Hydraulic oil for operation of the positioning mechanism is supplied by the hydraulic oil power module (HOPM) located at each reduction gear. Each HOPM delivers control oil and high-pressure oil to an oil distribution (OD) box on the forward end of each reduction gear. The OD box directs the high-pressure oil to the propeller hubs via a tube in the shaft. The control oil in the OD box is used to hydraulically actuate the propeller pitch changes. Pitch is normally controlled by the ECSS (engineering control and surveillance systems) but can be controlled manually at the OD box.

Electric Power Generation and Distribution System

Gas Turbine Generator Sets (GTGSs)

Ship's service electrical power is generated by three GTGSs. Each generator has continuous output power of 2000 kW, 450–V, three-phase 60-Hz at .8PF. Two GTGSs will normally supply all the electrical power needed. Number One and Two GTGSs are in Number One and Two enginerooms, respectively. Number Three GTGS is located in compartment 3–426–1–E (aft of engineroom Number Two). The GTGSs consist of:

1. Allison 501–K17 gas turbine engine
2. Reduction gear box
3. AC generator
4. Acoustical enclosure

Each GTGS has an independent L/O system and seawater cooling system. Fuel oil for Number One and Two GTGSs is taken from gravity-feed tanks in one F/O system in each engineroom. Number Three GTGS F/O is from the Number Two engineroom F/O service system. To go from a cold iron status to full load with HP air start takes 45 seconds; LP air start, 60 seconds.

Generator Gas Turbine Intake and Exhaust Ducts

They are basically the same as the main engine ducts except that they are small in cross-sectional area, and they do not allow turbine engine module removal. Associated with each GTGS is a waste heat boiler (WHB). The hot turbine gas goes through a waste heat boiler before entering the exhaust stacks. The boilers generate 100 psi steam for ship's service use in:

1. Heating
2. Cooking
3. Evaporation
4. Laundry
5. F/O and L/O heating

Electrical Distribution

There are three main switchboards; each one is located near the GTGS it serves. The generators can be connected in parallel (via bus tie breakers) or split. The switchboards supply main power to separate load centers for distribution or directly to certain vital electrical loads. Each switchboard has three panels, enclosing:

1. Main generator circuit breaker
2. Two bus tie breakers

3. Feeder breakers
4. Necessary instrumentation and local circuit breaker controls

Electrical Plant Control. Control is by means of:

1. CCS
2. Local manually at GTGS
3. Switchboard station

CCS and switchboard have start/stop and distribution control. GTGS local has only start/stop.

Engineering Plant Control Systems

Engineering Control and Surveillance System (ECSS)

This is an automated electronic control and monitoring system, using analog and digital circuitry. Its major control features are:

1. Auto sequencing for start/stop GTMs and auto control of MRGA clutch/brakes
2. Throttle and pitch control (auto schedules the rpm and pitch to the EOT order given)
3. Centralized control of the electronic plant (GTGS starting and auto sequencing of paralleling operations)
4. Auto sequencing and operation of selected prop support and auxiliary equipment

The monitoring features include:

1. Display of equipment status
2. Annunciation of abnormal conditions
3. A printed record of major engineering plant parameters

Central Control Station (CCS)

This is the main operating station from which a majority of the engineering plant machinery can be controlled and monitored. The ECSS equipment located in CCS includes:

1. Propulsion and auxiliary machinery control equipment (PAMCE)
2. Electric plant control system (EPCE)
3. Propulsion and auxiliary machinery information system equipment (PAMISE)

PAMCE. This equipment includes all control and displays necessary to operate both main propulsion plants and associated auxiliary equipment. It has integrated throttle/pitch control.

EPCE. This system contains logic circuitry to initiate start/stop of a GTGS and automatically sequence generator breakers and bus tie breakers. It provides auto paralleling of a selected combination of generators. An auto start-up feature is included in the event of a GTGS failure.

PAMISE. This equipment consists of a digital computer, signal conditioning equipment, and two printers. It receives, evaluates, and logs all engineering plant performance, status, and alarm status.

(Note: There are three sets of signal conditioning equipment, one in each main engineroom and one in the CCS. A signal conditioner is a central gathering point for sensory inputs; it processes inputs for computer use and to perform alarm-generation functions.)

Pilothouse

Ship control equipment (SCE) controls:

1. Throttle/pitch command to each shaft directly
2. EOT, rudder angle, and auto pilot control

Readouts obtained are:

1. Actual shaft rpm
2. Actual propeller pitch
3. Actual EOT settings

Main Engine Rooms

The ECSS equipment in each main engineroom includes:

1. PLOE (propulsion local operating equipment)
2. One signal condition (part of PAMISE)

The PLOE provides for local control and monitoring of the GTM and associated auxiliary equipment.

(Note: Throttle for each GTM and one pitch control for that shaft are available at the PLOE. The operators must schedule the throttle and pitch to the correct settings.)

Free-Standing Electronics Enclosure (FSEE)

One is located in each MER and provides a supporting electronic and control interface between the two GTMs and the ECSS.

Fuel System Control Console (FSCC)

Located in the CCS, the FSCC is used for monitoring and control of the fuel oil fill and transfer system. It operates in conjunction with the two F/O local control panels in the auxiliary machinery rooms and one JP–5 local control panel in the Number Two pump room.

Damage Control Console (DCC).

Located in CCS, it monitors hazardous conditions such as fire and high bilge levels. It also monitors the firemain and remotely operates six fire pumps.

Engineering Plant Auxiliary Systems

Ship's Service Steam System

Steam is generated by three waste heat boilers (WHB). As before, the WHB uses GTGS exhaust gast to heat feedwater into steam. The steam from the three boilers is fed into a common manifold and distributed throughout the ship. The system provides steam for the following:

1. FO heaters
2. Distillers (evaps)
3. Galley dishwashers

Steam condensates are collected in drain collecting tanks and returned to the feedwater system. Steam output is regulated by a control condenser which condenses excess steam from the boiler output and returns it to feedwater.

Fuel Oil Systems

Fuel oil is stored in seawater-compensated tanks* located in the ship's inner bottom. The F/O transfer system supplies purified F/O to service tanks and to transfer fuel oil among the storage tanks. Each engineroom contains its own independent F/O service system to supply the GTMs and GTGSs. Oil comes from the service tanks through a duplex strainer via an FOBP (fuel oil booster pump). After discharge from the FOBP it goes to a steam heater, then a filter/condenser, and then to the gas turbine. The GTGSs have gravity-feed F/O tanks. The tanks are supplied from a branch line off the F/O service line to the GTMs. Control and monitoring of the F/O transfer system is from one FSCC in the CCS. Control and monitoring of the two enginerooms' F/O service system is by either the PAMCE or FLOE.

Lube Oil System

The main function of this system is to provide lubrication and cooling for the main reduction gear and thrust bearing assemblies.

---

*Tanks are designed to allow seawater in as fuel oil is taken out, thereby keeping approximately the same load on the ship at all times, for stability reasons.

The secondary function is to cool the synthetic L/O from the gas turbines. Associated equipments include:

1. Settling tanks
2. Stowage tanks
3. Pumps
4. Filters
5. Coolers
6. Heaters
7. Purifiers

Control and monitoring is by either PAMCE or PLOE. The three GTGSs have their own separate L/O systems.

Compressed Air System

This system is composed of three systems:

1. Bleed air system
2. High pressure air system
3. Ship's service air system

Bleed Air System. This air is semicompressed air coming from the compressors of all gas turbines. Each engineroom has a bleed air header which can be connected to the other engineroom header to make one large manifold. The system provides:

1. Anti-icing—hot bleed air is mixed with intake air to prevent ice formation.
2. Prairie air—which is used to disguise the ship's signature radiated from the ship's propellers. Air is pumped down the propeller shafts and is emitted from small holes in the propeller blades.
3. Masker air—which helps mask the ship's noise radiated from the hull. Air is emitted from rings surrounding the hull, which coats the hull in air bubbles, thus reducing hull noise.
4. LP gas turbine starting—the bleed air is used to start additional gas turbines. It is mixed with cooler masker air and directs the starting air into the air-starting turbine.

High Pressure Air System (HPAS). This system consists of an HP loop supplied by two HP air compressors at 3000 psig and 27 storage flasks. Compressors are in auxiliary machinery Number One and aft compressor room at frame 464. The system is used in:

1. Emergency start of a gas turbine
2. Operation of the ASROC launcher
3. 5" guns (counterrecoil system)

4. Torpedo tubes
5. Helo services
6. Backup for ship's service air system

Ship's Service Air System (SSAS). This system supplies low pressure air at 150 psig and 100 psig. The 150 psig air is used for:

1. Radar electronic wave guides
2. HP air system dryer

The 100 psig air is used for LP pneumatic services such as:

1. Main reduction gear clutch/brake actuation
2. Waste heat boiler valve actuation
3. General service air

All of this is supplied by two 150 psig compressors, one in each engineroom.

10

Nuclear Power Plants

Nuclear reactors release nuclear energy by the fission process and transform this energy into thermal energy. While we are learning more daily about the phenomena that occur in nuclear reactions, the knowledge already gained has been put to use in both the submarine and the surface fleets. The Navy is now in the second decade of the utilization of nuclear energy for propulsion.

Nuclear engineering is in a stage of rapid development at present; therefore the discussion in this chapter is limited to the basic concepts of reactor principles. The discussion of nuclear physics is limited to the fission process, since all power reactors in operation at this time use the fissioning of a heavy element to release nuclear energy.[1]

Advantages of Nuclear Power

A major advantage of nuclear power for any naval ship is that less logistic support is required. For ships using conventional petroleum fuels as an energy source, the cruising range and strategic value are limited by the amount of fuel that can be stored in their hulls. A conventionally fueled ship of this type must either return to port to take on fuel or refuel from a tanker at sea—a time-consuming and hazardous operation.

Nuclear-powered ships have a virtually unlimited cruising range, since the refueling is done routinely as part of a regularly scheduled overhaul. On the first nuclear fuel load, the USS *Nautilus* steamed 62,562 miles, more than half of this distance fully submerged. The USS *Enterprise* steamed over 200,000 miles before being refueled. In 1963, Operation Sea Orbit, a 30,000-mile cruise around the world in 65 days, using all nuclear ships, completely without logistic support of any kind, proved conclusively the strategic and tactical flexibility of a nuclear-powered task force.

There are other (and perhaps less obvious) advantages of nuclear power for aircraft carriers. For one thing, tanks that would otherwise be used to store boiler fuels can be used in nuclear-powered carriers to store additional aircraft fuels, thus giving the ship a greater striking potential. Another advantage is the lack of stacks; since there are no stack gases to cause turbulence in the flight deck atmosphere, the operation of aircraft is less hazardous than on conventionally powered ships.

The fact that a nuclear-powered ship can be designed to require no outside source of oxygen from the earth's atmosphere means that the ship can be completely closed off, thereby reducing the hazards of any nuclear attack. The ship needs no oxygen for propulsion and recirculates the oxygen on board via CO_2 scrubbers for environmental use. This greatly increases the potential of the submarine fleet by giving it the capability of staying submerged for extended periods of time. In 1960, the nuclear-powered submarine USS *Triton* completed a submerged circumnavigation of the world, traveling a distance of 35,979 miles in 83 days and 10 hours.

Nuclear Fundamentals

There are now 103 known elements, of which the smallest particle that can be separated by chemical means is the *atom*. The Rutherford-Bohr theory of atomic structure (figure 10–1) describes the atom as being similar to our solar system. At the center of every atom is a nucleus which is comparable to the sun; moving in orbits around the

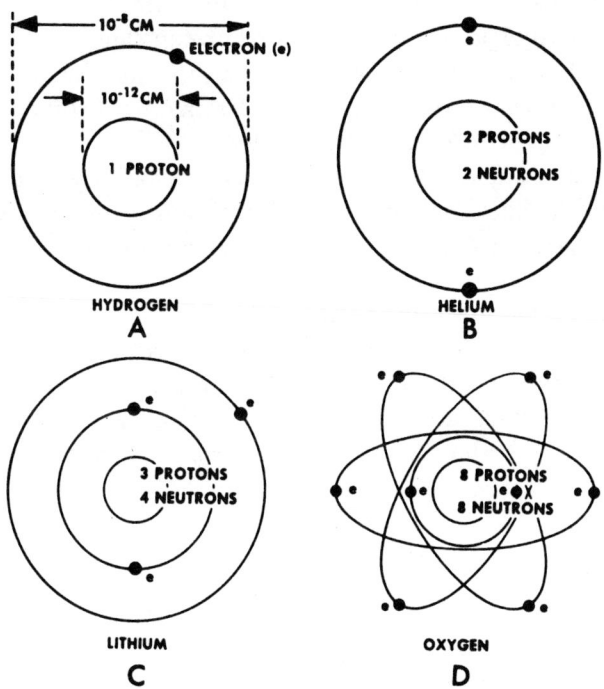

Figure 10–1. Rutherford-Bohr models of simple atoms.

nucleus are a number of particles called electrons. The electrons have a negative charge and are held in orbit by the attraction of the positively charged nucleus.

Two kinds of elementary particles, *protons* and *neutrons*, often referred to as *nucleons*, compose the atomic nucleus. The positive charge of atomic nuclei is attributed to the protons. A proton has an electrical charge equal and opposite to that of an electron. A neutron has no charge.

The number of electrons in an atom and their relative orbital positions predict how an element will react chemically, whereas the number of protons in an atom determines which element it is. An atom that is not ionized contains an equal number of protons and electrons; thus it is said to be neutral, since the total atomic charge is zero.

As shown in part A of figure 10–1, the hydrogen atom has a single proton in the nucleus and a single orbital electron. Hydrogen, the lightest element, is said to have a mass of approximately one. The next heavier atom, that of helium (part B of figure 10–1), has a mass of four relative to hydrogen and was at first expected to contain four protons. It was found, however, that the helium atom has only two protons instead of the four expected; the remainder of its mass is attributed to two neutrons located in the nucleus of the helium atom. The more complex atoms contain more protons and neutrons in the nucleus, with a corresponding increase in the number of planetary electrons. The planetary electrons are arranged in orbits or shells of definite energy levels outside the nucleus.

The characteristics of the elementary atomic particles are compiled in figure 10–2. Note that the mass of a proton is much greater than that of an electron; it takes about 1847 electrons to weigh as much as one hydrogen proton.

It is possible for atoms of the same element to have different numbers of neutrons, and therefore different masses. Atoms that have the same *atomic number* (number of protons in the nucleus of the atom) but different masses are called isotopes. Different isotopes of the same element are identified by the *atomic mass number*, which is

Particle	Charge	Mass (amu)
Proton	+1	1.00758
Neutron	0	1.00894
Electron	−1	0.00055

Figure 10–2. Characteristics of elementary atomic particles.

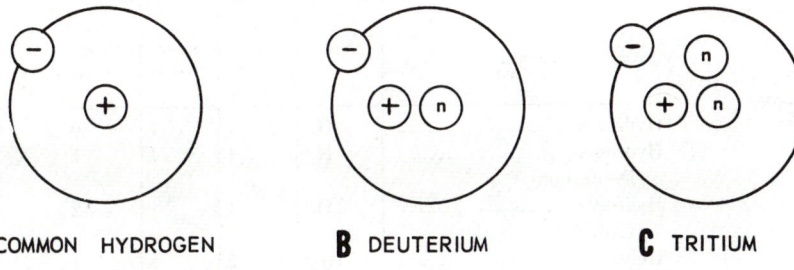

A COMMON HYDROGEN **B** DEUTERIUM **C** TRITIUM

Figure 10–3. Isotopes of hydrogen.

the sum of the number of neutrons and protons contained within the nucleus of the atom.

The element hydrogen has three known isotopes, as shown in figure 10–3. The simplest and most common known form of hydrogen consists of one proton, which is the nucleus, and one orbital electron. Another form of hydrogen, deuterium, consists of one proton and one neutron forming the nucleus, and one orbital electron. The third form, tritium, consists of one proton and two neutrons forming the nucleus, and one orbital electron.

In scientific notation, the three isotopes of hydrogen are written as follows:

  Common hydrogen H_1^1
  Deuterium H_1^2
  Tritium H_1^3

In this notation, the subscript following the symbol of the element indicates the atomic number of the element. The superscript following the symbol of the element is the atomic mass number; thus the superscript indicates which isotope of the element is being referred to.

The general symbol for any atom is thus:

$$X_Z^A$$

where X is the symbol of the element, z is the atomic number (number of protons in the nucleus), and A is the atomic mass number (sum of the number of protons and the number of neutrons).

Of the known 103 elements, there are approximately 1,000 isotopes, most of which are radioactive.[2] Figure 10–4 gives the nuclear composition of various isotopes.

Radioactivity

All isotopes with atomic number z greater than 83 are naturally radioactive, and many more isotopes can be made artificially radio-

Element	Symbol	Atomic no. or no. of protons	No. of electrons	No. of neutrons
Hydrogen	H_1^1	1	1	0
Hydrogen (Deuterium)	H_1^2	1	1	1
Hydrogen (Tritium)	H_1^3	1	1	2
Helium	He_2^3	2	2	1
Helium	He_2^4	2	2	2
Helium	He_2^5	2	2	3
Helium	He_2^6	2	2	4
Beryllium	Be_4^9	4	4	5
Cadmium	Cd_{48}^{113}	48	48	65
Polonium	Po_{84}^{210}	84	84	126
Radium	Ra_{88}^{226}	88	88	138
Uranium	U_{92}^{234}	92	92	142
Uranium	U_{92}^{235}	92	92	143
Uranium	U_{92}^{238}	92	92	146
Uranium	U_{92}^{239}	92	92	147
Neptunium	Np_{93}^{239}	93	93	146
Plutonium	Pu_{94}^{239}	94	94	145

Figure 10–4. Nuclear composition of various isotopes.

active by bombarding the nucleus with neutrons, which upset the neutron-proton ratio of the normally stable nucleus.

Naturally radioactive isotopes undergo *radioactive decomposition*, thereby forming lighter and more stable nuclei. Radioactive decomposition occurs through the emission of an *alpha particle* or a *beta particle*. One or more *gamma rays* may also be emitted with the alpha or beta particle.

An alpha particle (symbol a_2^4) is composed of two protons and two neutrons. It is the nucleus of a helium (He_2^4) atom, has an electrical charge of +2, and is very stable. In the decay process to a more stable element, many unstable nuclei emit an alpha particle. The results of alpha emission can be seen from the following equation:

$$U_{92}^{238} \rightarrow Th_{90}^{234} + a_2^4$$

In the above equation, the parent isotope of uranium (U_{92}^{238}) is a naturally occurring, radioactive isotope that decays by alpha emission. Since the A and z numbers must balance in a nuclear equation, and since an alpha particle contains two protons, we see that the uranium has changed to an entirely new element.

The radioactive isotope of thorium (Th_{90}^{234}) produced in the above reaction further decays by the emission of a beta particle (symbol e_{-1}^{0}) as indicated in the following equation:

$$Th_{90}^{234} \rightarrow Pa_{91}^{234} + e_{-1}^{0}$$

The beta particle has properties similar to those of an electron.[3] However, the origin of the beta particle is within the nucleus rather than the orbital shells of an atom. It is postulated that a beta particle is emitted at an extremely high energy level when a neutron within the nucleus decays to a proton and an electron (beta particle). When this phenomenon occurs, the proton stays within the nucleus, forming an isotope of a different element but having the same mass as the parent isotope.

A radioactive isotope may go through several transformations of the above types before reaching a stable state. In the case of U_{92}^{238} there are a total of eight alpha particles and six beta particles emitted before a stable isotope of lead (Pb_{82}^{206}) is obtained.

The third manner in which a naturally radioactive isotope may reach a more stable configuration is by the emission of gamma rays (symbol γ). The gamma ray is an electromagnetic type of radiation with high frequency, high energy, and a short wavelength. Gamma rays are similar to X-rays in that the properties are the same. The distinguishing factor between the two is the fact that gamma rays originate in the nucleus of an atom, whereas the X-ray originates with the orbital electrons. In general it can be said that a gamma ray is of higher energy, higher frequency, and shorter wavelength than an X-ray. Gamma decay does not alter the mass or atomic numbers.

Frequently an isotope that emits an alpha or beta particle in the decay process will emit one or more gamma rays at the same time, as in the case of Co_{27}^{60}, an isotope that decays by beta emission and at the same time emits two gamma rays of different energy levels. Some radioactive isotopes reach a stable state by the emission of gamma rays only. In the latter case, since gamma rays have neither mass nor electrical charge, the A and z numbers of the isotope remain unchanged, but the energy level of the nucleus is reduced.

An important property of any radioactive isotope is the time involved in radioactive decay. To understand the time element, it is necessary to understand the concept of *half-life*. Half-life may be defined as the time required for one-half of any given number of radioactive atoms to disintegrate, thus reducing the radiation intensity of that particular isotope by one-half. Half-lives may vary from microseconds to billions of years. At times an isotope may be said to be "short-lived" or "long-lived," depending upon its peculiar radioactive half-life. Some half-lives of typical elements are:

$$\text{Half-life of } U_{92}^{238} = 4.51 \times 10^9 \text{ years}$$
$$\text{Half-life of } U_{92}^{235} = 7.13 \times 10^8 \text{ years}$$
$$\text{Half-life of } Ra_{88}^{226} = 1620 \text{ years}$$
$$\text{Half-life of } I_{53}^{135} = 6.7 \text{ hours}$$
$$\text{Half-life of } Po_{84}^{214} = 16^{-6} \text{ second}$$

As stated previously, naturally radioactive isotopes decay by the emission of alpha particles, beta particles, gamma rays, or a combination thereof. In the case of induced nuclear reactions there are many other phenomena that may occur, including fission and the emission of neutrons, positrons, nutrinos, and other forms of energy.[4]

Conservation of Mass and Energy

The conservation of energy was discussed in Chapter 1 of this text. It now becomes necessary to consider mass and energy as two phases of the same principle. In so doing, the law of conservation becomes:

(mass + energy) before = (mass + energy) after

Fundamental to the above and to the entire subject of nuclear power is Einstein's mass-energy equation, where the following relation holds:

$$E = mc^2$$

where E = energy in ergs, m = mass in grams, and c = velocity of light (3×10^{10} cm/sec). Mass and energy are not conserved separately but can be converted into each other. The conversion of mass into energy is not seen in normal chemical and thermodynamic processes, and that is why quantities seem to be conserved separately.

Several units and conversion factors that have become conventional in the field of nuclear engineering are listed below.

1 ev (electron-volt) = the energy acquired by an electron as it moves through a potential difference of 1 volt

1 Mev (million electron-volts) = 10^6 ev
 = 1.517×10^{-16} Btu

1 amu (atomic mass unit) = $\frac{1}{16}$ of the mass of an oxygen atom (by definition)

1 amu = 1.49×10^3 erg
 = 1.66×10^{-24} g
 = 931 Mev
 = 1.415×10^{-13} Btu

Nuclear Energy Source

It was previously stated that the atomic mass number is the total number of nucleons within the nucleus. It can also be said that the atomic mass number is the nearest integer (as found by experiment) to the actual mass of an isotope. In nuclear equations, the entire mass must be accounted for; therefore the actual mass must be considered.

The atomic mass of any isotope is somewhat less than indicated by the sum of the individual masses of the protons, neutrons, and orbital electrons that are the components of that isotope. This difference is termed *mass defect*: it is equivalent to the binding energy of the nucleus. *Binding energy* may be defined as the amount of energy released when a nucleus is formed from its component parts.

The binding energy of any isotope may be found as in the following example of copper (Cu_{29}^{63}), which contains 34 neutrons, 29 protons, and 29 electrons. Using the values given in figure 10–2 we find:

$$34 \times 1.00894 = 34.30396 \text{ amu}$$
$$29 \times 1.00758 = 29.21982 \text{ amu}$$
$$29 \times 0.00055 = 0.01595 \text{ amu}$$

$$\text{Total of component masses} = 63.53973 \text{ amu}$$
$$\text{Less actual mass of atom} = 62.9298 \text{ amu}$$
$$\text{Mass defect} = 0.60993 \text{ amu}$$

Converting to energy, we find:

$$931 \text{ Mev/amu} \times 0.60993 \text{ amu} = 567.84483 \text{ Mev}$$

or:

$$567.8 \div 63 = 9.0 \text{ Mev/nucleon}$$

The relationship between mass number and the average binding energy per nucleon is shown in figure 10–5.

Because binding energy is released when a nucleus is formed from its component parts, it is necessary to add energy to separate a nucleus. In the fissioning of uranium-235, the additional energy is supplied by bombarding the fissionable fuel with neutrons. The fissionable material absorbs a neutron and is converted into a compound nucleus of uranium-236, which fissions instantaneously.

There are more than 40 different ways a uranium-235 nucleus may fission, resulting in more than 80 different fission products.[5] For the purpose of this discussion, let us consider the most probable fission of a uranium-235 nucleus. In slightly more than 6 percent of the fissions, the uranium-235 nucleus will split into fragments having mass numbers of 95 and 139. The following equation is typical:

$$n_0^1 + U_{92}^{235} \rightarrow Y_{39}^{95} + I_{53}^{139} + 2n_0^1$$

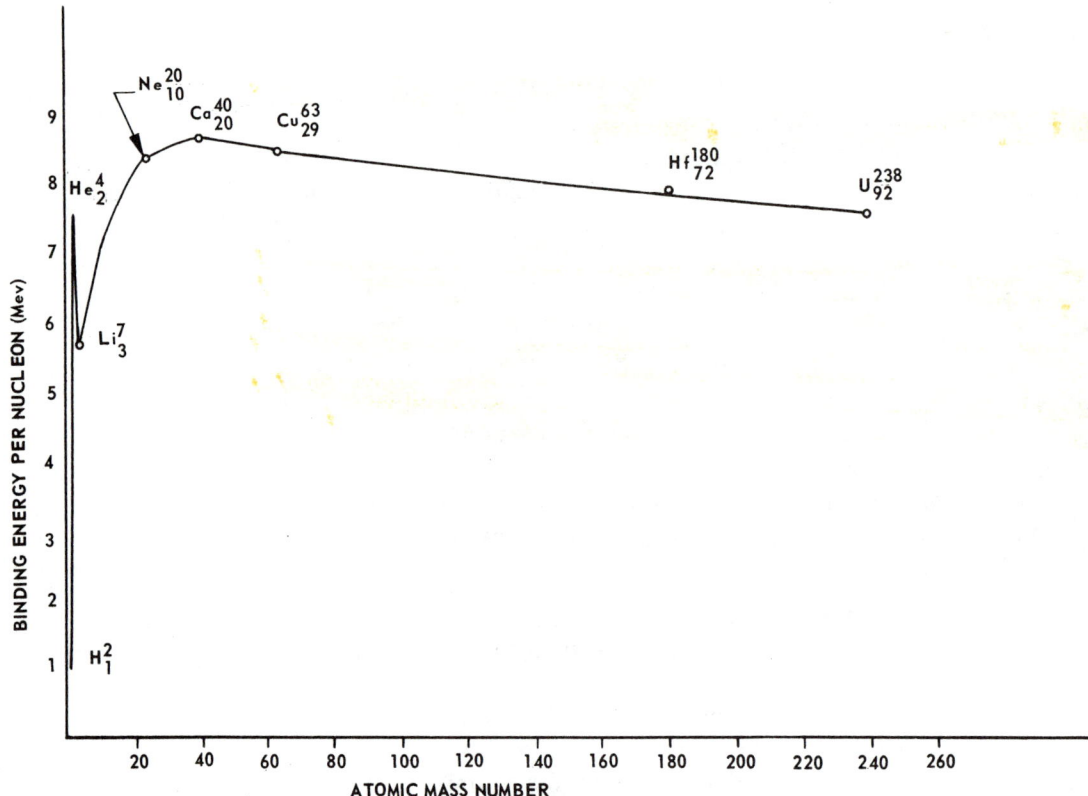

Figure 10–5. Relationship between atomic mass number and average binding energy per nucleon.

where the daughter products, yttrium and iodine, are both radioactive and decay through beta emission to the stable isotopes of molybdenum (Mo_{42}^{95}) and lanthanum (La_{57}^{139}), respectively.

One method of determining the energy released from the above reaction is to find the difference in atomic mass units of the daughter products and the original nucleus. It is also necessary that we account for the neutron used to bombard the uranium-235 atom and the two neutrons liberated in the fission process. In the investigation of energy released in this reaction we find:

Mass of uranium-235 atom	= 235.0439
Mass of neutron	= 1.00894
Original mass	= 236.05284 amu
Mass of molybdenum-95 atom	= 94.9058
Mass of lanthanum-139 atom	= 138.9061

Mass of two neutrons = 2.01788
Total mass of fission
  fragments = 235.82978
Mass defect = 236.05284 − 235.82978 = 0.22306 amu/fission

Hence,

0.22306 amu/fission × 931 Mev/amu = 207.7 Mev/fission

Thus we find that from each fission approximately 200 Mev of energy is released, most of which (about 80 percent) appears immediately as kinetic energy of the fission fragments. As the fission fragments slow down, they collide with other atoms and molecules; this results in a transfer of velocity to the surrounding particles. The increased molecular motion is manifested as sensible heat. The remaining energy is realized from the decay of fission fragments by beta particle and gamma ray emission, kinetic energy of fission neutrons, and instantaneous gamma ray energy.

In a nuclear reactor, the two neutrons liberated in the above reaction are available, under certain conditions, to fission other uranium atoms and assist in keeping the reactor *critical*. A nuclear reactor is said to be critical if the *neutron flux* remains constant. Neutron flux is defined as the number of neutrons passing through a unit area in a unit time. A neutron flux of 10^{13} neutrons per square centimeter per second is not uncommon. If the neutron flux is decreasing, the reactor is said to be subcritical; conversely, a reactor is supercritical if the neutron flux is increasing.

Neutron Reactions

Neutrons may be classified by their energy levels. A *fast neutron* has an energy level of greater than 0.1 Mev; an *intermediate neutron* in the process of slowing down possesses an energy level between 1 ev and 0.1 Mev; a *thermal neutron* is in thermal equilibrium with its surroundings and has an energy level of less than 1 ev.

Neutrons lose their kinetic energy by interacting with atoms in the surrounding area. The probability of a neutron's interacting with one atom is dependent upon the target area presented by that atom for a neutron reaction. This target area (which is the probability of a neutron reaction occurring) is called the *cross section*. The unit of cross section measurement is barns. The size of a barn is 10^{-24} square centimeter. Four of the different cross sections that an element may have for neutron processes are described as follows:

Scattering cross section is a measure of the probability of an elastic (billiard ball) collision with a neutron. In this type of collision part of the kinetic energy of the neutron is imparted to the atom, and the neutron rebounds after collision. Neutrons are thermalized (reduced to an energy level below 1 ev) by elastic collisions.

Capture cross section is a measure of the probability of the neutron's being captured without causing fission.

Fission cross section is a measure of the probability of fission of the atom after neutron capture.

Absorption cross section is a measure of the probability that an atom will absorb a neutron. The absorption cross section is the sum of the capture cross section and the fission cross section.

The cross section for any given element may vary with the energy level of the approaching neutron. In the case of uranium-235, the absorption cross section for a thermal neutron is 100 times the cross section for a fast neutron.

Reactor Principles

A nuclear reactor must contain a *critical mass*. A critical mass contains sufficient fissionable material to enable the reactor to maintain a self-sustaining chain reaction, thereby keeping the reactor critical. A critical mass is dependent upon the species of fissionable material, its concentration and purity, the geometry and size of the reactor, and the matter surrounding the fissionable material.[6]

Reactor Fuels

The form and composition of a reactor fuel may vary both in design and in the fissionable isotope used. Many commercial power reactors use a solid fuel element fabricated in plate form, with the fissionable material being enriched uranium in combination with aluminum, zirconium, or stainless steel. Fuel elements may be arranged in thin sandwich layers, as shown in figure 10–6. This construction provides a relatively large heat-transfer area between the fuel elements and the reactor coolant.

The outer cladding on the fuel elements confines the fission fragments within the fuel elements and serves as a heat-transfer surface. Cladding materials should be resistant to corrosion, should be able to withstand high temperatures, and should have a small cross section for neutron capture. Three common cladding materials are aluminum, zirconium, and stainless steel.

The fuel elements may be assembled in groups, some of which may contain control rods. Several groups of fuel elements placed within a reactor vessel make up the reactor core. It is not necessary that all fuel groups within the reactor contain control rods.

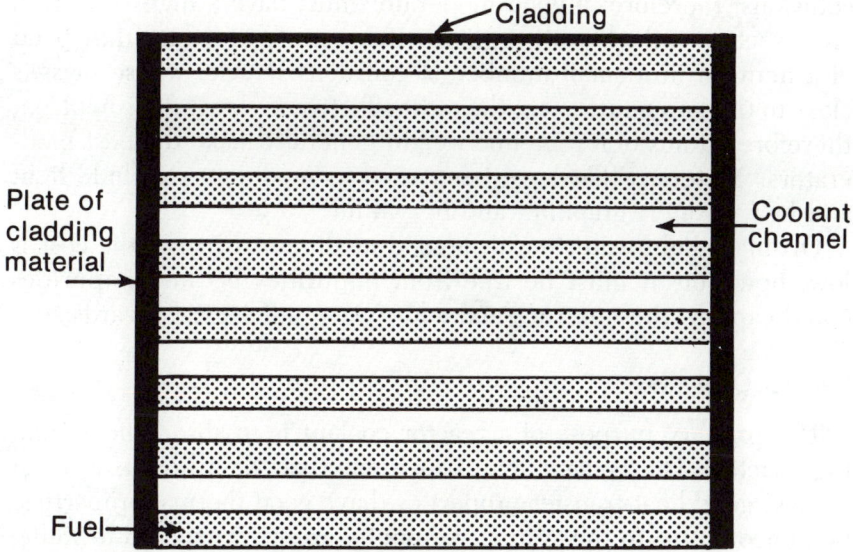

Figure 10-6. Drawing of a PWR fuel element.

Control Rods

Control rods serve a dual purpose in a reactor. They keep the neutron density (neutron flux) constant within a critical reactor, and they provide a means of shutting down the reactor.

The material for a control rod must have a high capture cross section for neutrons and a low fission cross section. Three materials suitable for control rod fabrication are cadmium, boron, and hafnium. Hafnium is particularly suitable for control rods because it has a relatively high capture cross section and because several daughter products after neutron capture are stable isotopes that also have good capture cross sections.

The control rods are withdrawn from the reactor core until criticality is obtained; thereafter very little movement is required. It is important to note at this point that after criticality is reached, movement of control rods does not control the power output of the reactor; it controls only the temperature of the reactor.

Control rod drive mechanisms are so designed that, should an emergency shutdown of the reactor be required, the control rods may be inserted in the core very rapidly. A shutdown of this type is called a *scram*.

Moderators

A moderator is the material used to thermalize the neutrons in a reactor. As previously stated, neutrons are thermalized by elastic collisions; therefore, a good moderator must have a high scattering cross section and a low absorption cross section, to reduce the speed of a neutron in a small number of collisions. Nuclei whose mass is close to that of a neutron are the most effective in slowing the neutron; therefore, atoms of low atomic weight generally make the best moderators. Materials that have been used as moderators include light and heavy water, graphite, and beryllium.

Ordinary light water makes a good moderator because the cost is low; however, it must be free from impurities because impurities could capture the neutrons and add to the radiological hazards.

Reactor Coolants

The primary purpose of a reactor coolant is to absorb heat from the reactor. The coolant may be either a gas or a liquid; it must possess good heat-transfer properties, have good thermal properties, be noncorrosive to the system, be nonhazardous if exposed to radiation, and be of low cost. Coolants that have been used in operational and experimental reactors include light and heavy water, liquid sodium, and carbon dioxide.

Reflectors

In a reactor of finite size, the leakage of neutrons from the core becomes somewhat of a problem. To minimize the leakage, a reflector is used to assist in keeping the neutrons in the reactor. The use of a reflector reduces both the required size of the reactor and the radiation hazards of escaping neutrons. The characteristics required for a reflector are essentially the same as those required for a moderator.

Because ordinary water of high purity is suitable for moderators, coolants, and reflectors, the inference is that it could serve all three functions in the same reactor. This is indeed the case in many nuclear reactors.

Shielding

The shielding of a nuclear reactor serves the dual purpose of (1) reducing the radiation so that it will not interfere with the necessary instrumentation, and (2) protecting operating personnel from radiation.

The type of shielding material used is dependent upon the purpose of the particular reactor and upon the nature of the radioactive particles being attenuated or absorbed.

Shielding against alpha particles is a relatively simple matter. Since an alpha particle has a positive electrical charge of 2, a few centimeters of air is all that is required for attenuation. Any light material such as aluminum or plastic makes a suitable shield for beta particles.

Neutrons and gamma rays have considerable penetrating power; therefore, shielding against them is more difficult. Since neutrons are best attenuated by elastic collisions, any hydrogenous material such as polyethylene or water is suitable as a neutron shield. Sometimes polyethylene with boron is used for neutron shields since boron has a high neutron capture cross section. Gamma rays are best attenuated by a dense material such as lead.

Types of Nuclear Reactors

The purpose of any power reactor is to provide thermal energy that can be converted to useful work. Several types of experimental and operational reactors have been designed. They include the pressurized water reactor (PWR), the sodium cooled reactor, the experimental boiling water reactor, the experimental breeder reactor, and the experimental gas cooled reactor.

The first full-scale nuclear-powered central station in the United States was the pressurized water reactor (PWR) at Shippingport, Pennsylvania. The Shippingport PWR is a thermal, heterogeneous reactor fueled with enriched uranium-235 "seed assemblies" arranged in a square in the center of the core, surrounded by "blanket assemblies" of uranium-238 fuel elements. Figure 10–7 shows a cross-sec-

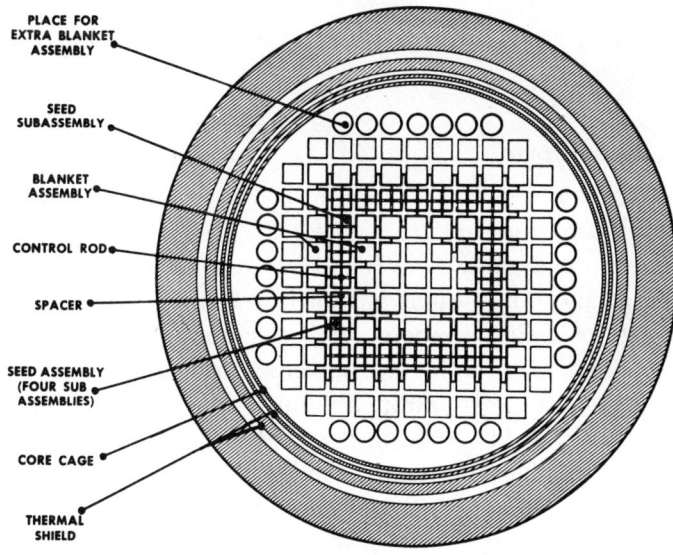

Figure 10–7. Cross-sectional view of a PWR reactor and core. (Courtesy Westinghouse Corp.)

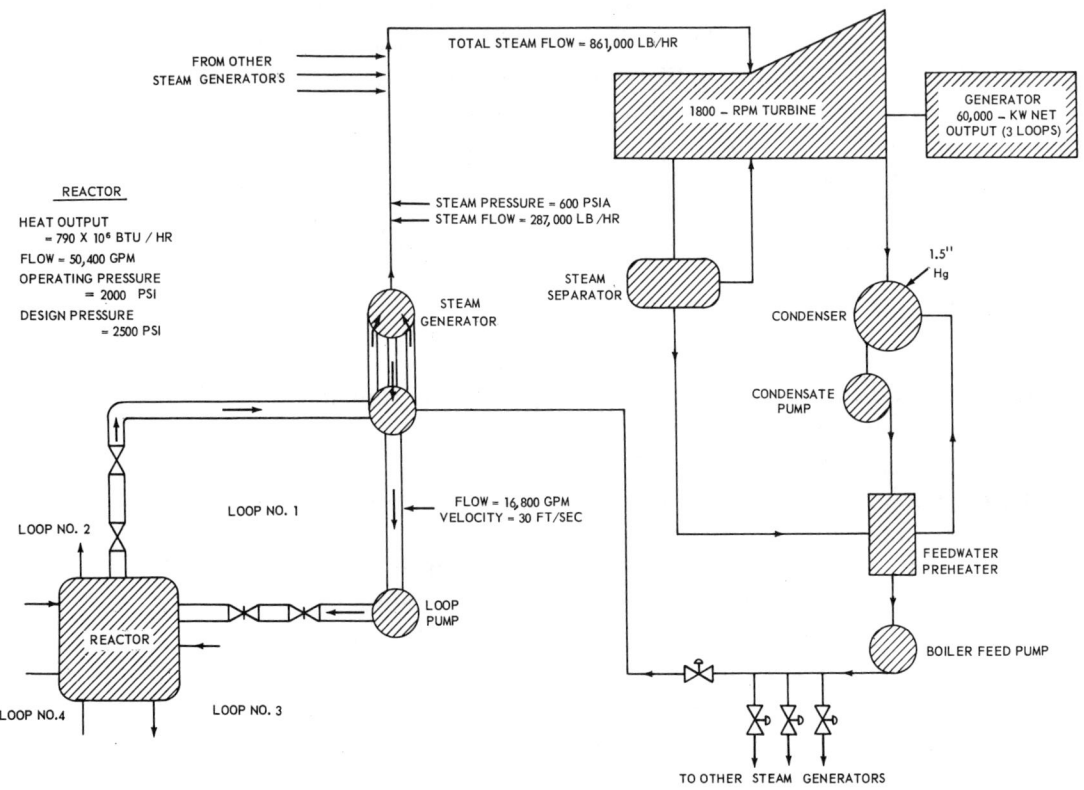

Figure 10–8. Schematic diagram of a PWR plant. (Courtesy Westinghouse Corp.)

tional view of the PWR reactor and core. This type of reactor can be called a *converter,* since the uranium-238 is converted into the fissionable fuel of plutonium-239.

A schematic diagram of a PWR and its associated steam plant with power output and flow ratings is shown in figure 10–8. The reactor plant consists of a single reactor with four main coolant loops; the plant is capable of maintaining full power on three loops. Each coolant loop contains a steam generator, a pump, and associated piping.

High-purity water at a pressure of 2000 psia serves as both moderator and coolant for the plant. At full power the inlet water temperature to the reactor is 508°F, and the outlet temperature is 542°F.

The coolant enters the bottom of the reactor vessel (figure 10–9) where 90 percent of the water flows upward between the fuel plates, with the remainder bypassing the core in order to cool the walls of

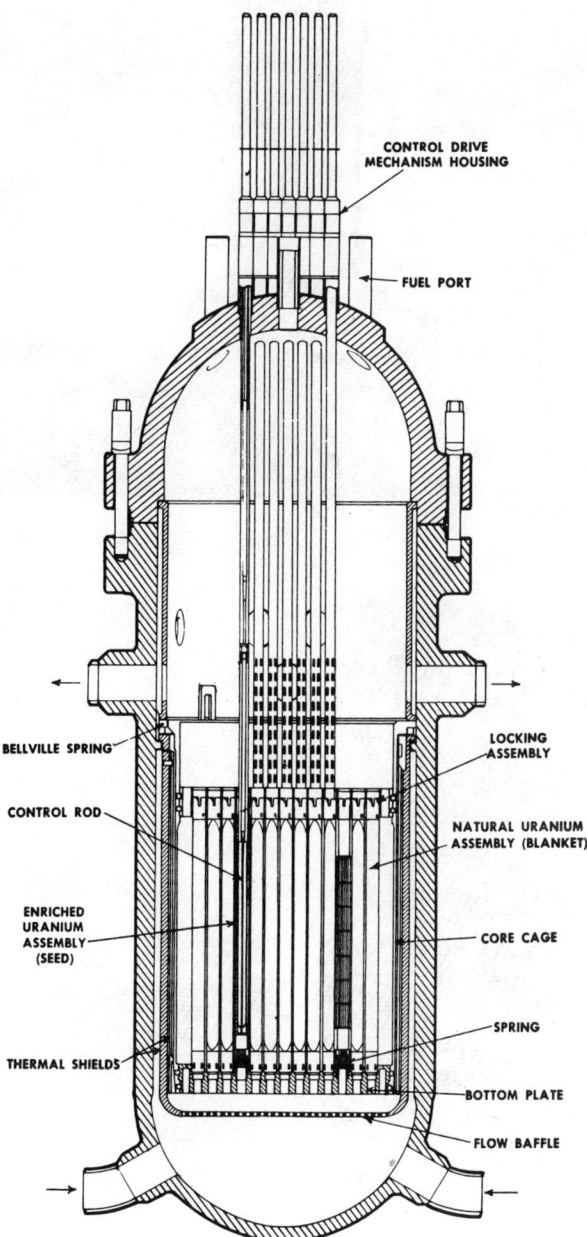

Figure 10-9. Longitudinal section of a PWR reactor. (Courtesy Westinghouse Corp.)

Nuclear Power Plants 181

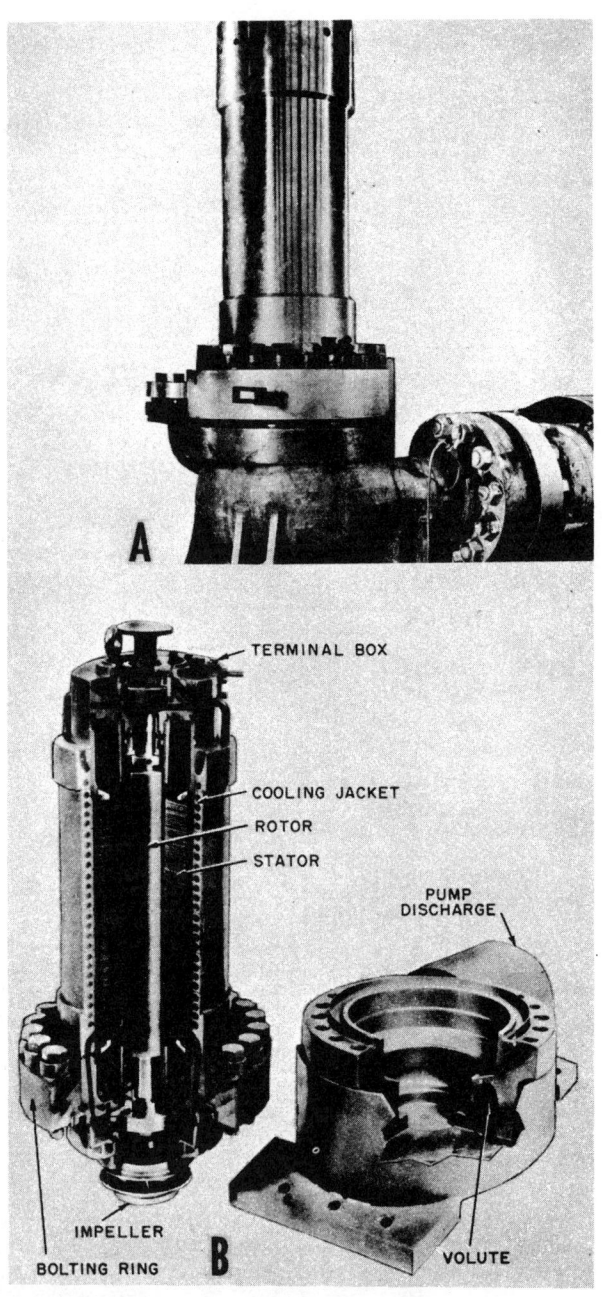

Figure 10–10. PWR main coolant pump: (A) external view, (B) cutaway view. (Courtesy Westinghouse Corp.)

the reactor vessel and the thermal shield. After having absorbed heat as it goes through the core, the water leaves the top of the reactor vessel through the outlet nozzles and flows through connecting piping to the steam generator.

The steam generator is a shell-and-tube type of heat exchanger with the primary coolant (reactor coolant) flowing through the tubes and the secondary water (boiler water) surrounding the tubes. Heat is transferred to the secondary water in the steam generator, producing high-quality saturated steam for use in the turbines.

The primary coolant flows from the steam generator to a hermetically sealed (canned rotor) pump (figure 10–10) and is pumped through connecting piping to the bottom of the reactor vessel to complete the primary coolant cycle.

The pressure on the reactor vessel and the main coolant loop is maintained by a pressurizing tank (figure 10–11), which operates under the saturation conditions of 636°F and 2000 psia. A second function of the pressurizing tank is to act as a surge tank for the primary system. Under no-load conditions the inlet, outlet, and average temperatures of the reactor coolant are nearly equal in value. As the power increases, the average temperature remains constant, but the inlet and outlet temperatures diverge. Since the colder leg of the primary coolant is the longer, the net effect in the pressurizer is a decrease in level to make up for the increase in density of the water in the primary loop. The reverse holds true with a decreasing power level. Electrical heaters and a spray valve with a supply of water from the cold leg of the primary coolant assist in maintaining a steam blanket in the upper part of the pressurizer, and also assist in maintaining saturation conditions of 2000 psia and 636°F.

Principles of Reactor Control

Reactor control principles[7] that are of particular interest to this discussion include the concepts of the negative temperature coefficient, delayed neutron action, and poisoning of fuel.

The term *negative temperature coefficient* is used to express the relationship between temperature and reactivity—as temperature decreases, reactivity increases. The negative temperature coefficient is a design requirement and is achieved by the proper ratio of elements in the reactor, the geometry of the reactor, and the physical size of the reactor. The negative temperature coefficient makes it possible to keep a power reactor critical with minimum movement of the control rods.

The concept of negative temperature coefficient may be most easily

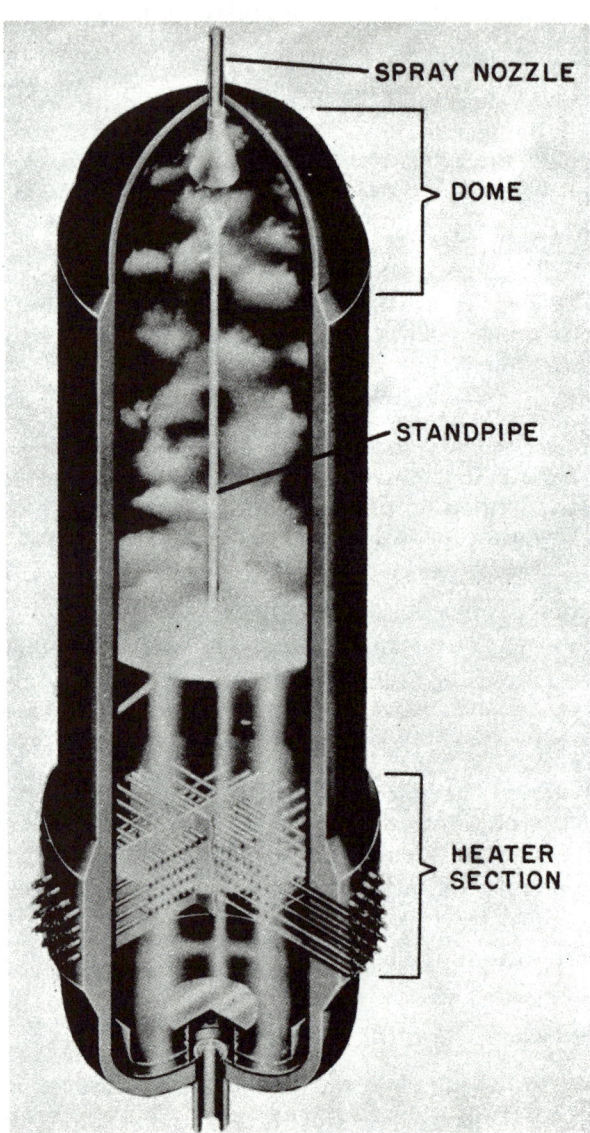

Figure 10–11. Cutaway view of a PWR pressurizing tank. (Courtesy Westinghouse Corp.)

understood by an example. Assume that, in the PWR plant shown in figure 10–8, the reactor is critical and the machinery is operating at a given power level. Now, if the valve is opened to increase the turbine speed, there will also be an increase in the rate of steam flow, and the power level of the reactor, and a resultant decrease in the temperature of the primary coolant leaving the steam generator. The decrease in temperature is small but significant in that it results in an increase in density of the coolant. As the density of the coolant increases, so does the magnitude of the neutron scattering cross section. The higher value of the scattering cross section allows the coolant, in its capacity as moderator, to thermalize neutrons at a faster rate, supplying more thermal neutrons to be absorbed in the fuel. As more neutrons are absorbed in the fuel, more fissions occur, resulting in a higher power level and more heat being generated by the reactor. The additional heat is removed by the reactor coolant to the secondary water in the steam generator to compensate for the increased steam demand by the turbine. The temperature of the primary coolant leaving the steam generator increases slightly, lowering the scattering cross section of the moderator, and the reactor stabilizes at a higher power level.

The phenomenon of *delayed neutron action* simplifies reactor control considerably. Each fission in a nuclear reactor releases on the average between two and three neutrons that either leak out of the reactor or are absorbed in reactor materials. If the reactor material that absorbs the neutron happens to be the fissionable fuel, and the neutron is of proper energy level, another fission is likely to result. The majority of the neutrons released in the fission process appear instantaneously and are termed prompt neutrons, but other neutrons are born after fission and are termed delayed neutrons. The delayed neutrons appear in a time range of seconds to 3 or more minutes after the fission takes place. The weighted mean lifetime of the delayed neutrons is approximately 12 seconds. About 0.75 percent of the neutrons produced in the fission process are delayed neutrons.

Should a reactor become prompt critical (critical on prompt neutrons), it would be very difficult to control, and any delayed neutrons would tend to make it supercritical. However, the delayed neutrons have the effect of increasing the reactor period sufficiently to permit reactor control. (Reactor period is the time required to change the power level by a factor of e, the base of the system of natural logarithms.)

A nuclear *poison* is material in the reactor that has a high absorption cross section for neutrons. Some poisons are classed as burnable poisons and are placed in the reactor for the purpose of extending the core life; other poisons are generated in the fission process and have a tendency to be a hindrance to reactor operation.

A burnable poison has a relatively high cross section for neutron absorption but is used primarily in the early part of the core life. If a burnable poison is added to the reactor, more fuel can be loaded into the core, thus extending the life of the core.

Most of the fission products produced in a reactor have a small absorption cross section. The most important one that does have a high absorption cross section for neutrons is xenon-135; it can become a problem near the end of core life. Xenon-135 is only a direct fission product a small percentage of the time, but is mostly produced in the decay of iodine-135, as indicated in the following decay chain:

$$I_{53}^{135} \xrightarrow{6.7 \text{ hr}} Xe_{54}^{135} \xrightarrow{9.2 \text{ hr}} Cs_{55}^{135} \xrightarrow{2.0 \times 10^6 \text{ yr}} Ba_{56}^{135}$$

Xenon-135 has a high neutron-absorption cross section. In normal operation of the reactor, xenon-135 absorbs a neutron and is transformed to the stable isotope of xenon-136, which presents no poison problem to the reactor. Xenon equilibrium is reached after about 40 hours of steady-state operation. At this point the same amount of xenon-135 is being "burned" by neutron absorption as is being produced by the fission process:

$$Xe_{54}^{135} + n_0^1 \rightarrow Xe_{54}^{136}$$

The second, and perhaps more serious, effect of xenon poisoning occurs near the end of core life. As indicated by the half-lives shown in the xenon decay chain, xenon-135 is produced at a rate faster than its decay rate. The buildup of xenon-135 in the reactor reaches a maximum about 11 hours after shutdown. Should a scram occur near the end of core life, the xenon buildup may make it impossible to make the reactor go critical until the xenon has decayed off. In a situation of this type, the reactor may have to sit idle for as much as two days before it is capable of overriding the poison buildup.

The Naval Nuclear Power Plant

Since many aspects of the design and operation of naval nuclear propulsion plants involve classified information, the information presented here is necessarily brief and general in nature.

In a nuclear power plant designed for ship propulsion, weight and space limitations and other factors must be taken into consideration, in addition to the factors involved in the design of a shore-based power plant.

The thermodynamic cycle of the shipboard nuclear propulsion plant is similar to that of the conventional steam turbine propulsion plant. Instead of a boiler, however, the nuclear propulsion plant utilizes a pressurized water reactor as the heat source and a steam generator as a heat exchanger to generate the steam used to drive the propulsion turbines.

The steam generator is a heat exchanger in which the primary coolant transfers heat to the secondary system (boiler water) by conduction. The water in the secondary side of the steam generator, being at lower pressure, changes from the physical state of water to the physical state of steam. This steam then flows through piping to the engineroom.

The engineroom equipment consists of propulsion turbines, turbogenerators, condensers, and associated auxiliaries.

Problems of Nuclear Power

Although many developmental and engineering problems associated with nuclear power have been solved to some extent, some problems remain. A few problems that are of particular importance in connection with the shipboard nuclear power plant are noted here briefly.

The remote possibility of radiological hazards exists even though the radiation is well contained in the shipboard nuclear reactor. To eliminate or minimize the radiological hazards, a high degree of quality control is essential in the design, construction, and operation of nuclear power plants. The high pressures and temperatures used in nuclear reactors, together with the prolonged periods of continuous operation, pose materials problems. For shipboard use, the great weight of the materials required for shielding presents still other problems.

Although many of these problems may be solved by further technological developments, the problems involved in the selection and training of personnel for nuclear ships appear to be continuing ones. The safe and efficient operation of a shipboard nuclear plant requires highly skilled, responsible personnel who have been thoroughly trained in both the academic and the practical aspects of nuclear propulsion. The selection and training of such personnel is inevitably costly in terms of time and money.

Notes

1. For a discussion of nuclear fusion, see M. O. Hagler and M. Kristiansen, *An Introduction to Controlled Nuclear Fusion* (Lexington, Mass.: Lexington Books, 1977).

2. For a detailed discussion of nuclear stability, see Francis W. Sears, Mark W. Zemansky, and Hugh D. Young, *University Physics* (5th ed.; Reading, Mass.: Addison-Wesley Publishing Co., 1976).

3. Sears, Zemansky, and Young, *University Physics*.

4. For detailed information on nuclear particles, refer to Edward J. Burge, *Atomic Nuclei and Their Particles* (Oxford: Clarendon Press, 1977).

5. For a detailed discussion on nuclear fission, see James J. Duderstandt and Louis J. Hamilton, *Nuclear Reactor Analysis* (New York: John Wiley & Sons, 1976).

6. For a discussion of the aspects of reactor design, see Duderstandt and Hamilton, *Nuclear Reactor Analysis*.

7. John F. Hogerton, *The Atomic Energy Deskbook* (New York: Reinhold Publishing Corp., 1963), p. 463.

11

Diesel and Gasoline Engine Theory

Much of the machinery and equipment discussed in the preceding chapters utilizes steam as the working fluid in the process of converting thermal energy to mechanical energy. This chapter deals with internal combustion engines, in which air (or a mixture of air and fuel) serves as the working fluid. The internal combustion engines considered are those to which the thermodynamic cycles of the open- and heated-engine types apply. In engines that operate on these cycles, the working fluid is taken into the engine, heat is added to the fluid, the energy available in the fluid is utilized, and then the fluid is discarded. During the process, thermal energy is converted to mechanical energy. The purpose of this chapter is to present the basic theory and the fundamental principles underlying the energy conversion in internal combustion engines. These internal combustion engines can be of the reciprocating type (diesel and gasoline), or a gas turbine providing thrust or turning a shaft.

Internal combustion engines are used extensively in the Navy, serving as propulsion units in a variety of installations such as ships, boats, airplanes, and automotive vehicles. Engines of the internal combustion type are also used as prime movers for auxiliary machinery such as emergency generators.

Principles of Operation

Most of the internal combustion engines in marine installations of the Navy are of the reciprocating type. This classification is based on the fact that the cylinders in which the energy conversion takes place are fitted with pistons, which employ a reciprocating motion. Internal combustion engines of the reciprocating type are commonly identified as diesel and gasoline engines.

The general trend in naval service is to install diesel engines rather than gasoline engines unless special conditions favor the use of the latter. Small boats used in conjunction with airplane facilities are frequently powered with gasoline engines, since the available fuel supply is gasoline. In addition, gasoline engines are used in many installations because of their small size or because of a lack of suitable diesel engines.

The gasoline engine and the diesel engine differ principally in that the former has a carburetor and a spark ignition system. The fuel and air for the *spark ignition (gasoline) engine* are mixed in the carburetor. This mixture is drawn into the cylinders where it is compressed and ignited by an electric spark.

The *compression ignition* type of engine is commonly known as a *diesel engine*. The diesel engine takes in atmospheric air, compresses it, and then injects the fuel into the combustion space. The heat generated by compression ignites the fuel; hence the term compression ignition is used for diesel engines.

The operation of an internal combustion engine of the reciprocating type involves the admission of fuel and air into a combustion space and the compression and ignition of the charge. The resulting combustion releases gases and increases the temperature within the space. As temperature increases, pressure increases and forces the piston to move, and this movement is transmitted through a chain of parts to a shaft. The resulting rotary motion of the shaft is utilized for work; thus, heat energy is transformed into mechanical energy. In order for the process to be continuous, the expanded gases must be removed from the combustion space, a new charge admitted, and then the process repeated.

In the study of engine operating principles, starting with the admission of air and fuel and following through to the removal of the expanded gases, it will be noted that a series of events takes place in the cylinder of an engine for each power impulse transmitted to the crankshaft. These events always occur in the same order each time the cycle is repeated. The number of events occurring in a cycle of operation will depend upon the engine type—diesel or gasoline. The difference in the events occurring in the cycle of operation for these engines is shown in Table 11–1.

The principal difference, as shown in the table, in the cycles of operation for diesel and gasoline engines involves the admission of fuel and air to the cylinder. Although this takes place as one event in the operating cycle of a gasoline engine, it involves two events in diesel engines. Thus, insofar as events are concerned, there are six main events taking place in the diesel cycle of operation and five in the cycle of a gasoline engine. This is pointed out in order to emphasize that the events that take place and the piston strokes that occur during a cycle of operation are not identical. Even though the events of a cycle are closely related to piston position and movement,

Table 11–1. Events and their sequence in a cycle operation for diesel and gasoline engines.

Diesel engine	Gasoline engine
Intake of air	*Intake* of fuel and air
Compression of air	*Compression* of fuel–air mixture
Injection of fuel	*Ignition* and *combustion* of charge
Ignition and *combustion* of charge	*Expansion* of gases
Expansion of gases	*Removal* of waste
Removal of waste	

Note: Actually expansion and combustion occur together, simultaneously.

all of the events will take place during the cycle regardless of the number of piston strokes involved.

Four-Stroke and Two-Stroke Cycles

All reciprocating internal combustion engines operate on either a two-stroke or a four-stroke cycle. A stroke is a single up or down movement of the piston, or the distance a piston moves between limits of travel. Each piston executes two strokes for each revolution of the crankshaft. The number of piston strokes occurring during any one series of operations required to perform a cycle determines whether it is a two-stroke (power stroke every crankshaft revolution) or four-stroke (power stroke for every two revolutions) cycle.

Let us use one cylinder of a gasoline engine to trace its operation through the four strokes that make up a cycle (see figure 11–1). The engine parts shown in this figure include only a cylinder, a crankshaft, a piston and connecting rod, and the inlet and exhaust valves. To simplify the diagrams (A through D), numerous engine components and accessories have been omitted.

In part A of the figure, the intake valve is open, and the exhaust valve is closed. The piston is moving downward and drawing a charge of air–fuel mixture from the carburetor into the cylinder through the open valve. This portion of the cycle, during which the piston is moving downward, is called the *intake* stroke.

When the crankshaft has rotated to the position shown in part B, the piston has moved upward, on the *compression* stroke, almost to the top of the cylinder. Both the intake and exhaust valves are closed during this stroke. The mixture that entered the cylinder during the intake stroke is compressed into the small space above the piston.

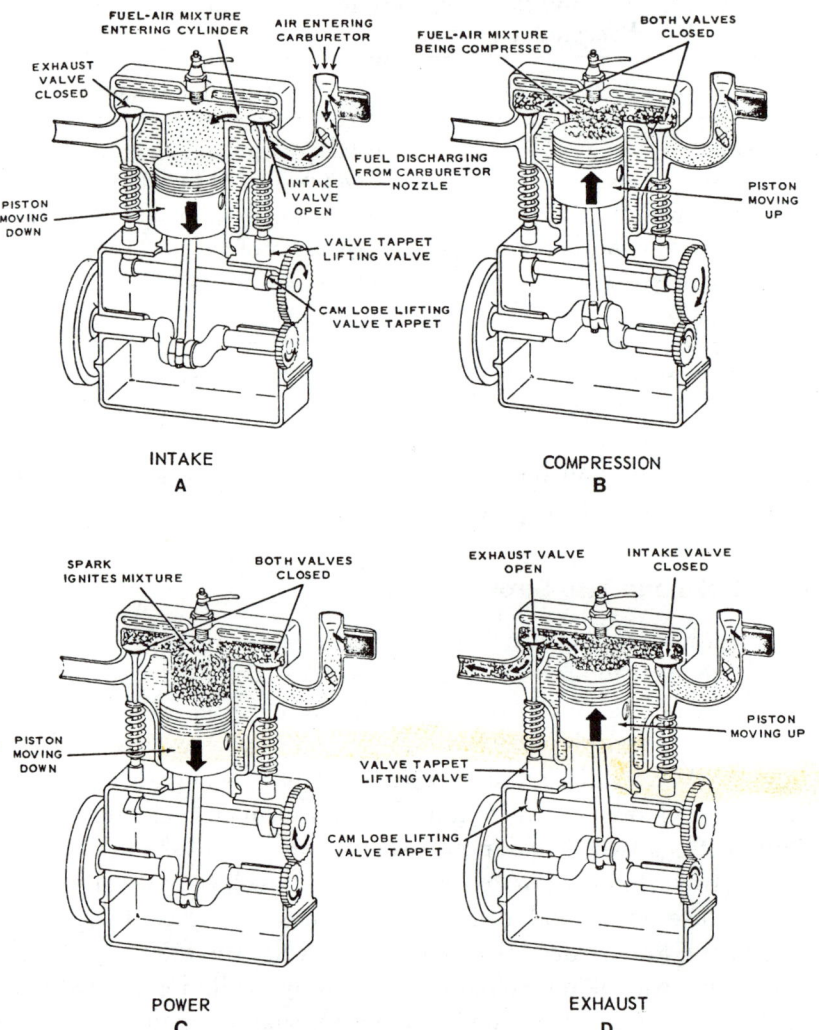

Figure 11-1. *Four-stroke cycle, spark ignition engine.*

The volume of this air may be reduced to less than one-eighth of what it was at the beginning of the stroke.

This compresssed mixture is now ready for ignition. A spark is introduced through a spark plug and ignites the air–fuel mixture.

During the *power* stroke, indicated in part C, the inlet and exhaust valves are both closed. The increase in temperature, resulting from the burning fuel, greatly increases the pressure on top of the piston. This increased pressure forces the piston downward and rotates the crankshaft. This is the only stroke in which power is furnished to the crankshaft by the piston.

During the *exhaust* stroke, shown in part D, the exhaust valve is open, and the intake valve remains closed. The piston moves upward, forcing the burned gases out of the combustion chamber through the exhaust valve. This stroke, which completes the cycle, is followed immediately by the intake stroke of the next cycle, and the sequence of events continues.

The four-stroke cycle diesel engine operates on the same mechanical cycle as the gasoline engine. In the diesel engine, air only is drawn into the cylinder during intake and is compressed. The high pressure that is a result of the great reduction in volume raises the temperature of the air far above the fire point of the fuel. When the piston reaches the top of the compression stroke, a shot of fuel is forced into the cylinder through the injector. The hot air which has been heated by compression ignites the fuel. The increased energy in the combustion gases forces the piston down, turning the crankshaft. As the piston approaches the bottom of the power stroke, the exhaust valves open. On the succeeding exhaust stroke, the piston forces the spent gas from the cyclinder.

Two-stroke cycle diesel engines are widely used in the Navy. Although some gasoline engines operate on the two-stroke cycle, their use is limited principally to small outboard motors, snowmobiles, and chain saws.

Every second stroke of a two-stroke cycle engine is a power stroke. The strokes between are compression strokes. The intake and exhaust functions take place rapidly at the bottom of each power stroke. With this arrangement there is one power stroke for each revolution of the crankshaft, or twice as many as in a four-stroke cycle engine.

The steps in the operation of a two-stroke diesel cycle engine are shown in figure 11–2. The cylinder has exhaust but no intake valves; instead it has holes or ports in the cylinder wall near the lowest point of the piston's travel. As the piston nears the bottom of the power stroke, shown in part A, it uncovers these intake ports. Air delivered under pressure by a blower (air pump) forces air in through the intake ports, and the burned gases are carried out through the exhaust valve. This scavenging operation takes place almost instantly and corresponds to the intake and exhaust strokes of the four-stroke cycle.

In part B of figure 11–2, the piston is moving upward on the compression stroke. The exhaust valves and the intake ports are now closed, and the piston is compressing the air trapped in the combustion chamber. At the top of the stroke, part C, fuel is sprayed into the cylinder where it is ignited by the hot (over 1000°F) compressed air. The power stroke has started.

In part D of the figure, the piston is moving downward, completing

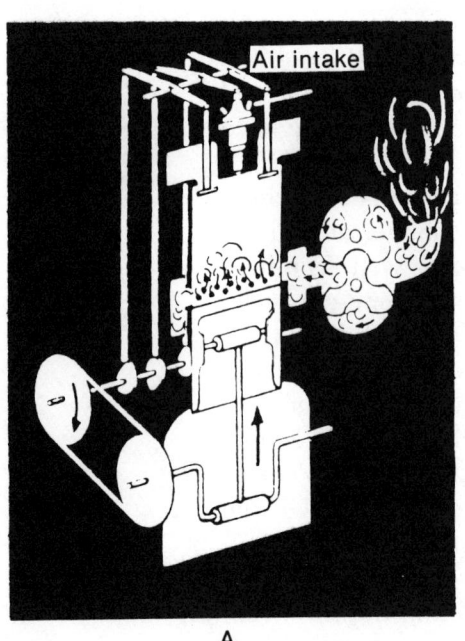

A

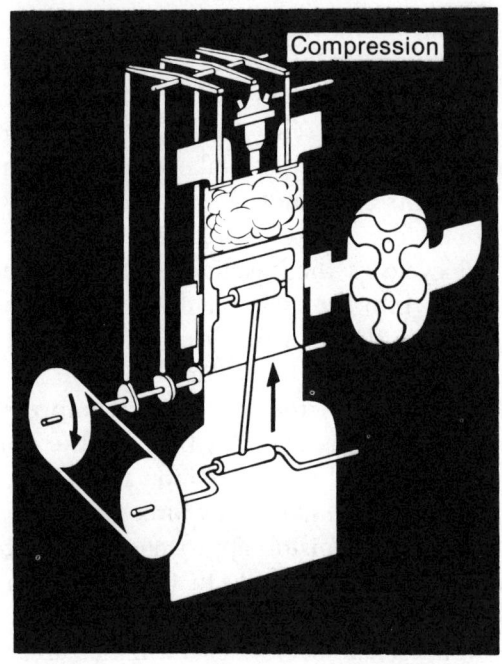

B

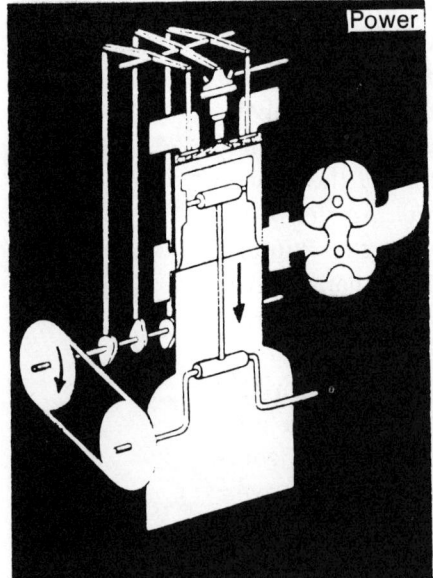

C

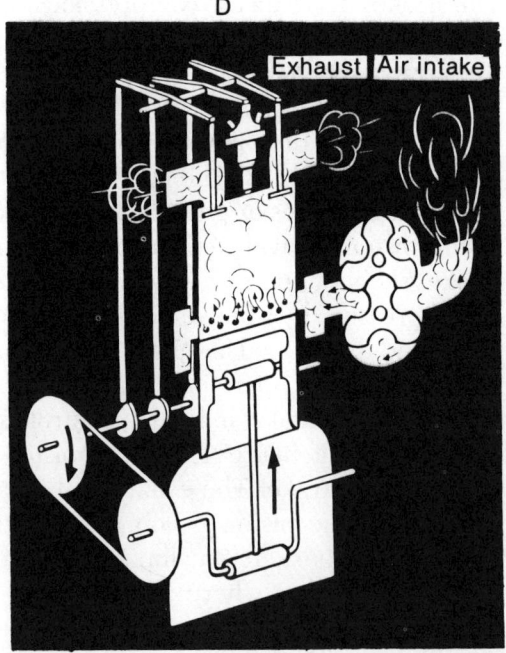

D

Figure 11-2. Steps in the operation of a two-stroke cycle engine.

the power sroke. The exhaust valve now opens, and the intake ports are uncovered, allowing the scavenging air to force the exhaust gas from the cylinder.

You might expect a two-stroke cycle engine to develop twice as much power as a four-stroke cycle engine. However, such is not the case because some of the engine's power is required to drive the blower, and some is lost because of the position of the intake ports at the cylinder bottom. Nevertheless, two-stroke cycle diesel engines give an increase in brake horsepower (BHP) and specific fuel consumption (SFC). Small gasoline engines operating on the two-stroke cycle principle operate satisfactorily, but in the larger sizes it is not practical because of SFC and unburned hydrocarbons.

Classification By Combustion-Gas Action

The classification of engines according to combustion-gas action is based upon a consideration of whether the pressure created by the combustion gases acts upon one or two surfaces of a single piston or against single surfaces of two separate and opposed pistons. The three types of engine under this classification are commonly referred to as single-acting, opposed-piston, and double-acting engines.

Single-Acting Engines

Engines of this type are those that have one piston per cylinder and in which the pressure of combustion gases acts only on one surface of the piston. This is a feature of design rather than principle, for the basic principles of operation apply whether an engine is single-acting, opposed-piston, or double-acting. The barrel or wall of a piston of this type has one end closed (crown) and one end open (skirt end). Only the crown of this piston serves as part of the combustion space surface; therefore, the pressure of combustion can act only against the crown. Thus, with respect to the surfaces of a piston, pressure is single-acting. Most modern gasoline engines, as well as many of the diesel engines used by the Navy, are single-acting.

Opposed-Piston Engines

With respect to combustion-gas action, the term opposed piston is used to identify those engines that have two pistons and one combustion space in each cylinder. The pistons are arranged in "opposed" positions—that is, crown to crown with the combustion space in between. When combustion takes place, the gases act against the crowns of both pistons, driving them in opposite directions. Thus, the term opposed not only signifies that, with respect to pressure and piston surfaces, the gases act in "opposite" directions, but also signifies piston arrangement within the cylinder.

In modern engines that have the opposed-piston arrangement, the crankshafts (upper and lower) are required for transmission of power. Both shafts contribute to the power output of the engine. They may be connected by chains or gears, the latter being more common.

The cylinders of opposed-piston engines have scavenging air ports located near the top. These ports are opened and closed by the upper piston. Exhaust ports located near the bottom of the cylinder are closed and opened by the lower piston. Opposed-piston engines used by the Navy operate on the two-stroke cycle.

Modern engines of the opposed-piston design have a number of advantages over single-acting engines of comparable rating. Some of these advantages are: less weight per horsepower developed, lack of cylinder heads and valve mechanism (and the cooling and lubricating problems connected with them), and fewer moving parts.

Double-Acting Engines

The term double-acting is used to identify two-stroke cycle engines in which the combustion gases in a cylinder act on both ends of a piston in alternating order. It should be noted that the Navy does not use this type of engine.

The pistons in a double-acting engine are usually shorter than the pistons common to single-acting and opposed-piston engines. Since combustion takes place in both ends of the cylinder, the cylinder must be sealed and both ends of the piston closed. A piston rod, fastened to the lower end of the piston, extends through a stuffing box in the lower cylinder head. The lower end of the rod is connected to a crosshead, which is attached by a pin to the connecting rod. As the engine operates, the flat bearing surface on the crosshead moves up and down in a crosshead guide, keeping the piston and rod in proper alignment in the cylinder.

If only one combustion space were considered, the cycle of operation in a double-acting engine would be similar to that of any two-stroke single-acting or opposed piston engine. However, since two combustion spaces per cylinder are involved, two series of events take place, alternately, above and below the piston during the cycle.

Diesel and Gasoline Engine Parts

The design of most internal combustion engines of the reciprocating type follows much the same general pattern. Though engines are not all exactly alike, there are certain features common to all, and the principal components of most engines are similarly arranged. Since the general structure of gasoline engines is basically the same as that of diesel engines, the following discussion of the engine components applies generally to both types of engines. However, differences do exist, which will be pointed out wherever applicable.

The principal components of an internal combustion engine may be divided into two principal groups—parts and systems. The main parts of an internal combustion engine may be further divided into structural parts and moving parts. Structural parts, for the purpose of this discussion, include those that, with respect to engine operation, do not involve motion—namely, the structural frame and its components and related parts. The other group of engine parts includes those that are mounted within the main structure of an engine and are moving parts. Moving parts are considered as those that convert the power developed by combustion in the cylinder to the mechanical energy that is available for useful work at the output shaft.

Main Structural Parts

The main purpose of the structural parts of an engine is to maintain the moving parts in their proper relative position. This is necessary if the gas pressure produced by combustion is to fulfill its function.

The term frame identifies several stationary parts fastened together to support most of the moving engine parts and engine accessories. As the load-carrying part of the engine, the frame of the modern engine may include such parts as the cylinder block, crankcase, bedplate or base, sump or oil pan, and end plates.

The part of the engine frame that supports the engine's cylinder liners and head or heads is generally referred to as the *cylinder block*, as seen in figure 11–3.

The engine frame which serves as a housing for the crankshaft is commonly called the *crankcase*. In some engines, the crankcase is an integral part of the cylinder block, requiring an oil pan, sump, or base to complete the housing. In others, the crankcase is a separate part and is bolted to the block.

Since lubrication is essential for proper engine operation, a reservoir for collecting and holding the engine's lubricating oil is a necessary part of the engine structure. The reservoir may be a *sump* or an *oil pan*, depending upon its design, and is usually attached directly to the engine. In most cases, an oil pan serves both as the lower portion of the crankshaft housing and as the oil reservoir.

The *cylinder assembly* completes the structural framework of an

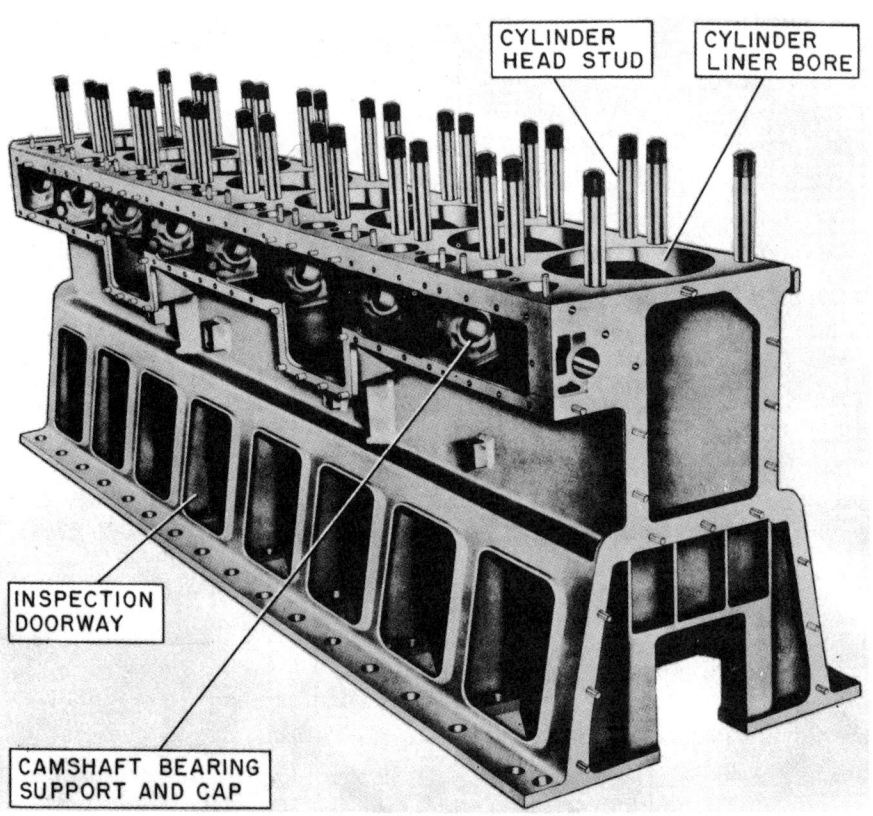

Figure 11–3. Cylinder block with in-line cylinder arrangement. (Courtesy Cooper-Bessemer Corp.)

engine. As one of the main stationary parts of an engine, the cylinder assembly, along with various related working parts, serves to confine and release the gases. For the purpose of this discusion, the cylinder assembly will be considered as consisting of the head (figure 11–4), the liner, and studs, and the gasket.

The design of the parts of the cylinder assembly varies considerably from one type of engine to another. Regardless of the differences in design, however, the basic components of all cylinder assemblies function, along with related moving parts, to provide a gas-and-liquid-tight space.

The barrel or bore in which an engine piston moves back and forth may be an integral part of the cylinder block, or it may be a separate sleeve or liner. The first type, common in gasoline engines, has the disadvantage of not being replaceable. Practically all diesel engines are constructed with replaceable cylinder liners.

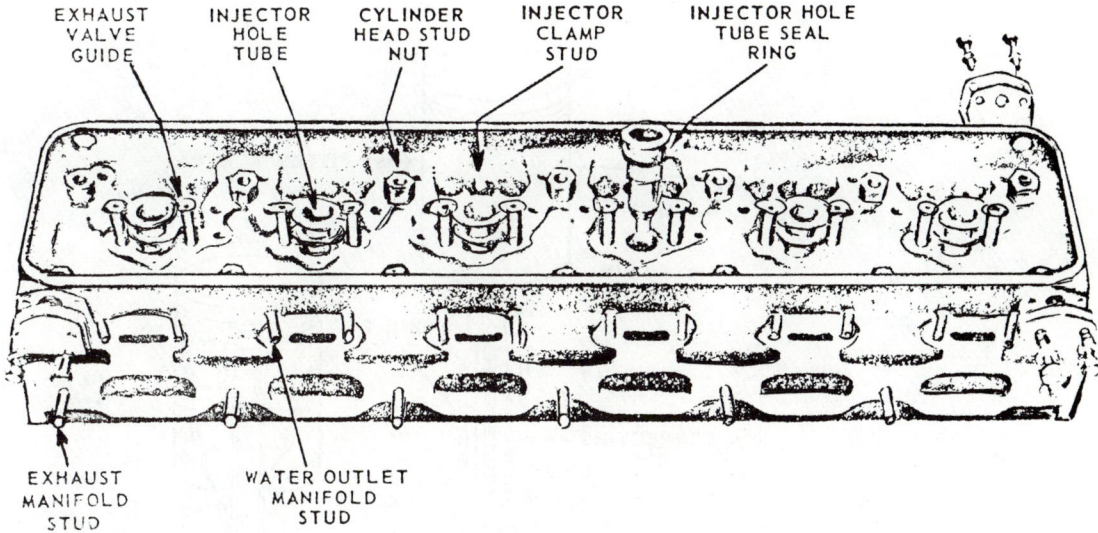

Figure 11–4. Cylinder head (GM 6-71).

The liners or bores of an internal combustion engine must be sealed tightly to form the combustion chambers. In most naval engines, except for engines of the opposed-piston type, the space at the combustion end of a cylinder is formed and sealed by a cylinder head which is a separate unit from the block. A number of engine parts that are essential to engine operation may be found in or attached to the cylinder head.

In most cases, the seal between the cylinder head and the block depends principally upon the studs and gaskets. The studs, or stud bolts, secure the cylinder head to the cylinder block. A gasket between the head and the block is compressed to form a seal when the head is properly tightened down.

Principal Moving Parts

In order that the power developed by combustion be converted to mechanical energy, it is necessary for reciprocating motion to be changed to rotating motion. The moving parts included in the conversion process, from combustion to energy output, may be divided into the following three major groups: (1) the parts that have only reciprocating motion (pistons), (2) the parts that have both reciprocating and rotating motion (connecting rods), and (3) the parts that have only rotating motion (crankshafts), as seen in figures 11–5 and 11–6.

The first two major groups of moving parts may be further grouped under the single heading of *piston and rod assemblies*. Such an as-

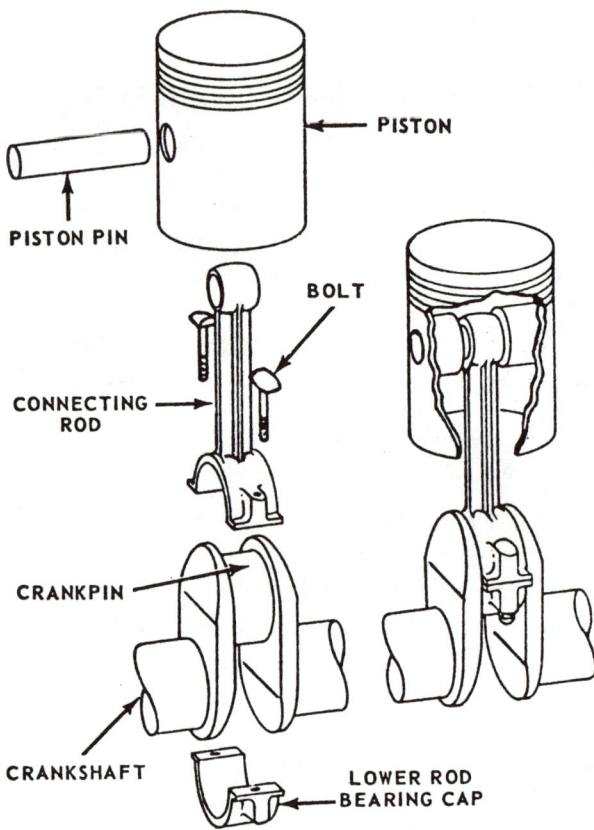

Figure 11-5. Piston assembly.

sembly may include a piston, piston rings, piston pin, connecting rod, related bearings, and, in some cases, a piston rod and crosshead assembly.

As one of the principal parts in the power-transmitting assembly, the piston must be so designed and must be made of such materials that it can withstand the extreme heat and pressure of combustion. Pistons must also be light enough to keep inertia loads on related parts to a minimum. The piston aids in the sealing of the cylinder to prevent the escape of gas and transmits some of the heat through the piston rings to the cylinder wall. In additon to serving as the unit that transmits the force of combustion to the connecting rod and conducts the heat of combustion to the cylinder wall, a piston serves as a valve in opening and closing the ports of a two-stroke cycle engine.

Piston rings are particularly vital to engine operation in that they must effectively perform three functions: seal the cylinder, distribute

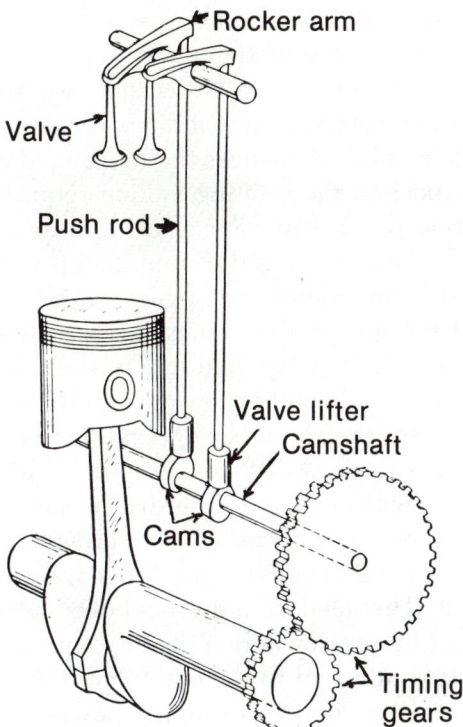

Figure 11-6. Valve and piston assembly arrangement.

and control lubricating oil on the cylinder wall, and transfer heat from the piston to the cylinder wall. All rings on a piston perform the latter function, but two general types of rings—compression and oil—are required to perform the first two functions. There are numerous types of rings in each of these groups, constructed in different ways for particular purposes.

On single-acting and opposed-piston engines the connection between the piston and connecting rod is usually the piston pin (sometimes referred to as the wrist pin) and its bearings. These parts must be of especially strong construction because the power developed in the cylinder is transmitted from the piston through the pin to the connecting rod. The pin is the pivot point where the straight line or reciprocating motion of the piston changes to the reciprocating and rotating motion of the connecting rod. Thus, the principal forces to which a pin is subjected are the forces created by combustion and the side thrust created by the change in direction of motion.

The connecting link between the piston and crankshaft is the connecting rod. In order that the forces created by combustion be transmitted to the crankshaft, the rod changes the reciprocating motion of the piston to the rotating motion of the crankshaft.

One of the principal engine parts that has only rotating motion is the crankshaft. As one of the largest and most important moving parts in an engine, the crankshaft changes the movement of the piston and the connecting rod into the rotating motion required to drive such items as reduction gears, propeller shafts, generators, pumps, and so on. As a result of its function, the crankshaft is subjected to all the forces developed in an engine. (See figure 11–7.)

The speed of rotation of the crankshaft increases each time the shaft receives a power impulse from one of the pistons; and it then gradually decreases until another power impulse is received. These fluctuations in speed (their number depending upon the number of cylinders firing in one crankshaft revolution) would result in an undesirable situation with respect to the driven mechanism as well as the engine; therefore, some means must be provided to stabilize shaft rotation. In some engines, this is accomplished by installing a flywheel on the crankshaft. The need for a flywheel decreases as the number of cylinders firing in one revolution of the crankshaft and the mass of the moving parts attached to the crankshaft increases.

A flywheel stores up energy during the power event and releases

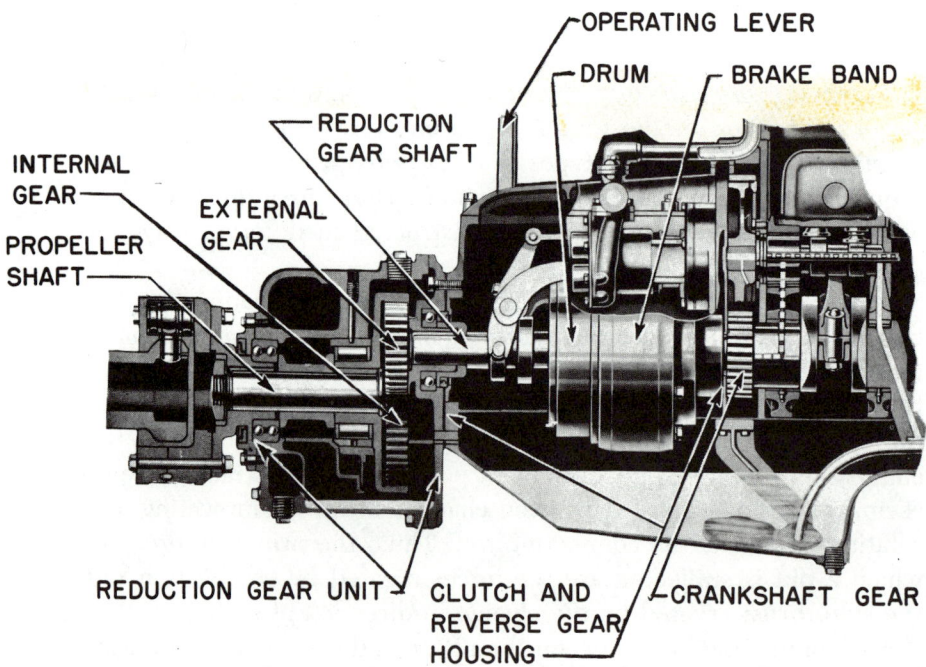

Figure 11–7. *Clutch and reverse gear assembly with attached reduction gear unit.*

it during the remaining events of the operating cycle. In other words, when the speed of the shaft tends to increase, the flywheel absorbs energy, and when the speed tends to decrease, the flywheel gives up energy to the shaft in an effort to keep shaft rotation uniform. In doing this, a flywheel (1) keeps variations in speed within desired limits at all loads; (2) limits the increase or decrease in speed during sudden changes of load; and (3) aids in forcing the piston through the compression event when an engine is running at low or idling speed.

Air Systems

Parts and accessories that supply the cylinders of an engine with air for combustion, and remove the waste gases after combustion and the power events are finished, are commonly referred to as the intake and exhaust systems. These systems are closely related and, in some cases, are referred to as the air systems of an engine.

The following information on air systems deals primarily with the systems' diesel engines; nevertheless, much of the information dealing with the parts of diesel-engine air systems is also applicable to most of the parts in similar systems of gasoline engines. However, the intake event in the cycle of operation of a gasoline engine includes the admission of air and fuel as a mixture to the cylinder. For this reason, the intake system of a gasoline engine differs, in some respects, from that of a diesel engine (see subsequent section on fuel systems).

A discussion of the air systems of diesel engines frequently involves the use of two terms that identify processes related to the functions of the intake and exhaust system. These terms—scavenging and supercharging—and the processes they identify are common to many modern diesel engines.

In the intake systems of all modern two-stroke cycle diesel engines and some four-stroke cycle engines, a device, usually a blower, is installed to increase the flow of air into the cylinders. This is accomplished by the blower compressing the air and forcing it into an air box or manifold that surrounds or is attached to the cylinders of an engine. Thus, a quantity of air under constant pressure is available as required during the cycle of operation.

The increased amount of air available as a result of blower action is used to fill the cylinder with a fresh charge of air, and, during the process, aids in clearing the cylinder of the gases of combustion. This process is called scavenging. Thus, the intake system of some engines, especially those operating on the two-stroke cycle, is sometimes called the scavenging system. The air forced into the cylinder is called scavenge air, and the ports through which it enters are called scavenge ports.

The process of scavenging must be accomplished in a relatively short portion of the operating cycle. The intake and exhaust openings are both open during this interval of time. The overlap of intake and exhaust permits the air from the blower to pass through the cylinder into the exhaust manifold, cleaning out the exhaust gases from the cylinder and, at the same time, cooling the hot engine parts.

An increase in air flow into cylinders of an engine can be used to increase power output, in addition to being used for scavenging. Since the power of an engine is developed by the burning of fuel, an increase of power requires more fuel; the increased fuel, in turn, requires more air, since each pound of fuel requires a certain amount of air for combustion. Supplying more air to the combustion spaces than can be supplied through the action of atmospheric pressure and piston action (in four-stroke cycle engines) or scavenging air (in two-stroke cycle engines) is called turbocharging. Turbocharging or scavenge blowers are the only types used in the Navy on diesel engines.

In some two-stroke cycle diesel engines, the cylinders are supercharged during the air intake simply by increasing the amount and pressure of scavenge air. The same blower is used for supercharging and scavenging. Whereas scavenging is accomplished by admitting air under low pressure into the cylinder while the exhaust valves or ports are open, supercharging is done with the exhaust ports or valves closed. This latter arrangement enables the blower to force air under pressure into the cylinder and thereby increase the amount of air available for combustion. The increase in pressure resulting from the compressing action of the blower will depend upon the engine involved, but it is usually low. With this increase in pressure, and the amount of air available for combustion, there is a corresponding increase in the air–fuel ratio and in combustion efficiency within the cylinder. In other words, a given-size engine that is supercharged can develop more power than the same-size engine that is not supercharged.

Even though the primary purpose of a diesel engine intake system is to supply the air required for combustion, the system generally has to perform one or more additional functions. In most cases, the system cleans the air and reduces the noise created by the air as it enters the engine.

The system that functions primarily to convey gases away from the cylinders of an engine is called the exhaust system. In additon to this principal function, an exhaust system may be designed to perform one or more of the following functions: muffle exhaust noise, quench sparks, remove solid material from exhaust gases, and furnish energy to a turbocharger.

Fuel Systems

The method of getting fuel into the cylinder is one of the major differences between gasoline and diesel engines. As pointed out earlier, fuel for gasoline engines is mixed with air outside the cylinder, and the mixture is then drawn into the cylinder and compressed. On the other hand, fuel for diesel engines is injected or sprayed into the combustion space after the air is already compressed. The equipment that supplies fuel to the cylinders of a gasoline engine would necessarily be different from that of a diesel engine.

A diesel engine has a fuel injection system, the primary function of which is to deliver fuel to the cylinders in accordance with the power requirements of the engine. The quantity of fuel injected determines the amount of energy available, through combustion, to the engine. Smooth engine operation and even distribution of the load between the cylinders depend upon the same volume of fuel being admitted to all cylinders of the engine. The measuring device of a fuel injection system must also be designed to vary the amount of fuel being delivered as changes in load and speed vary.

In addition to measuring the amount of fuel injected, the system must properly time injection to ensure efficient combustion so maximum energy can be obtained from the fuel. Early injection tends to decrease power output and high "peak" pressures causing engine degradation; and if extremely late, injection will cause incomplete combustion. In many engines, fuel injection equipment is designed to vary the time of injection as speed or load varies. A fuel system must also control the rate of injection.

Injection should proceed at such a rate that the rise in combustion pressure is not excessive, yet the rate of injection must be such that fuel is introduced as rapidly as is permissible in order to obtain a complete combustion. An incorrect rate of injection will affect engine operation in the same way as improper timing. If the rate of injection is too high, the results will be similar to those caused by an excessively early injection; if the rate is too low, the results will be similar to those caused by an excessively late injection.

A fuel injection system must increase the pressure of the fuel sufficiently to overcome compression pressures and to ensure proper distribution of the fuel injected into the combustion space. Proper distribution is essential if the fuel is to mix thoroughly with the air and burn efficiently. While pressure is a prime contributing factor, the distribution of the fuel is influenced in part, by "atomization" and "penetration" of the fuel, which again is influenced by the pressure. As used in connection with fuel injection, atomization means the breaking up of the fuel, as it enters the cylinder, into small particles that form a mistlike spray. Penetration is the distance through which the fuel particles are carried by the kinetic energy imparted to them as they leave the injector or nozzle.

The fuel system of a gasoline engine is basically similar to that of a diesel engine, except that a carburetor is used instead of injection equipment. Whereas injection equipment handles fuel only, the carburetor handles both air and fuel. The carburetor must meet requirements similar to those of an injection system except that in the carburetor air is also involved. In brief, the carburetor must accurately meter fuel and air, and in varying percentages, according to engine requirements. The carburetor also functions to vaporize the fuel charge and then mix it with the air, in the proper ratio. The amount of fuel mixed with the air must be carefully regulated, and must change with the engine's different speeds and loads. The amount of fuel required by an engine that is warming up is different from the amount required by an engine that has reached operating temperature. Special fuel adjustment is needed for rapid acceleration. All of these varying requirements are met automatically by the modern carburetor.

Ignition Systems

The methods by which the fuel mixture is ignited in the cylinders of diesel and gasoline engines differ as much as the methods of obtaining a combustible mixture in the cylinders of the two engines. An ignition system, as such, is not commonly associated with diesel engines. There is no one group of parts in a diesel engine that functions only to cause ignition, as there is in a gasoline engine. However, a diesel engine does have an "ignition system." Otherwise, combustion would not take place in the cylinders.

In a diesel engine, the parts that may be considered as forming the ignition system are the piston, the cylinder liner, and the cylinder head. These parts are not commonly thought of as forming an ignition system, since they are generally associated with other functions, such as forming the combustion space and transmitting power. Nevertheless, ignition in a diesel engine depends upon the piston, the cylinder, and the head. These parts not only form the space where combustion takes place but also provide the means by which the air is compressed to generate the heat necessary for self-ignition of the combustible mixture. In other words, both the source (air) of ignition heat and its generation (compression) are wholly within a diesel engine.

This is not true of a gasoline engine because the combustion cycles of the two types of engine are different. In a gasoline engine, even though the piston, the cylinder, and the head form the combustion space, as in a diesel engine, the energy necessary for ignition comes from a source external to the combustion space. The completion of the ignition process, involving the transformation of mechanical energy into electrical energy and then into heat energy, which is a catalyst for the heat process, requires several parts, each performing a specific function. The parts that make the transformation of energy and the system that they form are commonly thought of when reference is made to an ignition system.

The spark that causes the ignition of the explosive mixture in the cylinders of a gasoline engine is produced when electricity is forced across a gap formed by two electrodes in the combustion chamber. The electrical ignition system furnishes the spark periodically to each cylinder, at a predetermined position of piston travel. In order to accomplish this function, an electrical ignition system must have, first of all, either a source of electrical energy or a means of developing electrical energy. In some cases, a storage battery is used as the source of energy; in other cases, a magneto generates electricity for the ignition system. The voltage from either a battery or a magneto is not sufficiently high to overcome the resistance created by pressure in the combustion chamber and to cause the proper spark in the gap. Therefore, it is essential that an ignition system include a device that increases the voltage of the electricity supplied to the system sufficiently to cause a "hot" spark in the gap of the spark plug. The device that performs this function is a voltage transformer, generally called an ignition coil or induction coil.

Since a spark must occur momentarily in each cylinder at a specific time, an ignition system must include a device that controls the timing of the flow of electricity to each cylinder. This control is accomplished by interrupting the flow of electricity from the source to the voltage-increasing device. The interruption of the flow of electricity also plays an important part in the process of increasing voltage. The interrupting device is generally called the breaker assembly. A device that will distribute electricity to the different cylinders in the proper firing order also is necessary. The part that performs this function is called the distributor. Spark plugs to provide the gaps and wiring and switches to connect the parts of the system are essential to complete an ignition system.

All ignition systems are basically the same, except for the source

Diesel and Gasoline Engine Theory   207

of electrical energy. The source of energy is frequently used as a basis for classifying ignition systems—thus, the battery-ignition system and the magneto-ignition system.

Cooling Systems

A great amount of heat is generated within an engine during operation. Combustion produces the greater portion of this heat; however, compression of gases within the cylinders and friction between moving parts add to the total amount of heat developed within an engine. Since the temperature of combustion alone is about twice that at which iron melts, it is apparent that without some means of dissipating heat, an engine would operate for only a very limited time. Without proper temperature control, the lubricating-oil film between moving parts would be destroyed, proper clearance between parts could not be maintained, and metals would fail.

Of the total heat supplied to the cylinder of an engine by the burning fuel, only one-third approximately is transformed into useful work; an equal amount is lost to the exhaust gases. This leaves approximately 30 to 35 percent of the heat of combustion which must be removed in order to prevent damage to engine parts. The greater portion of the heat that may produce harmful results is transferred from the engine through the medium of water; lubricating oil, air, and fuel are also utilized to aid in the cooling of an engine. All methods of heat transfer are utilized in keeping engine parts and fluids (air, water, fuel, and lubricating oil) at safe operating temperatures.

In a marine engine, the cooling system may be of the open or closed type. In the open system, the engine is cooled directly by salt water. In the closed system, fresh water (or an antifreeze solution) is circulated through the engine. The fresh water is then cooled by salt water. In marine installations, the closed system is the type commonly used; however, some older marine installations use a system of the open type. The cooling systems of diesel and gasoline engines are similar mechanically and in function performed.

Lubricating Systems

It is essential to the operation of an engine that the contacting surfaces of all moving parts of an engine be kept free from abrasion and that there be a minimum of friction and wear. If sliding contact is made by two dry metal surfaces under pressure, excessive friction, heat, and wear result. Friction, heat, and wear can be greatly reduced if metal-to-metal contact is prevented by keeping a clean film of lubricant between the metal surfaces.

Lubrication and the system that supplies lubricating oil to engine parts that involve sliding or rolling contact are as important to successful engine operation as air, fuel, and heat are to combustion. It is important not only that the proper type of lubricant be used, but also that the lubricant be supplied to the engine parts in the proper quantities, at the proper temperature, and that provisions be made to remove any impurities that enter the system. The engine lubricating oil system is designed to fulfill the above requirements.

The lubricating system of an engine may be thought of as consisting of two main divisions, that external to the engine and that within the engine. The internal division, or engine part, of the system consists principally of passages and piping; the external part of the system includes several components that aid in supplying the oil in the proper quantity, at the proper temperature, and free of impurities. In order to meet these requirements, the lubricating systems of many engines include, external to the engine, such parts as tanks and sumps, pumps, coolers, strainers and filters, and purifiers.

The engine system that supplies the oil required to perform the functions of lubrication is of the pressure type in practically all modern internal combustion engines. Even though many variations exist in the details of the engine lubricating systems, the parts of such a system and its operation are basically the same; the difference in the lubrication systems of the two types of engines is generally due to differences in engine design and in opinions of manufacturers as to the best location of the component parts of the system. In many cases, similar types of components are used in the systems of diesel and gasoline engines.

12

Gas Turbine Engines

The gas turbine engine, long regarded as a promising but experimental prime mover, has in recent years been developed to the point where it is entirely practical for ship propulsion and for a number of auxiliary applications. Gas turbine engines are currently installed as primary power plants in *Spruance*-class destroyers, *Perry*-class frigates, minesweepers, landing craft, PT boats, air-sea rescue boats, hydrofoils, hydroskimmers, and other craft. In addition, the gas turbine engine is finding increasing application as the driving unit for ship's service generators, pumps, and other major auxiliary equipments.

Although the gas turbine engine, as a type, need no longer be regarded as experimental, many specific models of gas turbine engines are still at least partially experimental and subject to further change and development. The discussion in this chapter, therefore, deals primarily with the general principles of gas turbine engines rather than with specific models. Detailed information may be obtained from the manufacturer's technical manual furnished with the equipment. It is recommended that you refer to the plant layout portion for *Spruance* (Figure 12–20) when studying this chapter.

Basic Principles

All gas turbines, no matter how large or complex, are made up of three basic parts: a compressor, a combustor or combustion chamber, and a turbine (see figure 12–1). Although these parts will be discussed in greater detail later, you should be familiar at this point with the basic function of each component. The compressor draws in air and sends it under pressure to the burner, where air and fuel burn. The combustion gases then flow through the turbine blades, producing work to drive both the compressor and the load.

Although the gas turbine engine exhibits some resemblance to both an internal combustion engine of the reciprocating type and a steam turbine, a brief consideration of the basic principles of a gas turbine engine reveals several ways in which the gas turbine engine is quite different from either the reciprocating internal combustion engine or the steam turbine.

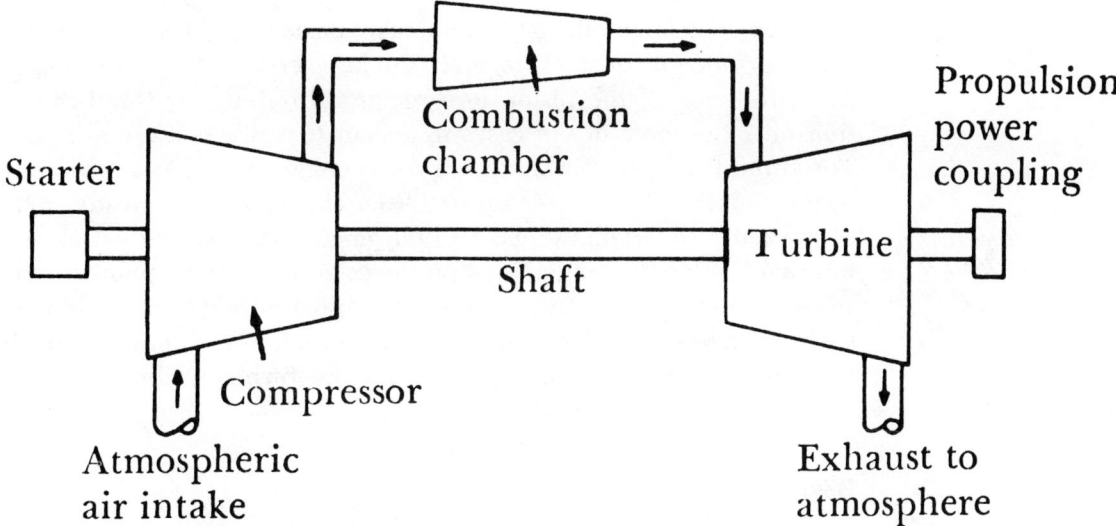

Figure 12–1. Basic parts of a gas turbine.

Let us look first at the thermodynamic cycles of the three engine types. The reciprocating internal combustion engine has an open, heated-engine cycle, and the steam turbine has a closed, unheated-engine cycle. In contrast, the gas turbine has an open, unheated-engine cycle—a combination we have not previously encountered in our study of naval machinery. The gas turbine cycle is open because it includes the atmosphere; it is an unheated-engine cycle because the working substance is heated in a device that is separate from the engine.

Another way in which the three types of engines differ is in the working substance. The working fluid in a steam turbine installation is steam. In both the reciprocating internal combustion engine and the gas turbine engine, the working fluid may be considered as being the hot gases of combustion that result from the burning of fuel in air. However, there are very important differences in the way the working fluid is used in the reciprocating internal combustion engine and in the gas turbine engine. The gas turbine is similar to reciprocating diesel engines in that air is compressed, a fuel–air mixture is burned, and the gases of combustion are expanded to produce useful power. Engines of the reciprocating type use one structure—the cylinder—for compression, combustion, and expansion. Since all three phases take place within one unit, the power impulse must occur intermittently or periodically, as the cycle is repeated. This is not the case in gas turbine engines. Instead, compression, combustion, and expansion take place in three separate components. Air is compressed in one unit, combustion takes place in an adjacent burner, and a turbine (or turbines) receives the force created by combustion. As in the case of the piston in a reciprocating engine, the turbine transmits the force of the gases to a shaft that drives a useful load. The three basic units of the gas turbine engine are so arranged and connected that the power output from the turbine is steady and continuous. In brief, the gas turbine engine can be defined as an internal combustion engine that produces power by a continuous and self-sustaining process. An air mass is compressed and then combined with atomized fuel. The resulting mixture then burns, and the combustion gases expand through one or more turbines which change some of the energy into useful power.

Basic Parts

Compressor. The compressor takes in atmospheric air and compresses it. Part of the compressed air, called *primary air*, enters directly into the combustion chamber where it is mixed with the atomized fuel so that the mixture can be ignited and burned. The remainder of the air, called *secondary air*, is mixed with the gases of combustion. The purpose of the secondary air is to cool the combustion gases down to the desired turbine inlet temperature. The division of air actually takes place in the combustor, not the compressor.

Both axial-flow compressors and centrifugal (radial-flow) compressors are currently used in gas turbine engines. There are several possible configurations of these basic types, some of which are in use and some of which are in experimental phases of development.

In the axial-flow compressor the air is compressed as it flows axially along the shaft. The centrifugal (radial-flow) compressor picks up the entering air and accelerates it outward by means of centrifugal force.

The advantages of the centrifugal compressor include a high pressure rise per stage, simplicity of manufacture, low initial cost, and relatively light weight. The disadvantages of the centrifugal compressor include the need for a relatively large frontal area for a given air flow and the losses that occur between the stages when two or more are used. This latter disadvantage is of concern mainly in aircraft engines.

Combustion Chamber. The combustion chamber is the component in which the fuel–air mixture is burned. The combustion chamber consists of a casing, a perforated inner shell, a fuel nozzle, and a device for initial ignition. The number of combustion chambers used in a gas turbine engine varies widely; as few as one and as many as 16 combustion chambers have been used in one gas turbine engine.

The combustion chamber is the most efficient component of a gas turbine engine. Efficiencies between 95 and 98 percent can be obtained over a wide operating range. To produce such efficiencies, combustion chambers are designed to operate with low pressure losses, high combustion efficiency, and good flame stability. Additional requirements for the combustion chamber include low rates of carbon formation, light weight, reliability, reasonable length of life, and the ability to mix cold air with the hot combustion gases in such a way as to give uniform temperature distribution to the turbine blades. Only a small part (perhaps one-fourth) of the air that enters the combustion chamber area is burned with the fuel (primary air). The remainder of the air is used to keep the temperature of the combustion gases low enough so that the turbine nozzles and blades will not be overheated and thereby damaged (secondary air).

Turbine. In theory, design, and operating characteristics, the turbines used in gas turbine engines are quite similar to the turbines used in a steam plant. The gas turbine differs from the steam turbine chiefly in the type of blading material used, the means provided for cooling the bearings and highly stressed parts, and the higher ratio of blade length to wheel diameter which is required to accommodate the large gas flow.

The turbine section of a gas turbine engine is located directly behind the combustion chamber outlet. The turbine consists of two basic elements, the stator and the rotor.

The stator element is referred to by various names, including turbine nozzle vanes and turbine guide vanes. The vanes of the stator element serve the same purpose as the nozzles in an impulse steam turbine or the stationary blading in a reaction steam turbine—they convert thermal energy into mechanical kinetic energy. The vanes of the stator element are contoured and set at such an angle that they form a number of small nozzles that discharge the gas as extremely high-speed jets. As in the case of the nozzles (or stationary blading) of steam turbines, the increase in velocity may be equated with the decrease in thermal energy. The vanes of the stator element direct the flow of gas to the rotor blades at the required angle while the turbine wheel is rotating.

The rotor element of the turbine consists of a shaft and a bladed wheel or disc. The wheel is attached to the main power transmitting shaft of the gas turbine engine. The jets of combustion gas leaving the vanes of the stator element act upon the turbine blades and cause the turbine wheel to rotate at a very high rate of speed. The high rotational speed imposes severe centrifugal loads on the turbine wheel, and at the same time the very high temperatures result in a lowering of the strength of the material. Consequently, the engine speed and temperature must be controlled to keep turbine operation within safe limits. Even so, the operating life of the turbine blading is accepted as the governing factor in determining the life of the gas turbine engine.

The turbine may be of the single-shaft type or the split-shaft type. Either single-shaft or split-shaft turbines may be used with either centrifugal or axial-flow compressors. In the single-shaft type of turbine, the power is developed by one rotor, and all engine-driven parts are driven by this single wheel. In the split-shaft type, the power is developed by two or more rotors. It is possible for one or more rotors to drive the compressor and the accessories, while one or more rotors are used for the power output. A single-shaft arrangement is shown in figure 12–1, and a split-shaft gas turbine is shown in figure 12–2.

Engine Systems

Fuel System. The fuel system supplies fuel (JP–5 or diesel fuel) for combustion.

The engine-driven pump receives filtered fuel from a motor-driven supply pump at a constant pressure. The engine-driven fuel pump increases the pressure and forces the fuel through a high-pressure

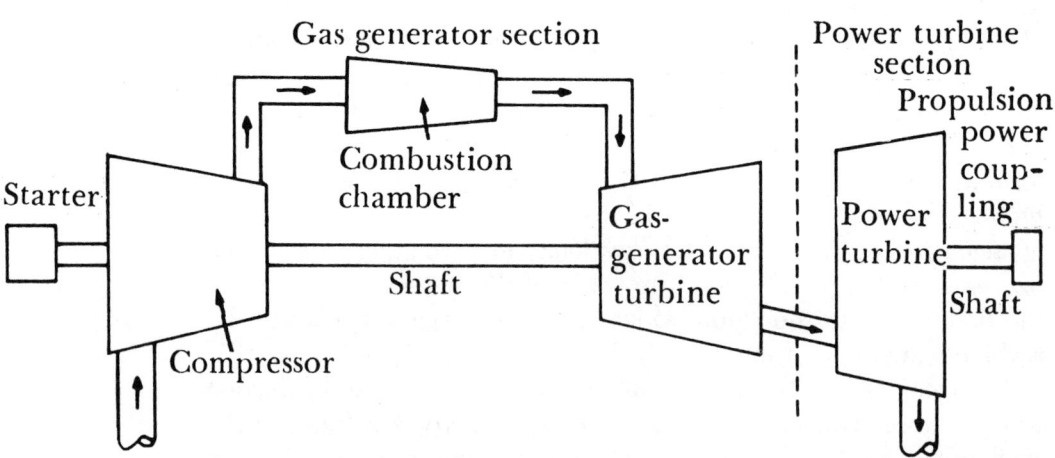

Figure 12–2. *Split-shaft gas turbine engine.*

filter to the fuel control governor in the fuel control assembly. The fuel control governor provides fuel to the nozzle at the pressure and volume required to maintain the desired engine performance. At the same time, the fuel control governor limits the fuel flow to maintain operating conditions within safe limits. The fuel nozzles serve to introduce the fuel into the combustion chamber. The fuel is sprayed into the combustion chamber under pressure, through small orifices in the nozzles. The fuel control assembly is the unit that regulates the turbine rpm by adjusting fuel flow from the high-pressure engine-driven pump to the nozzle.

Lubricating System. Because of the high operating rpm and the high operating temperatures encountered in gas turbine engines, proper lubrication is of vital importance. The lubricating system is designed to supply bearings and gears with clean lubricating oil at the desired pressures and temperatures. The main bearings in a gas turbine engine serve the critical function of supporting the compressor, the turbine, and the engine shaft. The number and position of main bearings required for proper support vary according to the length and stiffness of the shaft, with both length and stiffness being affected by the type of compressor used in the engine. Heat absorbed by the lubricating oil is transferred to the cooling medium in a lube oil cooler.

Starting System. A spark ignition system is used to provide the initial heat source in the combustion chamber. Once the engine is running, the heat in the chamber is sufficient to sustain combustion, and the igniter is no longer necessary.

To start a gas turbine engine, the compressor is rotated. Depending on engine design and construction, this is done manually, pneumatically, electrically, or hydraulically.

The air starter consists essentially of a turbine driven by pressurized air from an external source. The turbine output is geared to the compressor shaft and is designed to provide the required torque to rotate the compressor. The air supply is normally obtained from storage flasks which are replenished from ship's air compressors. When the compressor speed reaches a designated limit, ignition takes place. Following combustion, the compressor speed increases, and a centrifugal switch deactivates the starter. The ignition system is also deactivated at this time; it may be tied into the same switch or operated manually.

The electric starter provides the same function as the air starter. The unit consists of an electric motor energized by batteries. The motor output shaft is geared to the compressor shaft and provides the required torque to rotate the compressor. The starter can automatically be cut out along with the ignition system by a centrifugal switch, or the starter can be driven by the compressor shaft to, in effect, become a generator. As a generator it is used to recharge the batteries as well as supply current to a tachometer or other auxiliaries.

The hydraulic starter uses a closed, recirculating oil system with a pump supplying the power to a hydraulic motor to crank the gas turbine. It is deactivated in the same manner as the air starter. This type of starter provides the advantage of permitting extended cranking periods as compared to air and electric starters.

Power Transmission Systems

The process of transmitting engine power to a point where it can be used in performing useful work involves a number of factors, two of which are torque and speed. The gas turbine engine does not produce high torque, but it does produce high speed. Therefore, a gear train is used with most gas turbine engines to lower speed and increase torque. In the case of a gear drive installation, the gears are usually used between the gas turbine engine and the generator shaft, to reduce the rpm of the generator to a practicable operating value.

Many different types and models of gas turbine engines are in use. The gas turbine engine shown in figure 12–1 is called a single-shaft type because one shaft from the turbine rotor drives the compressor, and an extension of this same shaft drives the load.

The gas turbine engine shown in figure 12–2 is called a split-shaft type. This engine is considered to be split into two sections. In the gas-generator section, a stream of expanding gases is created as a result of continuous combustion; this section includes the compressor, the combustion chamber (or chambers), and the gas generator turbine. The power turbine section consists of a power turbine and the power output shaft. In this type of gas turbine engine, there is no mechanical connection between the gas-generator turbine and the power turbine. When the engine is operating, the two turbines produce basically the same effect as that produced by a hydraulic torque converter. The split-shaft gas turbine engine is well suited for use as a propulsion unit where loads vary because the gas generator section can be operated at a steady and continuous speed, while the power turbine is free to vary its speed with the load. Starting effort required for a split-shaft gas turbine engine is far less than that required for a single-shaft gas turbine engine connected to the reduction gear, propulsion shaft, and propeller.

Thus far, we have considered the gas turbine engine as a ship propulsion system delivering power in the form of torque on a shaft.

The student should also keep in mind that the gas turbine type engine also serves as the prime mover in the power plants of many military aircraft. When so adapted, the gas turbine engine develops power by converting thermal energy into mechanical kinetic energy in a high-velocity gas stream. The highly accelerated gas stream creates thrust which propels the aircraft. This method of creating thrust is called the direct reaction or jet propulsion method.

The concept of thrust is basic to an understanding of jet propulsion. This concept is based on Newton's third law of motion, which may be stated as follows: For every acting force there is an equal opposite reacting force. In the case of aircraft in flight, the acting force is the force the engine exerts on the air mass as it flows through the engine. The reacting force (thrust) is the force that the air mass exerts on the components of the engine as the heated air mass is discharged from the jet nozzle at the rear of the airplane. In other words, thrust is not produced by the ejected air mass's reacting against the atmosphere; rather, thrust is created within the engine as the air mass flowing through the engine is accelerated and discharged.

Engines that include the gas turbine and that create thrust by the direct reaction method are commonly identified as turbojet engines. Except for a diffuser and a different type of exhaust system in engines of the turbojet type, the basic components of the turbojet engine are similar in design and function to the components of any open-cycle gas turbine engine. The function of the diffuser is to decrease the velocity of the inlet air and to increase its pressure before the air enters the compressor. If this were not done, a supersonic shock wave could develop in the compressor, impairing its operation. The exhaust system of a turbine engine consists basically of a cone and a convergent nozzle. The exhaust cone is designed to channel to the nozzle the accelerated air mass that the other components have produced. When this air mass flows throught the convergent nozzle, its velocity is greatly increased, and thrust is created within the engine.

Regenerator

The largest single loss of energy in the basic gas turbine is the amount of heat rejected to the atmosphere. The use of a device called a regenerator allows us to recover some of this heat loss and thereby improve the efficiency of the gas turbine cycle.

The regenerator is a heat exchanger that resembles the steam generator's economizer both in construction and in theory. Like the economizer, the regenerator is placed in the path of the exhaust gases, as shown in figure 12–3. Air on its way from the compressor

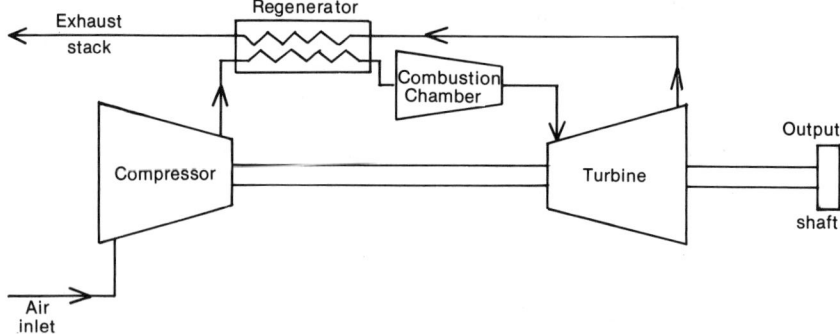

Figure 12–3. Inline regenerator.

to the combustion chamber passes through the regenerator and is preheated prior to entering the combustion chamber.

The use of preheated air in the combustion chamber increases the combustion rate and decreases fuel consumption. Regenerators, however, have not yet been installed on gas turbines used for propulsion units.

Basic Components

We will now briefly consider the engineering aspects of a gas turbine in a bit more detail.

The gas turbine power plant is not new; it has been used by the Navy as a prime mover for aircraft since 1948. Innovations in the gas turbine cycle have provided greater power and efficiency. Along with this greater power and efficiency have come a wider application and expanded use of these lightweight compact power plants in earthbound vehicles.

The U. S. Navy, from 1965 to 1969, greatly increased its use of gas turbines. The broader application of gas turbine power plants has been highlighted by the new DD 963 and PF classes of ships and the new YF 16 and 18 models selected by the Air Force and the Navy, respectively, as their fighter aircraft.

Since the Navy is utilizing gas turbines in increasing numbers and types for a variety of applications (surface and air), it is imperative that naval officers understand the basic concepts and principles of gas turbine design and operation.

The combination of processes requiring the intake and exhaust of the entire flow of working fluid from the atmosphere is called the open cycle. Figure 12–4 illustrates an open cycle gas turbine. A compressor is required to take ambient air and raise its pressure. In the combustion chamber, a portion of this pressurized air is mixed

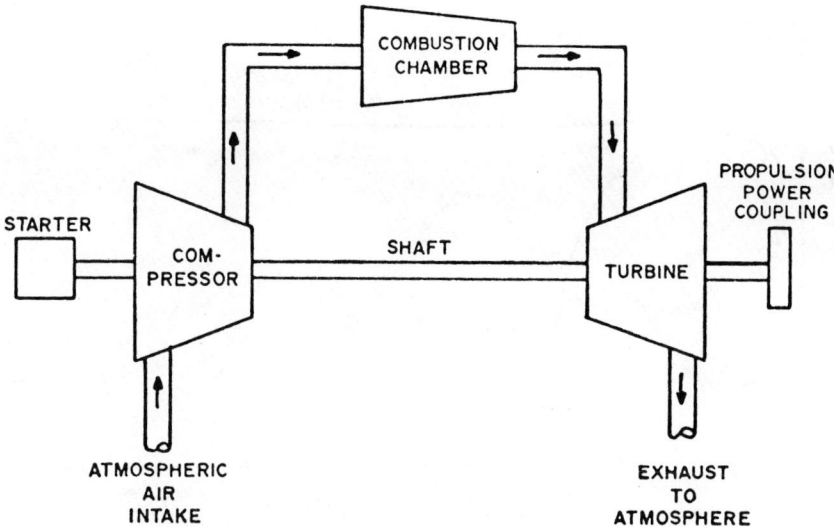

Figure 12–4. Single shaft gas turbine.

with fuel injected under pressure, which is burned, thus releasing heat. The heat added to the working fluid (air) from this burned fuel–air mixture raises its temperature and, hence, its energy level. This working fluid is now available, at a relatively high temperature and pressure, to be expanded through a turbine to develop mechanical or kinetic energy, or both. Because the atmosphere is the working fluid, it must be continuously replaced.

The compressor requires work input to turn a rotor in order to compress the working substance. The ancillaries require work in order to produce useful work. The propulsive power to drive the compressor and the ancillaries comes from the turbine through a connecting shaft. The turbine must be able to produce sufficient power to drive the compressor and the ancillaries and to provide enough remaining useful energy to perform useful work or thrust.

To be able to expand the hot gases in the turbine with a back pressure limited by atmospheric pressure, it is necessary first to compress the air prior to the heat-addition process. In order to provide the needed pressure with sufficient volume to burn the required amount of fuel and to control temperature, centrifugal and, subsequently, axial flow compressors have been developed for performing the compression function. Figure 12–5 shows the compression ratio/efficiency curves for these two types of compressors.

The efficiency of a single-stage centrifugal compressor is relatively low. The efficiency of a multistage centrifugal compressor is somewhat

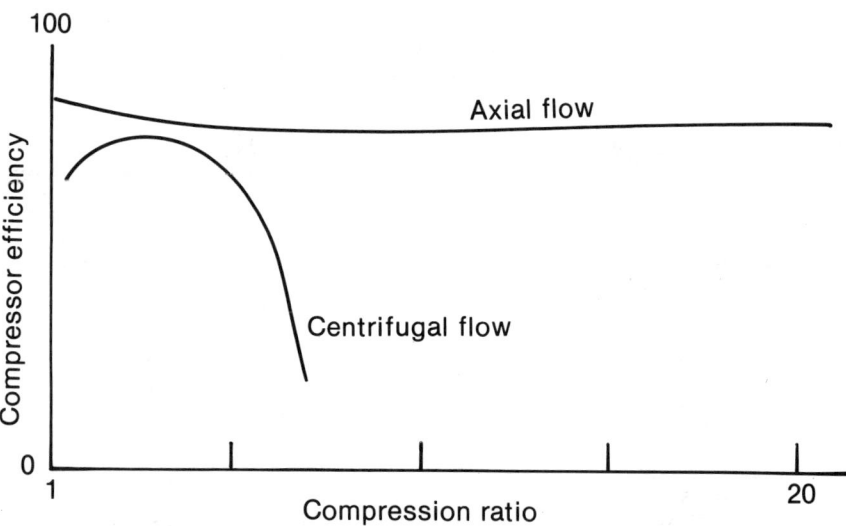

Figure 12–5. Compression ratio/efficiency of an axial flow vs. centrifugal flow compressor.

better, but still is far less than that of a multistage axial compressor. Figures 12–6 and 12–7 show the different stages of development that centrifugal compressors have undergone. A compression ratio of 4 or 5 to 1 is the maximum capability of a single-stage centrifugal compressor. These compressors are usually easier to design and produce than axial compressors; but higher pressure ratios are possible with multistage axial compressors (20 to 1), and therefore the latter are most frequently used in high-power engines. Figure 12–8 shows the components and the assembly of an axial flow compressor. Figure 12–9 shows the cross section of an axial flow compressor. The fluid in an axial compressor flows in an axial direction through a series of rotor blades and stator vanes, which are concentric with the axis of rotation.

The axial compressor, like the centrifugal compressor, has evolved through technology and a search for greater efficiency. A single axial compressor might theoretically be built to consist of as many stages as would be necessary to produce any required compression ratio. If such were the case, at some specific compressor speeds the rearmost stages of the compressor would operate inefficiently, and the foremost stages would be overloaded. Such a condition would produce compressor stall. That condition may be corrected, in part, by bleeding interstage compressed air overboard during part-throttle operation. However, excessive air bleeding is inefficient and wasteful. Greater flexibility for part-throttle conditions and for starting can be attained

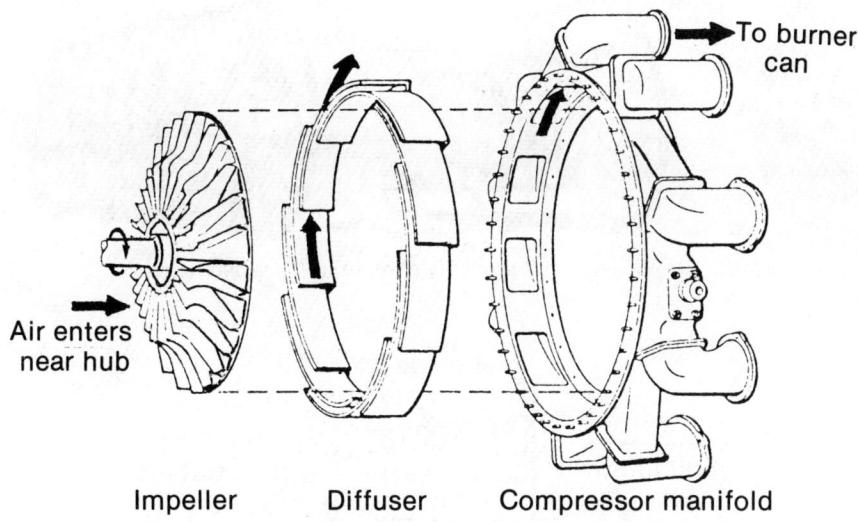

Figure 12–6. Single-stage centrifugal compressor components.

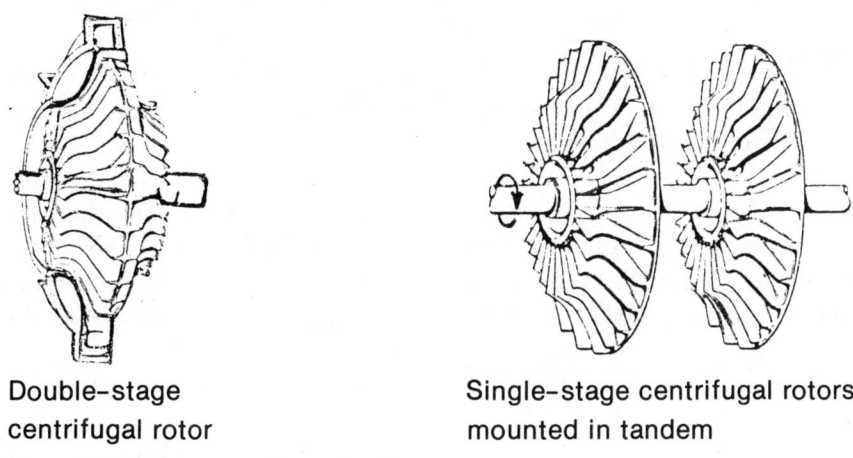

Figure 12–7. Two type of centrifugal compressors.

more efficiently by splitting the compressor into two mechanically independent rotor systems, as shown in figure 12–10. Each rotor system, high and low, is driven at its own best speed by its own separate turbine.

Combustion Chamber

Fuel is introduced into the working fluid at the front of the combustion chamber in spray form, suitable for rapid mixing with the

Gas Turbine Engines   **221**

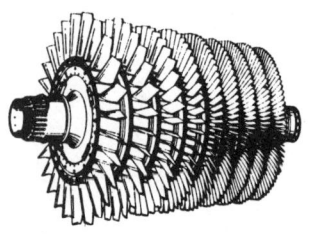 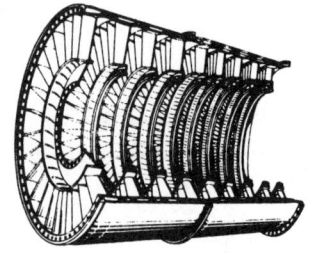 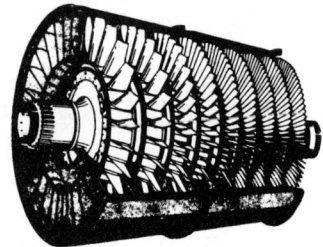

Figure 12–8. *Components and assembly of an axial flow compressor.*

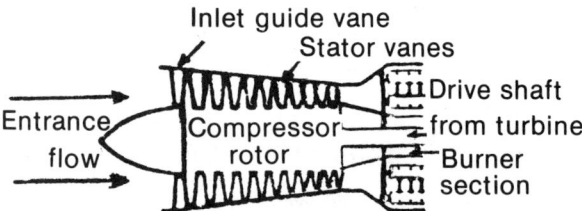

Figure 12–9. *Cross section of an axial flow compressor.*

fluid for combustion. The burner section, which contains the combustion chamber, is designed to burn the mixture of fuel and air, and to deliver the resulting gases to the turbine at a temperature that will not exceed allowable limits at the turbine inlet. The burners, within a very limited space, must add enough energy to the gases (passing through the engine) to produce sufficient turbine power to drive the compressor as well as to produce the desired thrust for the jet engines or mechanical power for shaft output engines and power for the turbine.

The criterion for an acceptable combustion chamber is that the pressure loss for the working fluid passing through the burner must be held to a minimum, while combustion efficiency is kept at a high level. There must be no tendency for the burner to blow out, and no burning should occur after the gases leave the burner.

The three types of combustion chamber are: (1) the can type, (2) the annular, and (3) the can annular. The can type is most frequently employed on centrifugal compressor engines. The annular type is employed on larger jet engines using axial flow compressors, as in figure 12–11. The can annular type is employed on some small gas turbines; figure 12–12 shows a cutaway view of the can annular combustion chamber.

In larger gas turbine engines, the compressor used is the axial flow type. Figure 12–13 shows the elements of an axial flow turbine. It

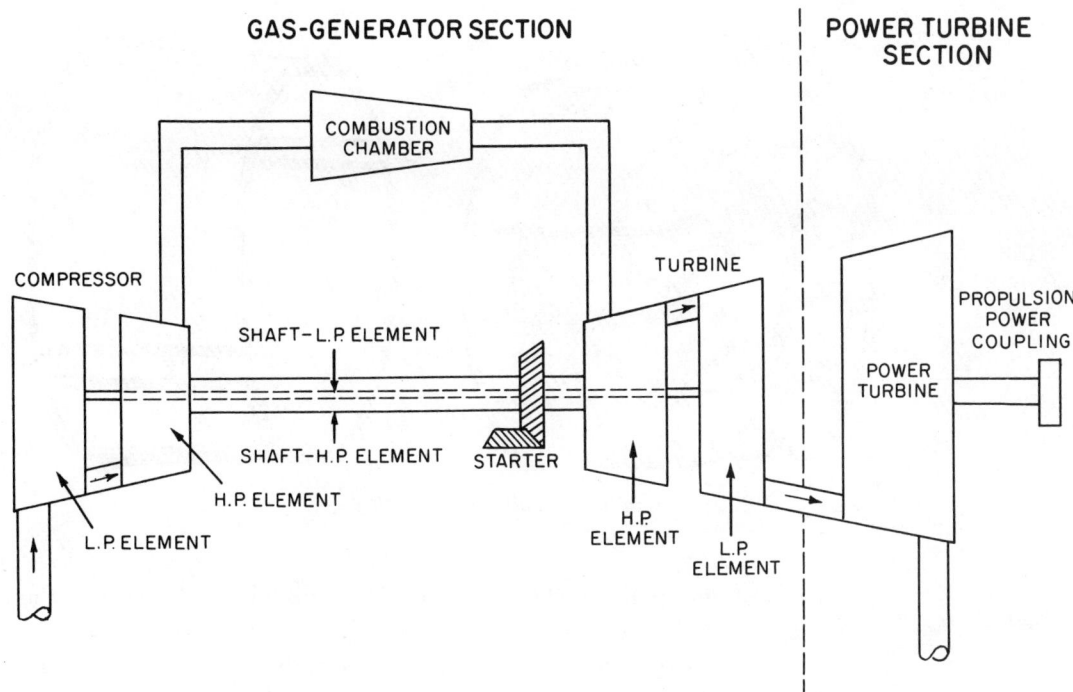

Figure 12–10. Relationship of parts in a dual axial gas turbine engine.

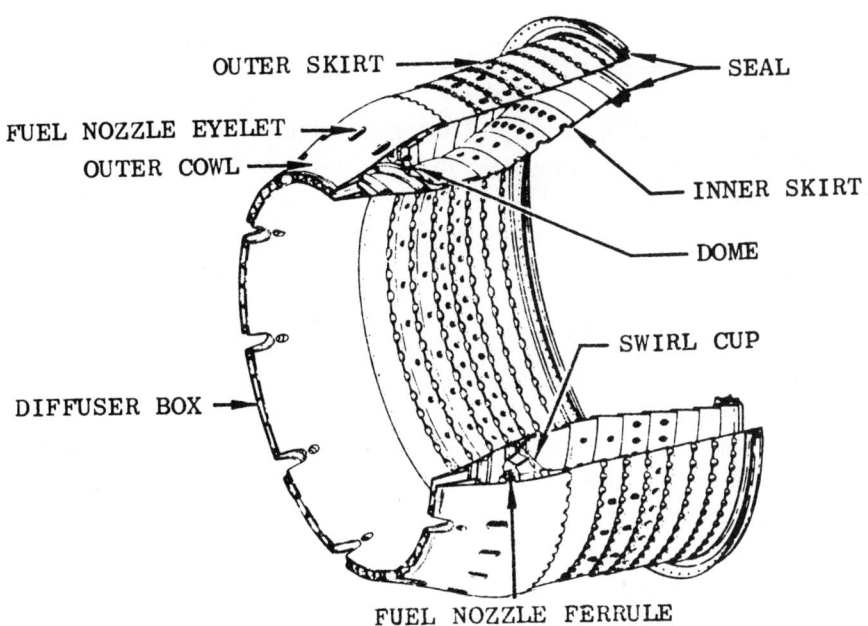

Figure 12–11. Annular combustion chamber.

Gas Turbine Engines

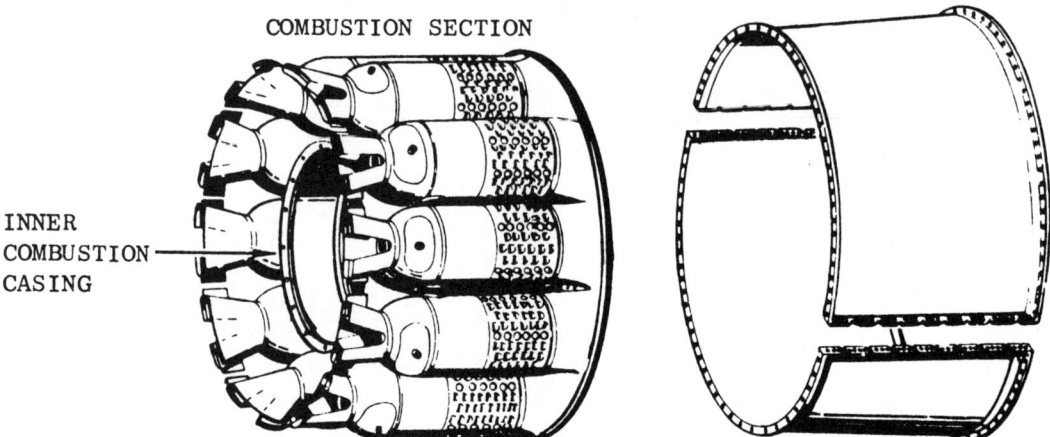

Figure 12–12. Can annular combustion chamber.

consists of one or more stages located immediately to the rear of the engine combustion chamber section. The turbine extracts kinetic energy from the gases after they come through the nozzles from the individual burner cans. The turbine converts this energy into shaft horsepower to drive the compressor and engine accessories. The remaining useful energy may be used to perform other work (thrust or mechanical work). Figure 12–14 shows the turbine location.

The axial flow turbine is comprised of two main elements, a set of turbine rotor blades, and a set of stationary vanes or blades (figure 12–13).

We have now discussed the basic gas turbine components. Figure 12–15 shows these components installed in a modern jet aircraft engine. This section view clearly indicates the various components and their relative locations within the jet aircraft power plant.

Power Plant Variations

Having discussed the basic gas turbine, it is appropriate that we review some of the variations that are found in gas turbine power plants. Again starting with the single-shaft gas turbine unit, we will examine its variations as shown in figure 12–16. This power plant can be used for either aircraft propulsion or to perform mechanical work. If it is used for aircraft, fewer turbine stages are present, and a nozzle is used (as shown in figure 12–16B). If it is used to perform mechanical work (as in driving a shaft), more turbine stages are used to drive the propulsion power coupling (as shown in figure 12–16A).

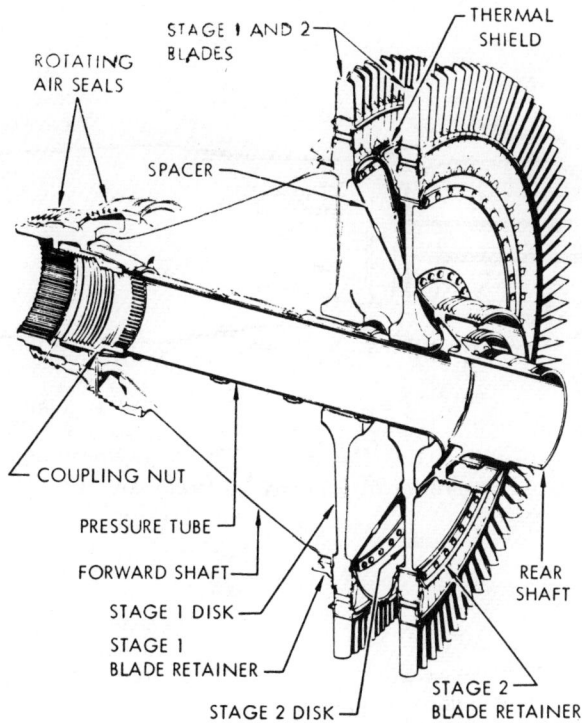

Figure 12–13. Elements of a high pressure turbine rotor.

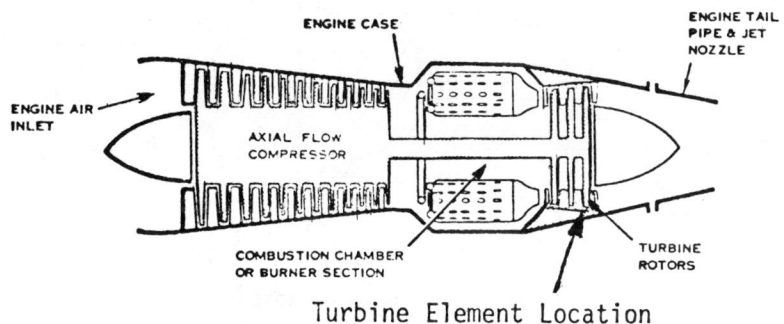

Figure 12–14. Turbine element location.

The gas turbine engine shown in figure 12–17 is considered to be split into two sections. The gas-generator section, in which a continuous stream of high temperature gas is created, includes the compressor, combustion chamber, and the gas-generator turbine. The power turbine section consists of a power turbine and the power output shaft. There is no mechanical link between the gas-generator

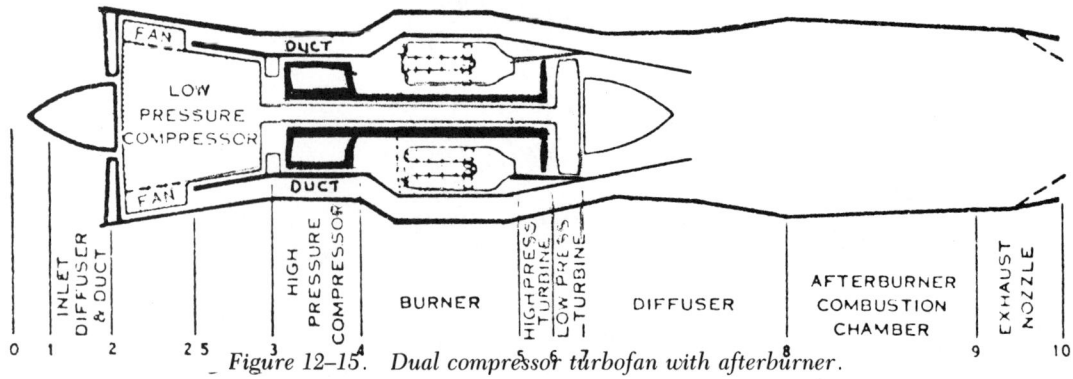

Figure 12–15. Dual compressor turbofan with afterburner.

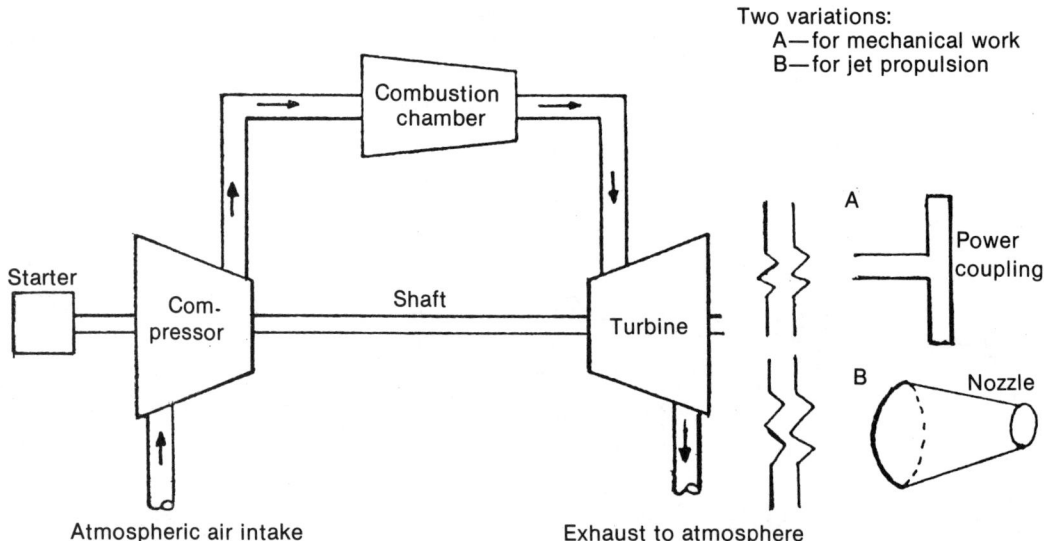

Figure 12–16. Single-shaft gas turbine.

turbine and the power turbine. The effect of this design equates basically to the effect created by a fluid torque converter. The benefit of this type of turbine is that it allows for varying speeds and loads, since the gas-generating section can be operated at a steady continuous speed, while the power turbine section is free to vary its speed with load.

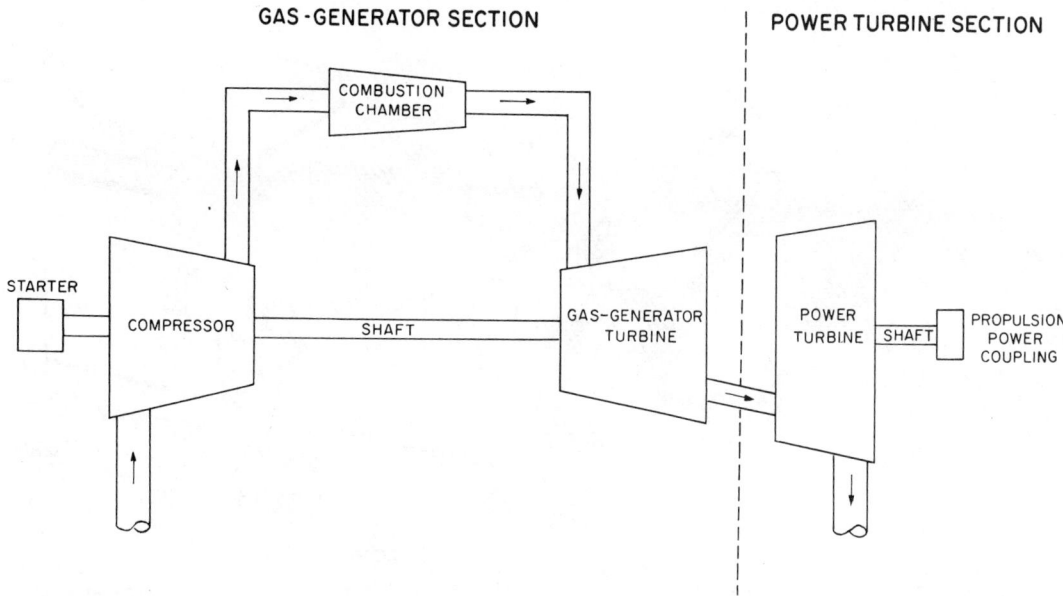

Figure 12–17. Split-shaft gas turbine.

The LM 2500 Gas Turbine Engine

In 1974, for the first time, large combatant ships in the U. S. Navy began using gas turbines as the sole source of propulsion. Until then the only use of gas turbines in the Navy had been in combination with diesels for propulsion of patrol gunboat class ships. The best current example of how gas turbine propulsion is applied to a naval vessel is the LM 2500 engine and its associated systems in the *Spruance*-class (DD–963) destroyer, in which four of these engines are used. The *Oliver H. Perry*–class frigates also use two LM 2500s for their main propulsin.

The LM 2500 gas turbine engine made gas turbine propulsion of large U. S. combatant ships economically feasible. This engine was developed by General Electric from the TF–39 aircraft engine that powers the C–5A Galaxy transport, and from the CF–6 used in the DC–10. The four engines in the *Spruance* are each mounted in a gas turbine module (GTM) (figure 12–18) which incorporates the following features:

- Engine mounting
- NBC protection
- Engine cooling
- Noise reduction
- An enclosed automatic CO_2 fire extinguishing system

Gas Turbine Engines   **227**

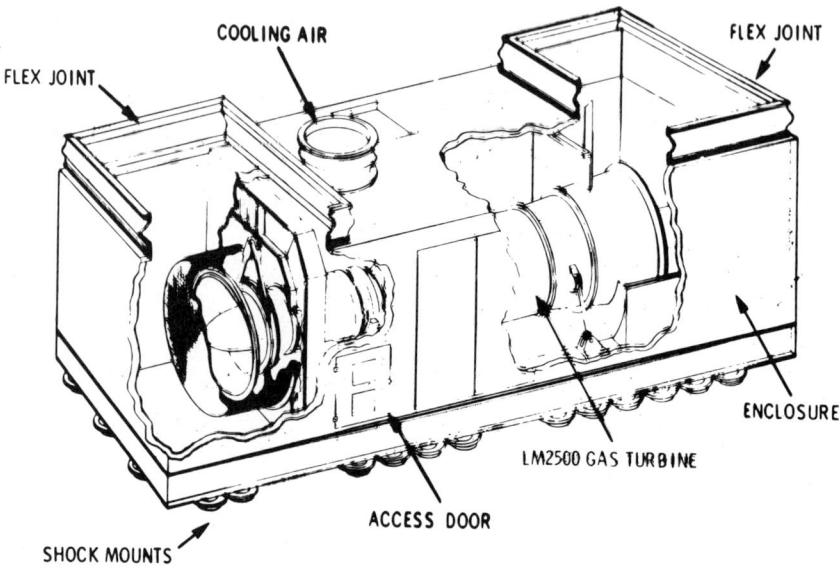

Figure 12–18. LM 2500 gas turbine module.

The engines are shock-mounted, both to minimize the amount of noise and vibration which is transmitted to the ship's hull, and to protect the engine from any sudden hull shocks which might damage the compressor.

The LM 2500 gas turbine engine has several advantages over a conventional steam plant. It is more compact and lighter, taking up less space altogether than only the main condenser of a steam plant of equal horsepower. It is easier to maintain and repair. And, it can go from the "cold iron" shut-down state to fully "ready for load" in roughly sixty seconds, in contrast to the several hours required to bring a steam engine on line from "cold iron."

The LM 2500 consists of four major components: the compressor, the combustion chamber, the gas generator turbine, and the power turbine. Each of these is described below.

The LM 2500 Compressor

The LM 2500 compressor is a 16-stage, axial-flow compressor, as shown in figure 12–19. The inlet guide vanes and first six stages of stator vanes are controllable in pitch, so as to aerodynamically match the low-pressure stages of compression to the high-pressure stages. Since axial-flow compressors are very sensitive to variations in rpm and air flow, controllable-pitch vanes ensure sufficient flow of air at all operating speeds and help to prevent compressor stall.

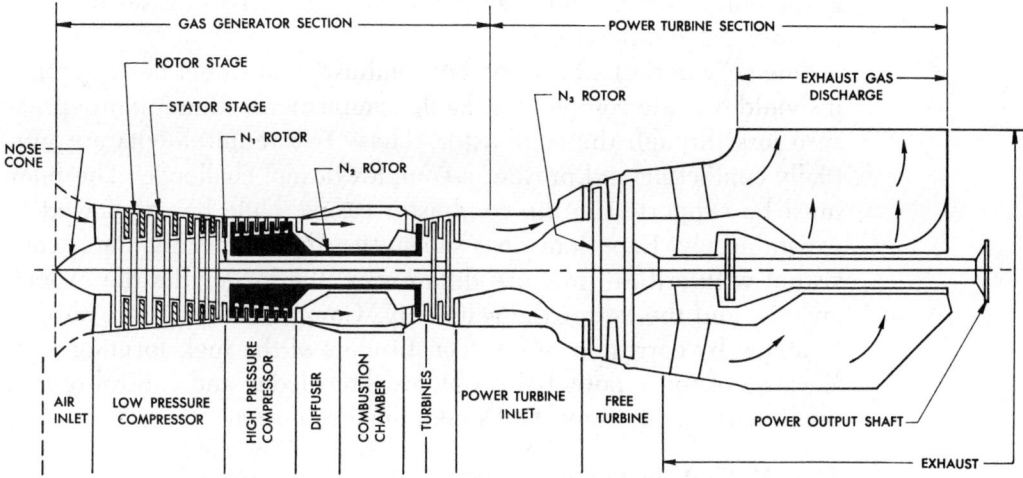

Figure 12–19. LM 2500 gas turbine.

The angle of the vanes is varied by two hydraulic activators which are operated by fuel pressure from the main fuel control. The stator vanes and inlet guide vanes are controlled by a single system. The function of the variable stator vanes is to maintain the velocity of air within acceptable limits for low air flow conditions and to permit high air flow at higher power rates with a minimum of restrictions. The function of the inlet guide vanes is to throttle and direct air flow to the first-stage rotor blades.

Compressor discharge air (the air leaving the compressor) splits and is used for three purposes. *Primary air* (25%) and *secondary air* (70%) go to the combustion chamber as described earlier. The other 5% is used for film cooling of the gas generator turbine (described later), and as a secondary source of compressed air to supplement that supplied by the regular ship's air compressors.

The LM 2500 Combustion Chamber

The combustion section of the LM 2500 (figure 12–19) consists of an annular combustor design with 30 fuel nozzles. The fuel–air mixture is ignited by two spark plugs. The inlet diffuser to the combustor is designed to provide uniform flow to the combustor despite large variations in inlet flow conditions caused by shifts in compressor discharge characteristics. Each nozzle is provided with a vortex-inducing swirl cup to provide flame stabilization and mixing of primary air with the fuel. The inner skirt or liner of the combustor is provided with scoop-type primary air entry holes. Thimble holes downstream in the outer skirt promote additional mixing of air and gases in order to lower gas temperatures.

One very important criterion of combustion chamber design is that it should provide complete, reliable combustion with minimum pressure loss through the combustor. These two requirements are mutually conflicting and present a complex design challenge. The liner must be supported in the combustor casing while being allowed to expand freely. Experience has shown that the thinner the liner, consistent with critical pressure differences, the longer the life expectancy is, and the better the reliability. Combustion parts are subject to attack by corrosion due to constituents of the fuel, forming H_2S. Because of these potential problems, knowledge and control of fuel quality are highly important operational requirements.

The LM 2500 Generator Turbine

The gas generator turbine in figure 12–19 is a two-stage, high-inlet-temperature, air-cooled turbine. Because of the high inlet temperature, secondary air is used to cool the first-stage nozzles and the first-stage blades. The rotor is cooled by a continuous flow of compressor discharge air, which provides a relatively cool layer of air at the metal surface of the blades. This method of cooling is called film cooling.

The LM 2500 Power Turbine

The hot gases leave the gas generator turbine and pass through the midframe section on the way to the power turbine; the midframe section houses the bearings and mounts to support the gas generator and power turbines. The power turbine (figure 12–19) consists of six stages of turbine nozzles and blading. The turbine blading is very similar in appearance to that used in steam turbines. Each power turbine can develop up to 20,000 shaft horsepower.

Spruance Engineering Plant

As indicated in figure 12–20, each propeller shaft is driven by two LM 2500 gas turbine engines delivering a total of 40,000 shaft horsepower to each propeller. The two LM 2500s are connected to a double reduction, double helical locked train reduction gear set through a pneumatic clutch, which allows an engine to be disconnected when not required. The two main engine rooms and auxiliary machinery spaces are operated from a central control station (figure 12–20).

To generate electricity, three 2000-kW gas turbine generating sets (GTGS) are installed, any two of which can supply the full electrical requirements of the ship. Each is powered by a single-shaft Allison

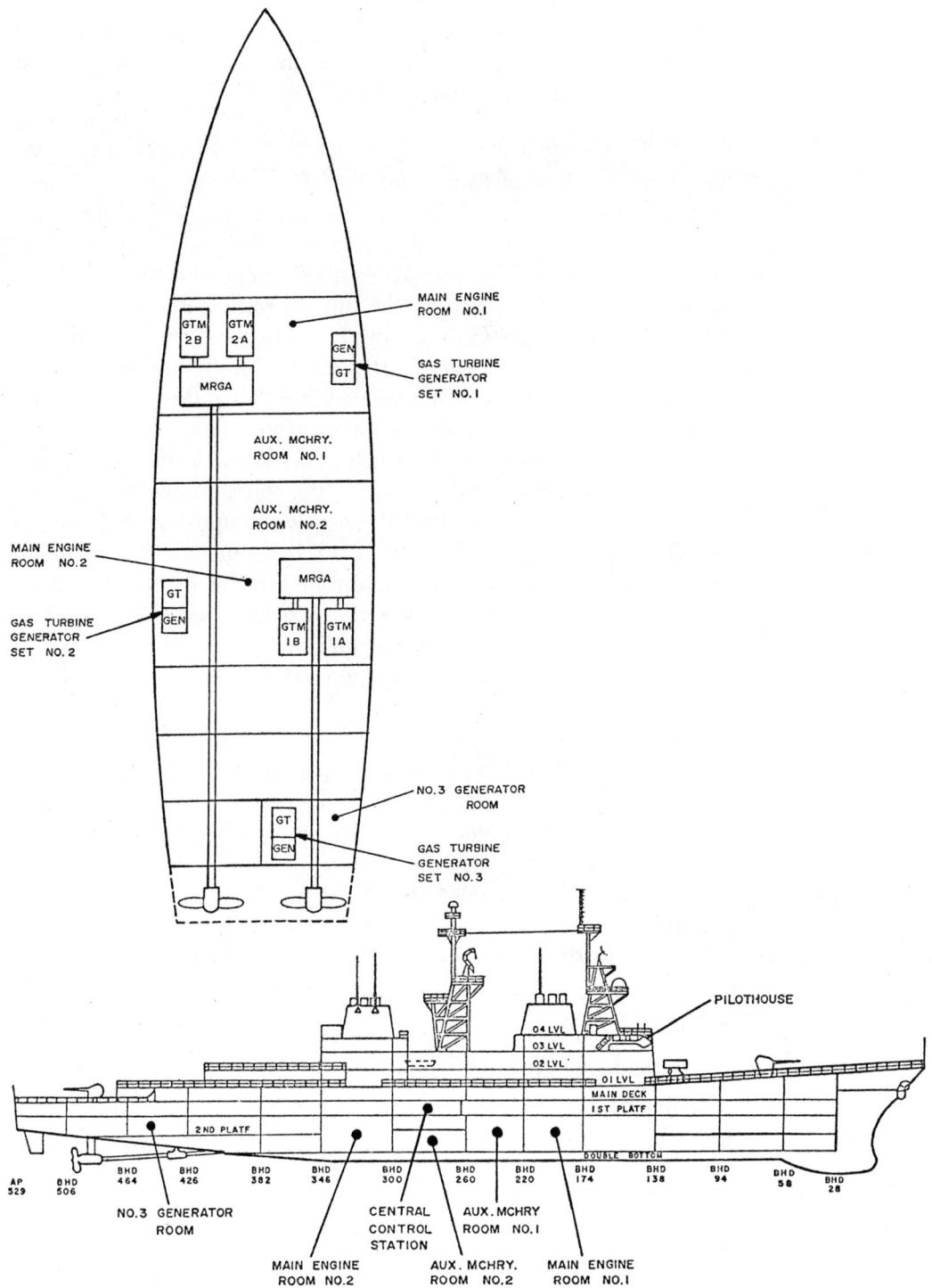

Figure 12–20. Engineering plant layout, Spruance-class.

Gas Turbine Engines 231

501 gas turbine engine, similar to those on P–3 and C–130 aircraft. Each GTGS is separated from the others by at least three water-tight bulkheads to minimize potential battle-damage losses of power. Each can be brought on line in only 30–45 seconds.

LM 2500 Engine Control

The basic method of controlling the speed of the LM 2500, and thus the power output, is by controlling the flow of fuel to the combustion chamber nozzles. On the *Spruance*, this is done by an electronic control system which regulates the gas generator turbine speed in accordance with the setting of the ship's throttles. For each throttle setting, the control system will supply enough fuel to keep the gas generator section running at a particular predetermined speed. The electronic control system will alter that fuel flow if the gas generator accelerates or decelerates improperly, gets too hot (overtemperature), or overspeeds, in which case it will shut down the fuel flow, and hence the turbine itself. As mentioned previously, split-shaft gas turbines are well-suited for shipboard use. Since compressors will quickly lose efficiency when placed under a load, the split-shaft design allows the gas generator section to run continuously at its set rpm even while the power turbine slows under a heavy load. The speed at which the power turbine will then run is determined by the amount of hot gases it receives from the gas generator, versus the load placed on it by the propeller. This is termed "power control" rather than "speed control" of the power turbine. Since the rpm is not directly controlled, there must be a means to prevent a sudden "run-away" such as would occur if the propeller left the water momentarily during rough seas. Therefore, a *topping* (or speed-limiting) *governor* is provided to cut back the power turbine speed if it reaches 104 percent of the preset level. It does this by reducing fuel flow in the gas generator section, thus reducing the power delivered to the power turbine. To further protect the power turbine, if it should reach 108 percent of the preset speed, the entire LM 2500 will be shut down by an *overspeed trip* feature of the governor.

Inlet and Exhaust Ducts

Shipboard gas turbine engine installations require the passage of large inlet and exhaust volumetric flows which must be accommodated within the confines of the ship structure with minimum amounts of pressure drop.

It is important that shipboard ducts be as small as possible to reduce the demands on valuable topside deck area. Furthermore, because the ducts are high in the ship, minimizing their weight significantly contributes to ship stability. In conflict with these considerations are the requirements for large ducts to minimize pressure losses and flow distortion and provide access for engine removal.

As previously mentioned, it is highly important to keep the air inlet pressure losses to a minimum, since a small inlet pressure drop may cause a considerable loss in performance. In addition, unequal air flow distribution to the compressor can reduce efficiency and cause vibrations leading to blade failure. The inlet ducts, as shown in figure 12–21, must be provided with suitable screens to protect the compressor from any foreign matter which may be drawn into the engine. It is very important to see that the screens remain in place and are maintained in good shape so that pieces of deteriorated screen are not drawn into the unit. Gas turbines require clean air at all times; otherwise the compressor blades will become coated with ingested material which would reduce the capacity and efficiency of the engine. In the marine environment the prevention of any form of salt water from entering the compressor is a necessity. Air inlets must be placed as high above the water as possible and equipped with baffles to prevent the entry of solid water. In addition, de-misters may be installed in the ducts to remove spray. They may consist of filter pads and/or inertial type separators. De-mister pads are effective in stopping the ingestion of salt spray and other foreign particles when they are kept wet, even with seawater.

The high volume of exhaust gases from the LM 2500 engines presents the problems of high noise levels and very hot exhaust gases (1200°F). Gas turbines generate large amounts of noise over a wide frequency spectrum. Most of the noise is in the high frequency range, and emanates aerodynamically from the blades of the compressor. Airborne noise in both the inlet and exhaust is attenuated by the use of suitable silencers. However, the greater the reduction in noise through attenuation, the greater the pressure losses with attending losses in power and efficiency. Therefore, noise reduction is limited to that level required by the particular ship's mission.

The high temperature of the exhaust gases poses two problems in the *Spruance* class: possible damage to personnel or equipment, and the increased possibility of detection and weapons guidance by the enemy through the use of infrared (IR) sensors. To cool the gases, each LM 2500 engine's exhaust is routed through an "air ejector" nozzle at the top of the stack that cools the exhaust to 450°F. Should the possibility of IR detection be a concern, an infrared suppression system can be actuated which sprays water from the fire main into the exhaust duct, further cooling the gases to 350°F.

GTGS Exhaust System

Each gas turbine generating set exhausts to a waste-heat boiler

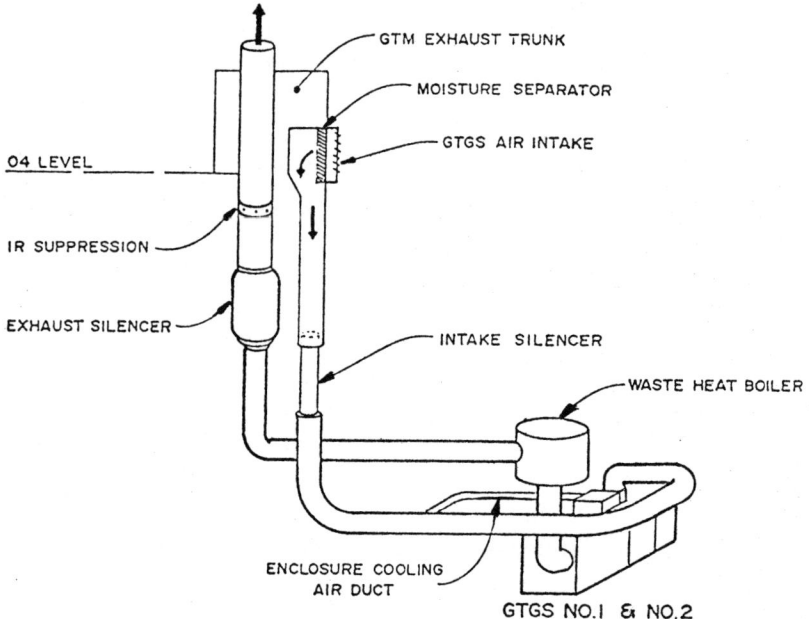

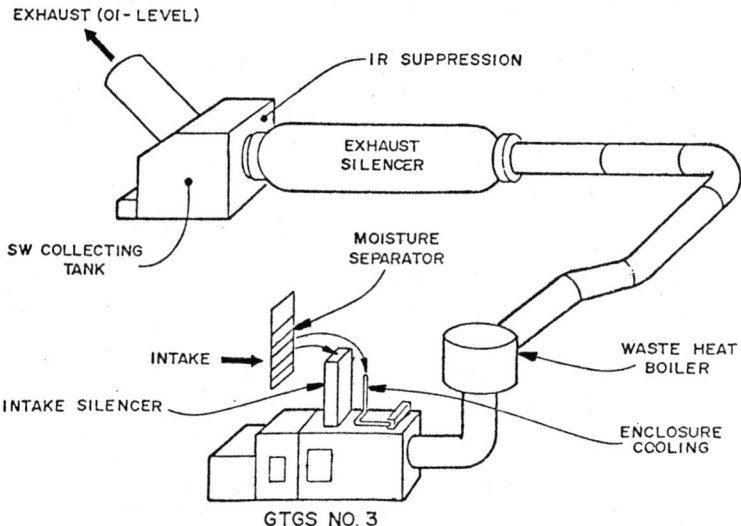

Figure 12–21. GTGS intake and exhaust systems.

(figure 12–21) which cools the exhaust gases to about 400°F before discharge. The waste heat boilers (WHB) are controlled circulation boilers, generating 100 psig 650°F ship's service steam for use in the following:

- Fuel oil heaters
- Lube oil heaters

- Hot water heaters
- Distilling plant
- Galley
- Laundry
- Ship's heating
- ASROC heaters

Lube Oil and Fuel Oil Systems

Lube Oil System (figure 12–22). As with all engines, the lube oil must be cooled to remove heat absorbed from the bearings and other heated engine areas. In order to safeguard against seawater contaminating the synthetic lube oil used in the gas turbine engines, an oil-to-oil heat exchanger is provided, utilizing the lubricating oil in the reduction gear sump to cool the synthetic turbine oil.

An attached shaft-driven lube oil pump is provided to deliver oil to the gear system down to 10 percent below cruising speed. At full power this would require bypassing of approximately 70 percent of the oil because of the relatively flat lube oil demand curves of recent naval gear designs. Two two-speed motor-driven pumps are used to back up the attached pump at high speed and to provide all the lubrication at speeds below the cruising speed. The two-speed feature minimizes the amount of lube oil that must be bypassed as the speed is increased, and the attached pump contributes to the overall flow.

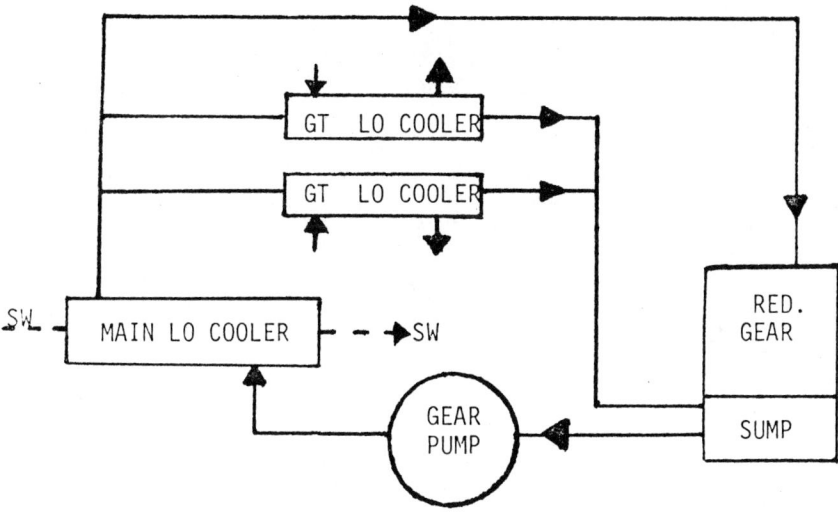

Figure 12–22. Schematic diagram of a DD 963 lube oil system.

Two pumps are provided for redundancy and reliability, and each pump is designed to supply the entire lube oil requirement.

Fuel Oil System (figure 12–23). An important function of the fuel oil system is to ensure that the fuel oil is delivered to the gas turbines clean and at the proper viscosity. The DD 963 is the first gas turbine ship to be designed specifically for operation on Diesel Fuel Marine (DFM), which is a relatively low-quality distillate oil. Because this fuel oil has higher concentrations of contaminants and is more viscous than the diesel oil or JP fuels commonly used for gas turbines, it is imperative that every precaution be taken to maximize the purity of the fuel oil delivered to the engines. To achieve this the following features are incorporated in the fuel oil system:

1. A centrifugal purifier is placed between the stowage tanks and the service tanks to reduce the impurity content of the fuel to less than 500 ppm.

2. A duplex filter-coalescer element is placed between the service tanks and the engines to further reduce the water content to 20 ppm and to trap any particle matter greater than 10 microns in size.

3. All of the fuel oil piping between the service tank and the content engine is made of stainless steel to minimize any corrosion.

4. Steam heaters are also included to heat the fuel oil to a temperature of 130°F before it reaches the gas turbines. The heaters are used only to improve centrifuge and filter performance.

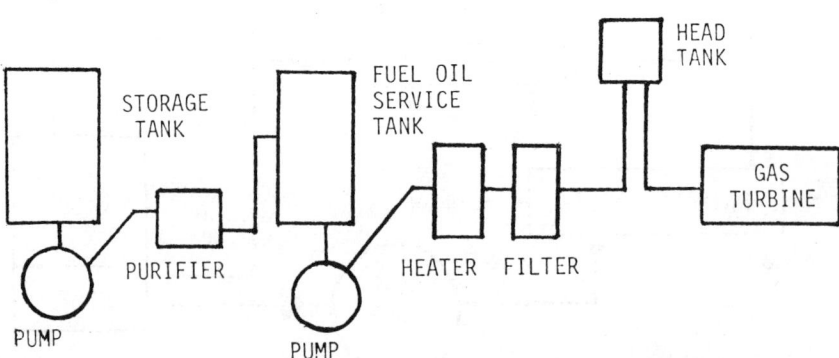

Figure 12–23. *Schematic diagram of a DD 963 fuel oil system.*

Controllable Reversible Pitch Propeller and Propulsion Shafting

Power from the LM 2500 engines is transmitted to the reduction gears and then via a hollow propeller shaft to a 5-bladed variable pitch propeller. The shaft is hollow to provide for a flow of hydraulic oil to and from the propeller. Since gas turbine engines run in only one direction, gas turbine–powered vessels of the *Spruance* class have controllable reversible pitch (CRP) propellers. In order to reverse thrust, the propeller blades rotate in the hub until they are pushing water ahead instead of astern.

However, there is another reason for the controllable pitch propeller. Unlike a steam turbine, which can be stopped without shutting down the boilers, the LM 2500 gas generator cannot operate at less than 5,000 rpm without "flaming out" (stalling) because of air starvation. At this minimum speed, the turbine delivers enough horsepower to drive the *Spruance* at 11 knots. In order to travel at slower speeds, some of this horsepower must be "wasted." Therefore, the pitch of the propeller blades is reduced, making the propeller less efficient, to drive the ship at slower speeds (Table 12–1). Speeds above 11 knots are attained by speeding up the propeller shaft, and setting the propeller to operate at a fixed extreme pitch, as can be seen in Table 12–1.

The pitch of the blades is controlled hydraulically through a system of pumps, pistons, and gears. Normally, the electronic control system automatically selects and sets the appropriate pitch as it controls the LM 2500. If necessary, however, the pitch setting can be manually selected. Finally, if hydraulic control is lost, the blades can be hand-pumped to the full ahead or full astern positions, and locked in place.

The two shafts are counter-rotating; the port shaft rotates clockwise, and the starboard shaft rotates counterclockwise, as viewed from astern.

The CRP propeller system for each propulsion shaft consists of the propeller blades and hub assembly, the propulsion shafting, hydraulic and pneumatic piping and control rod contained in the shaft, the hydraulic oil power module, the OD box, the control valve manifold block, the electro-hydraulic controls, the hydraulic oil sump tank, and the head tank. A block diagram of the system is shown in figure 12–24.

The CRP propeller has five blades, is 17 feet in diameter, and develops 40,000 shaft horsepower at 168 rpm. The blade pitch control hydraulic servo motor, mechanical linkage, and hydraulic oil regulating valve are housed in the propeller hub. High-pressure hydraulic control oil is provided for each propeller by an HOPM (hydraulic oil

Table 12–1. *Spruance*-class propeller rpm, pitch, and ship speed.

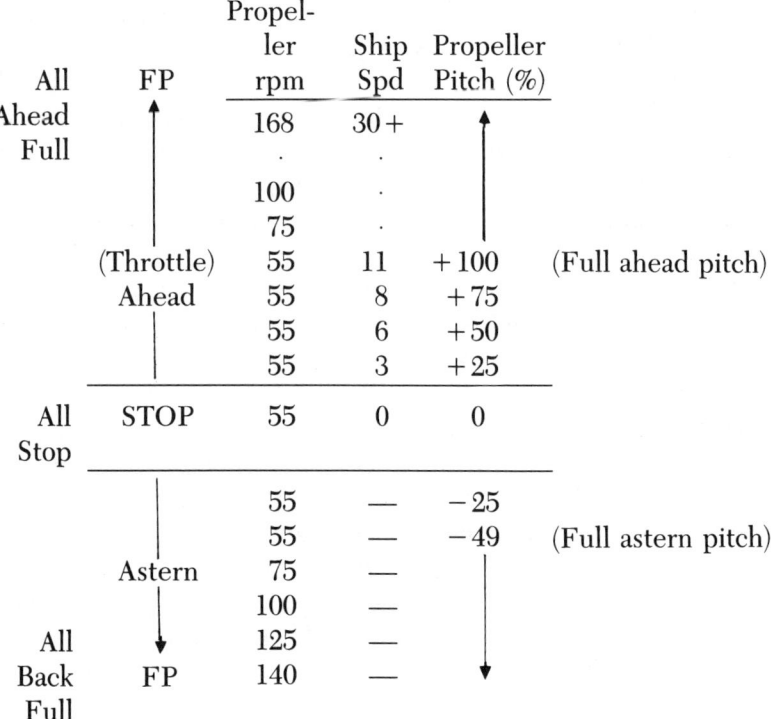

power module), which is located adjacent to the reduction gear in each engine room. An OD (oil distribution) box, mounted on the forward end of the reduction gear, is mechanically connected to the hydraulic oil regulating valve by a valve control rod. The OD box contains the hydraulic servo mechanism which positions the regulating valve rod. The OD box also provides the flow path connection between the hub servo motor and the HOPM. The regulating valve control rod, prairie masker air piping, and flow path for hydraulic oil supply and return are contained in the hollow propulsion shafting.

A control valve manifold block assembly, mounted on the side of the OD box, contains control valves for both manual and automatic pitch control. The manual control valves consist of a manual pitch control valve and two manual changeover valves. Automatic pitch control is accomplished through an electro-hydraulic control oil servo valve. This valve responds to an electrical signal generated from an electronic enclosure located in the engine room. The electronic enclosure signal is controlled by the pitch setting ordered through the ECSS (engineering control and surveillance system).

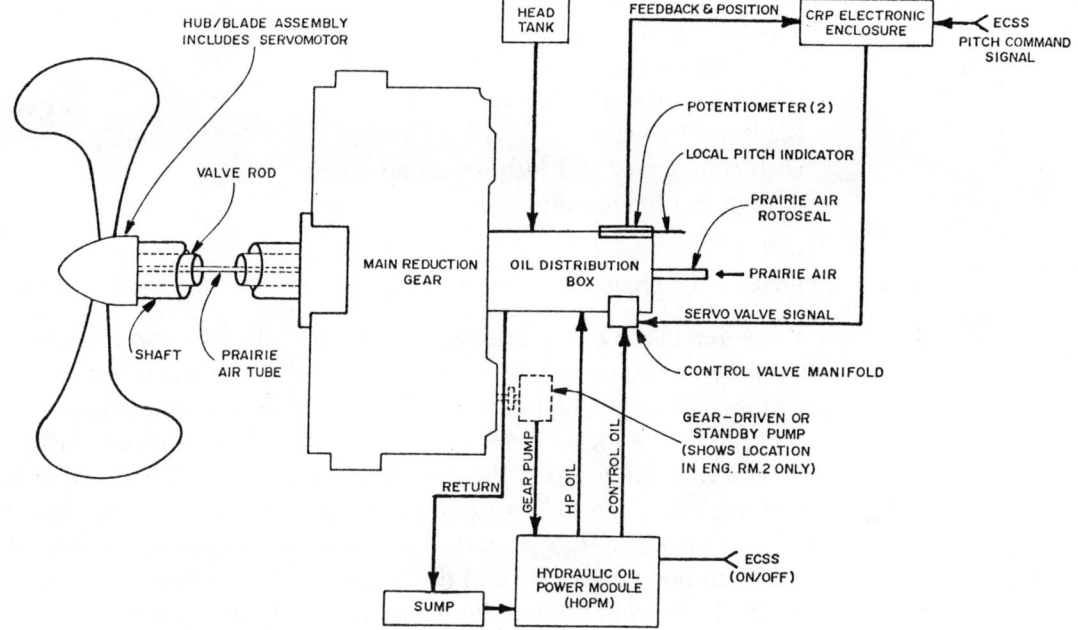

Figure 12–24. CRP propeller system block diagram.

List of Equipment

Each ship requires two independent CRP propeller systems, one for each propulsion shaft. Many components are identical, but some assemblies must have port and starboard configurations, such as the propeller hub and blade assembly, because of the counter-rotating shafts. Table 12–2 lists the major components and assemblies of each propeller system.

Hub-Blade Assembly

The hub-blade assembly provides the mounting for the five propeller blades, contains the hydraulic servo motor mechanism for blade pitch control, and transforms propulsion shaft rotational torque into axial thrust for ship propulsion. The hub body is secured to the tailshaft flange by fifteen bolts and five dowel pins. The rotational torque of the propulsion shaft is transmitted to the hub by the five dowel pins. The hub-blade assembly consists of the following major components:

Hub body assembly
Piston and piston rod assembly
Crosshead and sliding blocks

Gas Turbine Engines   **239**

Regulating valve
Tailshaft spigot
Crankpin rings
Blade seal base rings
Blade port covers
Hub cone cover and hub cone end cover
Blades and blade bolts

Principle of Operation

Propeller blade pitch is hydraulically controlled by the hydraulic servo motor. The servo motor, consisting of the piston and piston rod assembly and the regulating valve, is contained within the hub body. A one-piece crosshead is secured to the forward end of the piston rod. The hub body forms the crosshead chamber with the tailshaft spigot enclosing the forward end of the chamber and the hub body end plate enclosing the aft end. The servo motor cylinder is formed by the hub body end plate and the hub cone. The piston rod extends through the bore in the hub body end plate into the crosshead chamber.

Five sliding blocks fit into slots machined in the crosshead. The hub body contains five blade ports with a center post in each port; the center posts are integral parts of the hub body. A crankpin ring fits over each center post in the blade ports, and an eccentric crankpin on the underside of each crankpin ring fits into a hole in each sliding block. When the servo motor piston is translated, as a result of differential pressure across the piston, axial movement of the crosshead produces rotary movement of the crankpin rings through the sliding blocks and eccentric crankpins. The propeller blades, which are secured to the crankpin rings, are thus rotated, resulting in the desired pitch change.

Table 12–2. CRP propeller system equipment.

Hub-blade assembly
  Hub assembly
  Set of five blades

Valve rod assembly
  Piping, couplings, and guides

Oil distribution box
  OD box assembly
  Manifold valve block mounting

Hydraulic system
  Hydraulic oil power module
    Suction strainer
    Hydraulic oil pump and motor
    Duplex discharge filters (3)
    Air bleed valves (2)
    Pressure control assembly
    Manual bypass valve
    Gage panels (2)
  Local pump motor controller
  Manifold block assembly (on OD box)
    Electro-hydraulic servo valve
    Manual pitch control valve
    Manual changeover valves
  Standby hydraulic oil pump
    Suction strainer
  Return check valve

Electronic pitch control system
  Electronic control assembly
  Linear potentiometer—Feedback
  Potentiometer—Pitch readouts

Local pitch indicator
  Mechanical pitch indicator

Emergency pitch setting
  Emergency hand pump assembly

Prairie air system
  Rotoseal
  Check valves, hub and OD box
  Air tube assembly
  Connectors, hub to blades

13

Comparison of Marine Propulsion Plants

Assuming that for any given ship type the approximate amount of propulsion power required for satisfactory operation is known, the next logical requirement is to determine which particular type of marine propulsion plant can best meet our needs.

General Requirements

In general, marine propulsion plants must meet certain basic requirements. They must be of a minimum size and weight, be highly reliable, be economically efficient, and be reasonably easy to maintain. In addition, warship propulsion plants must be quiet, quick-reacting to changes in load, and resistant to shock, and must demonstrate efficiency through a wide range of power levels. With the present spiraling rise in operating costs, maintenance costs, and personnel costs, added emphasis must be placed on reliability, efficiency, and reducing the number of personnel required to operate and maintain our propulsion plant safely.

Experience has shown that five general types of propulsion systems have satisfied the broad requirements stated above for main propulsion of naval vessels. The five are: fossil-fueled steam power plants, nuclear-fueled steam power plants, turbo-electric power plants, diesel engine power plants, and gas turbine power plants. Table 13–1 compares four of these types of propulsion plants.

Steam Power Plant, Fossil-Fueled

Fossil-fueled steam power plants are the most common propulsion systems used in modern warships. They can be found in carriers, cruisers, destroyer types, amphibious warfare vessels, and fleet auxiliaries. The size and design of these systems vary, but the basic principles of operation are the same. The oil-fired steam turbine plant has been proved in large numbers of naval and merchant marine applications. Large quantities of oil are carried for conversion to thermal energy by combustion in a marine boiler. The thermal energy in the boiler is transferred to high-pressure steam, the working substance. Mechanical energy, to turn the ship's propeller, is provided

Table 13–1. Comparison of four propulsion plants (1973 dollars).

	Diesel	Steam	Nuclear	Gas turbine
lb/hp	3	15	35	.28
Cost	$18/BHP[a]	$34/BHP	$51/BHP	$24/BHP
SFC[b]	.3 lb/hr-shp[c]	.55 lb/hr-shp	NA	.4 lb/hr-shp
Comments	20,000 shp limit; too large and bulky above 20,000 for ship use.	280,000 shp; no limit; heavy and expensive to overhaul.	Initial cost and maintenance high; high state of training necessary.	Very light; rivals cost of diesel.

[a] BHP (brake horsepower)—a measurement of actual power produced by an engine.
[b] SFC—specific fuel consumption.
[c] shp (shaft horsepower)—a measurement of available power to the propeller.

by a high-speed turbine through which the steam passes. Reduction gears are necessary to connect the turbine to the propeller shaft, since the turbine operates most efficiently at high rotational speeds, while the propeller is most efficient at low rotational speeds.

Long and extensive usage has resulted in an efficient, reliable, well-tested propulsion system. The overall efficiency of the system is well above the minimum requirements at all load levels, and the steam propulsion system has the capability of partial load control, leading to good maneuvering characteristics. In addition to its mechanical advantages, the system's extensive service and usage have resulted in the availability of a large reservoir of qualified operators, technicians, and mechanics.

An additional indirect benefit of a steam propulsion plant, often overlooked, is that the same steam used for propulsion can also be used for a variety of auxiliary functions. Through the use of various valves and reducers, steam is used to generate electricity, produce fresh and feed water, heat water and food, operate the laundry, provide central heating throughout the ship, and operate the steam catapult for aircraft launch.

Many of the advantages of the fossil-fueled steam generation system are offset by associated disadvantages. Although the system operates at all loads, a rapid drop in efficiency is experienced at low partial load, resulting in poor fuel economy. A large percentage of full load displacement (as high as 20 percent), therefore, is devoted to fuel storage. Even with this large fuel load, the fossil-fueled plant has relatively low endurance at very high power levels, as shown in figure 13–1.

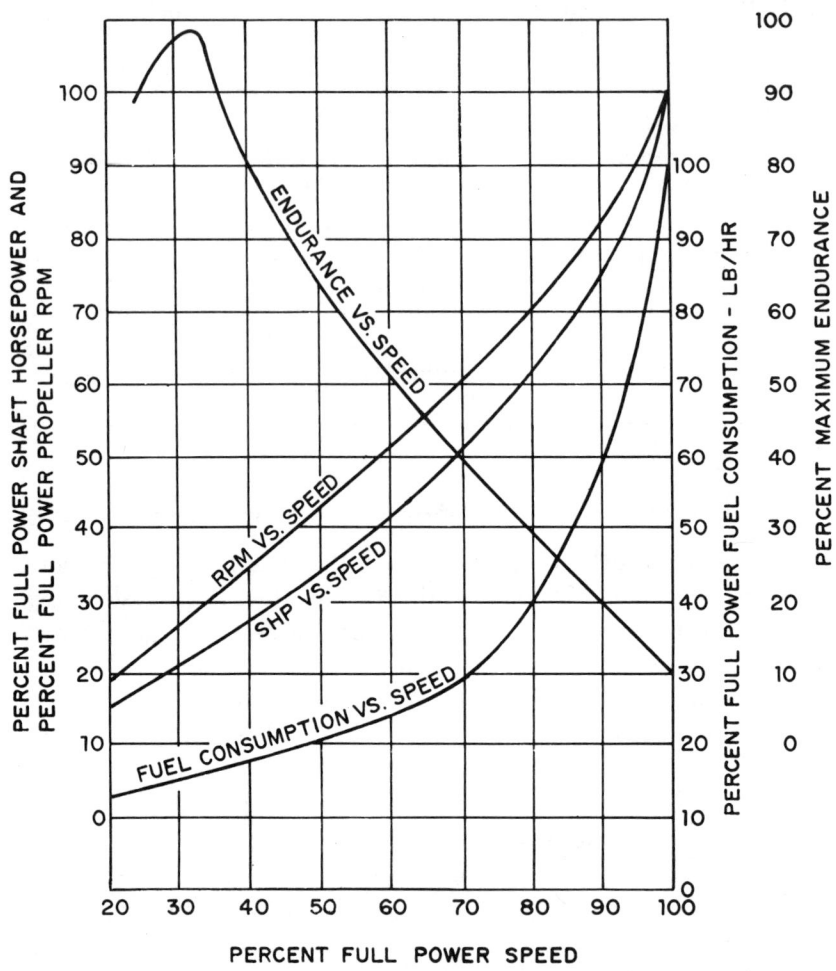

Figure 13–1. *Fossil-fueled engineering plant statistics.*

In general, steam generators are large and bulky, requiring excessive space. These large boilers, along with the extensive piping, valves, and insulation required for safe operation, add undesired weight to the ship. In addition, the slow combustion rate, coupled with the requirement to heat all elements of the steam system slowly, results in long light-off times, the minimum average light-off to underway time being four hours. This can be very costly in wartime or an emergency.

Other limitations and disadvantages are manpower-related. Because of the large number of monitoring and operating stations required and the extensive auxiliary machinery, manpower requirements are relatively high for operation and maintenance. Qualified-personnel costs are at a record high. In addition, more complicated repair-part support is required, owing to the large number of components of different manufacturers in any ship or even classes of ships.

Pressure-Fired Steam Generators

Some of the disadvantages of the conventional fossil-fueled steam generators have been offset by the development of the pressure-fired steam generator. The pressure-fired steam generator is approximately one-half the physical size of a conventional one of the same power output. Its operation differs from the conventional boiler in that air is provided by an eleven-stage compressor designed as a supercharger. The air from the supercharger is forced into the boiler under 60 pounds of pressure, increasing the combustion rate. This increased combustion rate, coupled with the reduced size and water capacity of the steam generator, significantly decreases the light-off and response time of the boiler. The average, safe light-off time for a pressure-fired boiler is one hour. In addition, response time for the pressure-fired boiler from 0 to 100 percent boiler load is 20 seconds.

The improvements, however, are not without disadvantages. The increased power-to-weight value of the pressure-fired steam generator is offset somewhat by the requirement for cleaner, higher-grade, and more expensive fuel. In addition, the operation, maintenance, and calibration of the pressure-fired steam generator and its associated automatic combustion control system vary greatly from what is required of the coventional steam generator, necessitating additional schooling for personnel, supply support, and maintenance procedures.

Steam Power Plant, Nuclear-Fueled

The nuclear steam plant utilizes the energy released by nuclear fission of reactor fuel, usually an alloy of uranium. The liberated heat is transferred from the reactor to a steam generator by a primary coolant fluid (such as pressurized water) where it is transferred to high-pressure steam. The steam is utilized in steam turbines in a fashion identical to its use in the fossil-fueled plant.

The major advantage of the nuclear-fueled plant is the increased endurance over the fossil-fueled steam plant. The USS *Enterprise* (CVN–65) steamed 207,000 miles prior to her first refueling. Compare this to a typical destroyer endurance of 5000 miles at most economical speed or about 1500 miles at top speed.

The related benefit of eliminating the requirement for propulsion fuel in advanced areas is equally important. Because huge quantities of fuel would no longer have to be transported, handled, and stored for an *all*-nuclear fleet, huge savings in ships and manpower would result. The tactical advantage of elimination of the highly vulnerable refueling evolution every two days in a war zone is apparent. However, requirements for aircraft fuel and rearming still limit CVN operations.

The nuclear plant is not dependent on air for combustion. For surface ships this permits simplified plant arrangement, since air inlet, stacks, and blowers are not required. It also provides inherent NBC warfare defense because of elimination of contaminants brought into the ship with combustion air. The advantages from an air pollution aspect are obvious. Large changes in navigational draft due to fuel oil consumption and ballasting are also eliminated.

Limitations of nuclear steam plants on board surface ships include the use of large amounts of shielding, which tends to eliminate the advantage of not carrying fuel oil. Construction and maintenance costs of nuclear steam plants are at premium rates because of the quality assurance and care necessary to ensure reliability and safety of the nuclear plant. This factor, coupled with lengthy training requirements for operational personnel, makes the nuclear plant installation about 50 percent more expensive than the fossil-fueled steam plant. However, with the ever increasing cost of fossil fuel, and the political and diplomatic difficulties in obtaining adequate quantities, this 50 percent increase in installation costs has been somewhat reduced.

As indicated by figure 13–2, the excessive weight of nuclear plants makes them burdensome for small and medium-sized combatants. In capital ships, however, the weight-to-shaft-horsepower ratio is comparable to that in fossil-fueled plants. This realization has resulted in the construction of two additional CVNs (the *Eisenhower* and the *Nimitz*) and their associated nuclear-powered escorts.

Turbo-Electric Power Plant

The turbo-electric plant uses a conventional fossil-fueled boiler to produce steam for turbine power. However, the heavy and bulky reduction gear is replaced by an equally cumbersome AC or DC generator/motor combination which produces electric power for propulsion through electric motors.

Overall weight and space needs of the turbo-electric plant are comparable to requirements for the geared turbine plant. Power transmission efficiency for the turbo-electric plant is about 90 percent (97 percent for the generator times 93 percent for the motor). In contrast, the reduction gears of the geared turbine plant operate with

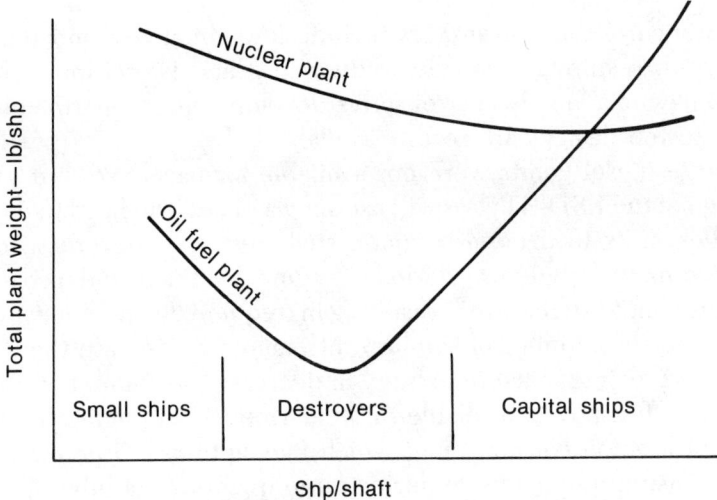

Figure 13-2. Comparison of nuclear plants and oil fuel plants; weight vs. shaft horsepower of various ship types.

a transmission efficiency of 98 to 99 percent. However, the turbo-electric plant allows the turbine to run at constant speed, thus eliminating the decreasing turbine efficiency at low partial load experienced in the geared turbine plant. The turbo-electric plant offers good economy and ease of maneuvering (speed control) and reversing. These were important advantages prior to World War II when the reduction gear of the geared turbine was still being developed. The turbo-electric drive offered little advantage over the geared turbine once the reliability of the reduction gears and astern turbine was proved.

Turbo-electric drive was used in many merchant ships as well as some World War II era escort types in use by the Navy and Coast Guard as late as 1969. The direct-current drive is especially well adapted to rapid maneuvering and is found in tugs, towing vessels, and ice breakers.

Diesel Engine Plant

Diesel engines have long been used for medium and light power applications, in both general marine and naval service. For high power naval applications, several engines have been connected to a common shaft or used to generate power in a diesel-electric plant in some surface ships as well as submarines.

Diesel engines afford relatively high efficiency at partial load, and distinctively higher efficiency at very low partial load than steam turbines. The diesel thus requires the least weight of fuel for a given endurance. Other advantages include low initial cost and relatively low rpm, resulting in smaller reduction gears. Diesel units are reliable, having a long history of active development for marine use, and fully tested designs are readily available.

Large diesel plants were not available for naval use, but current design of the LSD–41 class will use such a plant. The need to combine smaller units to drive a common shaft results in severe space and arrangement problems. Periodic engine overhaul and progressive maintenance are required, resulting in frequent down periods, which, owing to the number of similar units, may not increase the amount of inport maintenance time but will decrease the amount of time the ship has full power available. The marine diesel engine has a high rate of lube oil consumption, which may approach 5 percent of the fuel consumption, thus requiring large quantities of lube oil. At the present time, lighter-weight, higher-powered diesel engines are being developed.

Diesel plants have been widely used in the Navy for propulsion of small boats, service craft, and even larger surface ships including some destroyer escorts. All ocean minesweepers and some special ship types such as large submarine tenders, ice breakers, auxiliary tugs, fleet tugs, and salvage vessels are diesel-driven.

Figure 13–3 shows the increase in specific power (horsepower produced per pound of machinery plus fuel) over the years. With naval designers requiring ever more power in a smaller space, specific

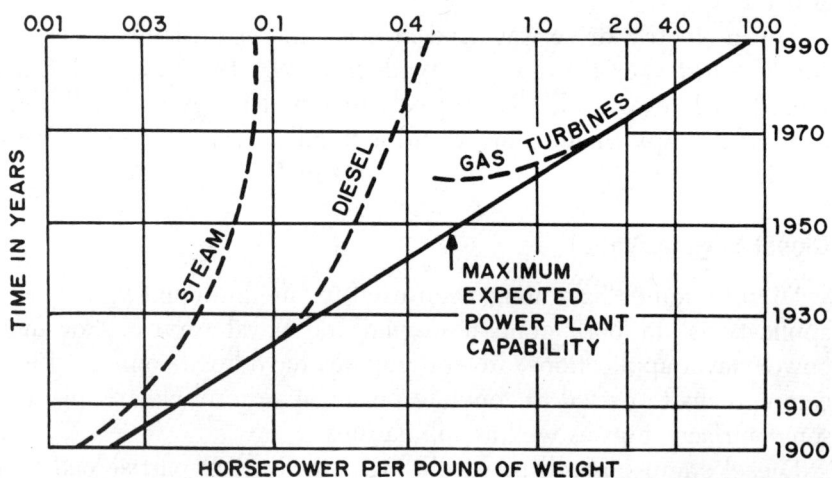

Figure 13–3. Comparison in specific weight for type propulsion over the time frame 1900–1990.

248   Introduction to Naval Engineering

weight of the propulsion plant assumes great importance. Figure 13–3 shows that specific weight of steam and diesel plants is peaking out, leaving the gas turbine as prime contender for high-powered marine plants in the near future.

Gas Turbine Plant

The gas turbine engine, long regarded as a promising but experimental prime mover, has in recent years been developed to the point where it is entirely practicable for ship propulsion and for a number of auxiliary applications. Gas turbine engines are presently in use in the PG–84 class high-speed gunboats, the DD 963 (*Spruance*)-class destroyers, minesweepers, landing craft, hydrofoils, hydro-skimmers, and other craft. Foreign navies and the Coast Guard have already built, tested, and efficiently operated gas-turbine-powered ships up through destroyer size. In addition, the gas turbine engine is finding increasing application as the driving unit for ship's service generators, pumps, and other auxiliary units.

The marine gas turbine plant has been using standard aircraft jet engines that have been converted to marine use by utilizing the jet exhaust to run a separate turbine in the same engine enclosure. This free turbine drives an external shaft, which in turn can be used to transmit mechanical work.

All gas turbines, no matter how large or complex, are composed of three basic parts: a compressor, a burner or combustion chamber, and a turbine. The compressor draws in air and sends it to the burner, where air and fuel burn. The combustion gas then flows through the turbine blades, producing work to drive the compressor and the load.

Advantages of marine gas turbine power include very low specific weight (shp/lb), compactness which greatly eases machinery space design problems, good efficiency at or near full load, little or no cooling-water requirement, full power available in less than five minutes from cold iron, and capability of removing the engine for shore overhaul while a new engine is installed for continuous shipboard use. Additionally, the engine emits much less hullborne vibration than other designs. Intensive aircraft industry development and subsequent marine development have led to gas turbine prime movers with a high degree of reliability. The vast majority of problems with marine gas turbine propulsion plants today are in associated clutching and reversing machinery external to the turbine.

Despite their vast potential, gas turbines are not without disadvantages and limitations. Because of the large air requirements (up to four times as much as in steam or diesel plants) external air ducting for inlet and exhaust can cause an arrangement problem as well as a possible pollution problem. This vast ducting decreases some of the saved space gained by discarding steam piping and insulation. In addition, this requirement for external air is unsatisfactory for NBC defense.

A second disadvantage is that the gas turbine is very inefficient at low partial loads. To be efficient the gas turbine must be run at or near full load, greatly increasing fuel consumption. Unlike the case of nuclear power, therefore, a large amount of the ship's displacement must be allocated for propulsion-fuel storage.

The greatest limitation placed on gas turbine plants is the need for external reversing arrangements owing to the use of a high-speed unidirectional engine. The recent development and improvement of the controllable pitch propeller has greatly alleviated this limitation; but, as previously mentioned, most gas turbine problems are still associated with clutching and reversing machinery. Future developments and improvements should remove this limitation completely.

Hybrid Propulsion Plants

In addition to research and development on the five previously mentioned basic propulsion systems, some attention is being given to hybrid propulsion plants combining favorable characteristics from each of the basic types. A combined diesel and gas turbine (CODAG) plant is presently in use in the PG–84 class gunboat and the Coast Guard *Hamilton*-class high-endurance cutters. A combination gas turbine/gas turbine (COGOG) plant utilizing a small gas turbine for low-speed operations and a large gas turbine for high-speed operations is installed in the Royal Canadian Navy DDH–280 class helicopter destroyers. A combination gas turbine/steam system (COSOG) has been used by the British Navy with good success. All these systems have proved to be reasonably efficient and reliable.

Since combatants spend very little of their at-sea time operating at or near top speeds, many feasible propulsion combinations exist that could meet the desired results. Such a feasible combination would be a steam plant for low-speed operations combined with a gas turbine for the short high-speed operating periods.

14

Distilling Plants

Naval ships must be self-sustaining as far as the production of fresh water is concerned. The large quantities of fresh water required on board ship for boiler feed, drinking, cooking, bathing, and washing make it impracticable to provide storage tanks large enough for more than a few days' supply. Therefore, all naval ships depend upon distilling plants to meet the requirements for large quantities of fresh water of extremely high chemical and biological purity.

Principles of Distillation

All shipboard distilling plants not only perform the same basic function but also perform this function in much the same way. The distillation process consists of heating seawater to the boiling point and condensing the vapor to obtain fresh water (distillate). The distillation process for a shipboard plant is illustrated very simply in figure 14–1.

At a given pressure, the rate at which seawater is evaporated in a distilling plant is dependent upon the rate at which heat is transmitted to the water. The rate of heat transfer to the water is dependent upon a number of factors. Of major importance are the temperature difference between the substance giving up heat and the substance receiving heat, the available surface area through which heat may flow, and the coefficient of heat transfer of the substances and materials involved in the various heat exchangers that constitute the distilling plant. Additional factors such as the velocity of flow of the fluids and the cleanliness of the heat-transfer surfaces also have a marked effect upon heat transfer in a distilling plant.

Since a shipboard distilling plant consists of a number of heat exchangers, each serving one or more specified purposes, the plant as a whole provides an excellent illustration of many thermodynamic processes and concepts. Practical manifestations of heat transfer—including heating, cooling, and change of phase—abound in the distilling plant, and the significance of the pressure–temperature relationships of liquids and their vapors is clearly evident.

The seawater which is the raw material of the distilling plant is an

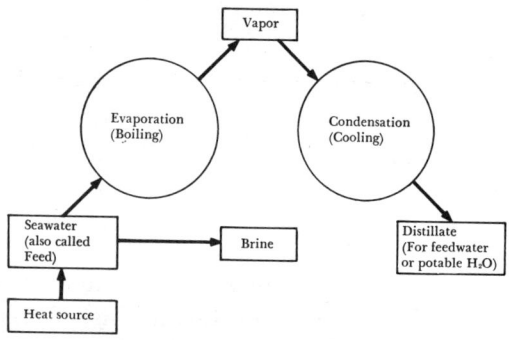

Figure 14-1. Basic block diagram of the distillation process.

aqueous solution of various minerals and salts. In addition to the dissolved material, seawater also contains suspended matter such as vegetable and animal growths and bacteria and other microorganisms. Under proper operating conditions, naval distilling plants are capable of producing fresh water that contains only minute traces of the chemical and biological contaminants that are found naturally in seawater. It should be noted that distilling plants are not effective in removing volatile gases or liquids that have a lower boiling point than water, nor are they effective in killing all microorganisms. These points are of particular importance when a ship is operating in contaminated or polluted waters.

One of the problems that arises in the distillation of seawater occurs because some of the salts present in seawater are negatively soluble—that is, they are less soluble in hot water than they are in cold water. A negatively soluble salt remains in solution at low temperatures but precipitates out of solution at higher temperatures. The crystalline precipitation of various sea salts forms scale on heat-transfer surfaces and thereby interferes with heat transfer. In naval distilling plants, this problem is partially avoided by designing the plants to operate under vacuum or (in the case of one type of plant) at approximately atmospheric pressure.

The use of low pressures (and therefore low boiling temperatures) has the additional advantage of greater thermodynamic efficiency than can be achieved when higher pressures and temperatures are used. With low pressures and temperatures, less heat is required to make the seawater boil, and less heat is lost overboard through the circulating water that cools and condenses the vapor.

Definition of Terms

The manner in which the various kinds of distilling plants accomplish the distilling process can be best understood if we first become familiar with certain terms related to the process. The terms defined here relate basically to all types of distilling plants now in naval use. Additional terms that apply specifically to a particular type of distilling unit are defined as necessary in subsequent discussion.

Distillation: The process of boiling seawater and then cooling and condensing the resulting vapor to produce fresh water.

Evaporation: The first part of the process of distillation. Evaporation is the process of boiling seawater in order to separate it into fresh water vapor and brine.

Condensation: The latter part of the process of distillation. Condensation is the process of cooling the vapor to produce usable fresh water.

Vapor: The product of the evaporation of seawater. The terms *vapor* and *fresh water vapor* are used interchangeably.

Distillate: The product resulting from the condensation of the fresh water vapor produced by the evaporation of seawater. Distillate is also referred to as *condensate,* as *fresh water,* as *fresh water condensate,* and as *seawater distillate.* However, the use of the term *condensate* should be avoided whenever there is any possibility of confusion between the condensate of the distilling plant and the condensate that results from the condensation of steam in the main and auxiliary condensers. In general, it is best to use the term distillate when referring to the product resulting from the condensation of vapor in the distilling plant.

Salinity: The concentration of salt in water.

Brine: Water in which the concentration of salt is higher than it is in seawater.

Steady operating conditions are beneficial to the satisfactory operation of a distilling unit. Fluctuations in the pressure and temperature of the generating steam will cause fluctuations of pressure and temperature throughout the entire unit. Such fluctuations may cause increased salinity of the distillate, and may also cause erratic operation of the feed and brine pump.

To achieve satisfactory operation of a distilling unit, it is necessary to maintain the designed vacuum. When the unit is operated at less than the designed vacuum, the heat level rises throughout the unit, and there is an increased tendency toward scale formation.

Scale formation is highly undesirable, since scale interferes with heat transfer and thus reduces the capacity of the unit. Excessive scale formation may also impair the quality of the distillate.

To reduce the rate of scale formation in distilling plants a chemical compound called ameroyal is used as the standard compound for evaporator feed treatment. In addition to ameroyal, other compounds that retard scale formation are also available for naval use. These chemical compounds do not, however, successfully kill bacteria or remove pollutants from the water.

Special restrictions are placed upon the operation of distilling units when the ship is operating in contaminated waters. Because most distilling plants operate at low pressures (and therefore at low temperatures), the distillate is not sterilized by the boiling process in the evaporators and may contain dangerous microorganisms or other matter harmful to health. All water in harbors, rivers, inlets, bays, landlocked waters, and the open sea within 10 miles of the entrance to such waters must be considered contaminated unless a specific determination to the contrary is made. In other areas, contamination may be declared to exist by the fleet surgeon or his representatives, as local conditions may warrant. When the ship is operating in contaminated waters, the distilling units must be operated in strict accordance with special procedures established by the Naval Sea Systems Command.

Flash-Type Distilling Units

Most new ships are equipped with flash-type distilling units. A typical two-stage shipboard installation is shown schematically in figure 14–2. The flash-type unit depends upon pressure differentials between the stages (or effects) to generate vapor from the seawater feed.

Flash-type units consist of two or more stages. Each stage of a flash-type unit has a flash chamber, a feed box, a vapor separator, and a distiller condenser. A two-stage or three-stage air ejector, a distillate cooler, and a feedwater heater are also provided. Feedwater passes through the tubes of the distillate cooler, the stage distiller condensers, and the air ejector condenser. In each of these heat exchangers the feed picks up heat. The final heating is done by low-pressure steam admitted to the shell of the feedwater heater. From this heater the feedwater enters the first-stage feed box and comes out through orifices into the flash chamber. As the heated feedwater enters the chamber, a portion flashes or vaporizes because the pressure in the chamber is lower than the saturation pressure corresponding to the temperature of the hot feed. The vapor condenses on the tubes of the first-stage distiller condenser. The feed that does not vaporize in the first chamber passes to the second chamber. The process is repeated in each stage, and the brine remaining in the last stage is removed by the brine overboard pump. Vapor formed in each

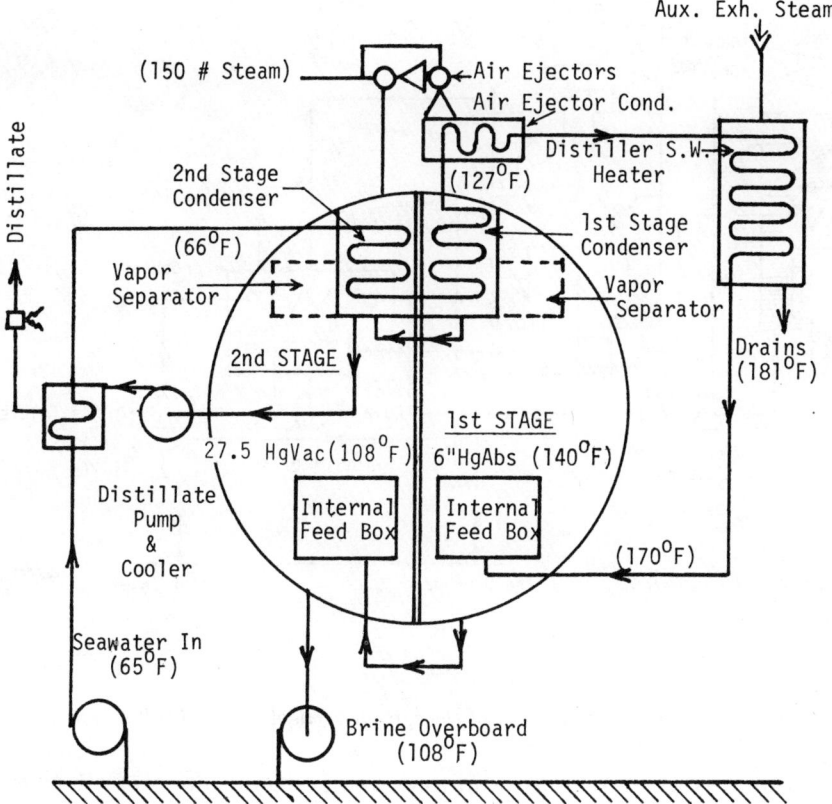

Figure 14–2. Schematic diagram of the typical flash-type distilling unit.

stage passes through a vapor separator and into the stage distiller condenser, where it is condensed into distillate. The distillate passes through a loop seal on its way to the distiller condenser of the next stage. The distillate pump removes the distillate from the last stage and discharges it through the distillate cooler to the test tanks.

Vertical Basket Distilling Units

Some ships, such as submarines, are equipped with vertical basket distilling units. A unit of this type is shown in figure 14–3. The unit shown has two effects; however, some units of this type may have more.

The vertical basket unit consists of two or more evaporators, a distiller condenser, vapor feed heaters, a distillate cooler, and air ejectors. In the vertical basket unit, each evaporator consists of a vertical shell in which a deeply corrugated vertical basket is installed.

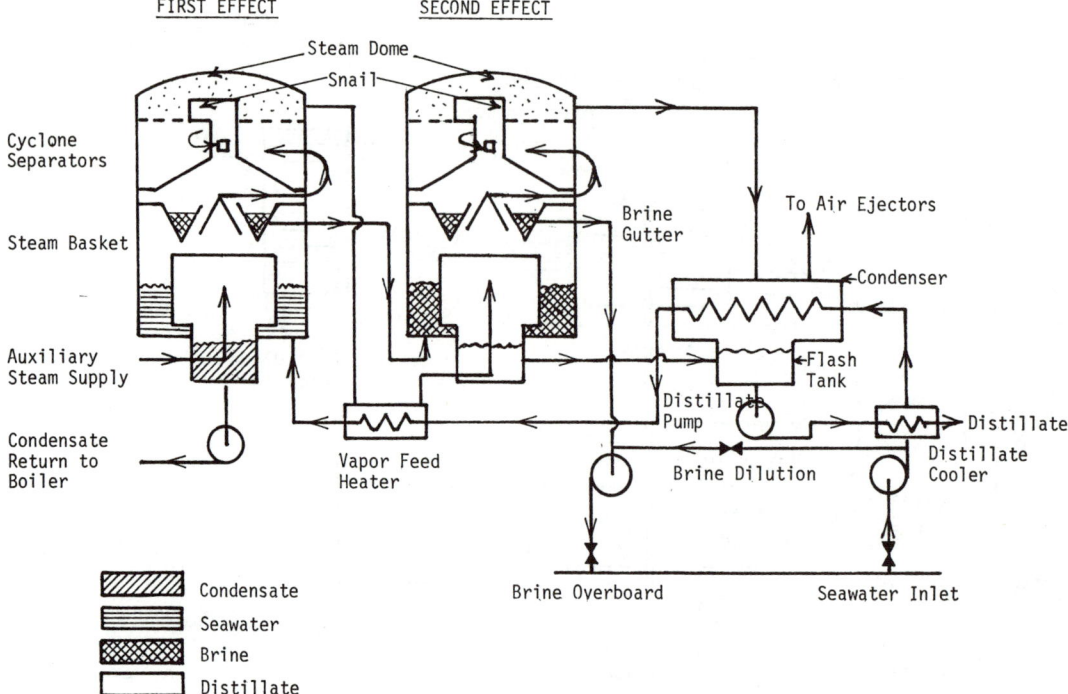

Figure 14–3. Schematic diagram of the typical vertical basket distilling plant.

Low-pressure steam is admitted to the inside of the first-effect basket. This steam boils the feedwater in the space between the outside of the basket and the shell of the evaporator. The condensate resulting from the condensation of steam drains downward and is returned to the boiler feed system. The vapor generated from the boiling seawater feed passes through the cyclonic separator above the evaporation section, where most of the entrained liquid particles are removed from the vapor by centrifugal force. The vapor continues on through the second vapor separator (called the snail), where the remaining water droplets are separated from the vapor. The liquid particles from both of these separators drain downward and become part of the brine drains.

The vapor generated in the first-effect shell passes from the steam dome of the first-effect shell. The first-effect vapor boils the second-effect feed and thus causes the generation of second-effect vapor. The second-effect vapor goes through the cyclonic separator and the snail in the second-effect shell. From the steam dome, this vapor then goes to the distilling condenser, where the vapor is condensed on the outside of the tubes. The second-effect distillate drains down and collects in the flash tank.

As the first-effect vapor is being used to boil the second-effect feed, some of the vapor condenses. This distillate drains downward into the second-effect steam chest and is discharged to the flash tank at the bottom of the distilling condenser, where it mixes with the distillate formed from the second-effect vapor. The distillate is removed from the flash tank by the distillate pump and is discharged through the distillate cooler and the solenoid-operated dump valve to the ship's tanks. Should the salinity of the distillate exceed 0.065 epm the dump valve would automatically dump the distillate to the bilges.

Seawater flows through the tubes of the distillate cooler and the distilling condenser, creating a suction for the brine pump and maintaining a back pressure for the feed system. About 25 percent of the seawater passes through supplementary heating sections in the distilling condenser to the air ejector condenser, and feeds the evaporator shells in parallel. As the seawater passes through the air ejector condenser, it condenses the air ejector steam; the resulting condensate drains to the fresh water drain collecting tank.

15

Air Compressors

Compressed air serves many purposes on board ship, and air outlets are installed in various suitable locations throughout the ship. The uses of compressed air include (but are not limited to) the operation of pneumatic tools and equipment, diesel engine starting and control, torpedo charging, aircraft starting and cooling, air deballasting, and the operation of pneumatic control systems. Compressed air represents a storage of energy. Work is done *on* the working fluid (air) so that work can later be done *by* the working fluid.

Compressor Classifications

The compressors that provide air at the pressures required for shipboard compressed air systems are classified in various ways according to a number of design and operational features. Perhaps the basic classification has to do with the type of compressor element. Compressor elements may be reciprocating, rotary, centrifugal, or—as in the case of one compressor discussed later in this chapter—a combination of rotary and centrifugal. Most air compressors used in the Navy have reciprocating elements.

Compressors may also be classified according to the type of drive. Most air compressors in naval service are driven by electric motors, but some are driven by steam turbines, reciprocating steam engines, or internal combustion engines.

One common way of classifying compressors is by pressure. Low pressure compressors are those that have a discharge pressure of 150 psig or less. Medium pressure compressors are those that have a discharge pressure between 151 and 1000 psig. High pressure compressors are those that have a discharge pressure above 1000 psig.

Reciprocating compressors may be classified according to the number of stages. Two-stage, three-stage, and four-stage reciprocating compressors are in common naval use. Five-stage and six-stage compressors are used in some high-capacity air systems.

Reciprocating Air Compressor Operating Cycle

Let us consider first the operating cycle that occurs during one stage of compression in a single-acting reciprocating air compressor. The operating cycle consists of two strokes of the piston: a suction stroke and a compression stroke.

The suction stroke begins when the piston moves away from top dead center (TDC). The air under pressure in the clearance space (above the piston) expands rapidly until the pressure falls below the pressure on the opposite side of the air inlet valve. At this point, the difference in pressure causes the inlet valve to open, and air is admitted to the cylinder. Air continues to flow into the cylinder until the piston reaches bottom dead center (BDC).

The compression stroke starts as the piston moves away from BDC, and compression of the air begins. When the pressure in the cylinder equals the pressure on the opposite side of the air inlet valve, the inlet valve closes. Air is increasingly compressed as the piston moves toward TDC; the pressure in the cylinder finally becomes great enough to force the discharge valve open against the discharge line pressure and the pressure of the valve springs. During the balance of the compression stroke, the air that has been compressed in the cylinder is discharged, at almost constant pressure, through the open discharge valve.

The basic operating cycle just described is repeated a number of times in single-acting compressors and in other stages of multistage compressors. In a double-acting compressor, each stroke of the piston is a suction stroke in relation to one end of the cylinder and a compression stroke in relation to the other end of the cylinder. In a double-acting compressor, therefore, two basic compression cycles are always in process when the compressor is operating; but each cycle, considered separately, is simply one suction stroke and one compression stroke.

Reciprocating Air Compressor Components

Compressing Element

The compressing element of a reciprocating air compressor consists of the air valves, the cylinder, and the piston. (See figure 15–1.)

The valves of modern compressors are of the automatic type. The opening and closing of these valves is caused solely by the difference between the pressure of the air in the cylinder and the pressure of the external air on the intake valve or the pressure of the discharged air on the discharge valve.

Pistons may be of various types. The most common is the trunk type, which is driven directly by a connecting rod from a crankshaft.

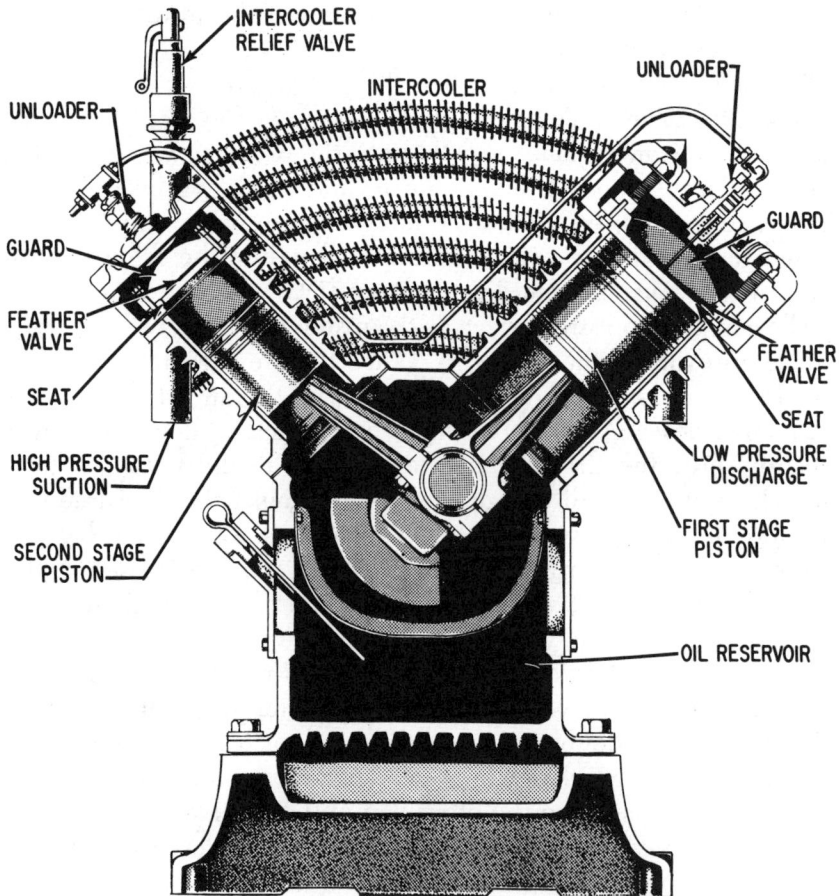

Figure 15–1. Cutaway of a reciprocating air compressor.

Double-acting pistons require a crosshead assembly and a piston rod. Differential pistons are modified trunk pistons having two or more different diameters. These pistons are fitted into special cylinders that are arranged so that more than one stage of compression is served by one piston. The compression for one stage takes place over the piston crown; compression for the other stage or stages takes place in the annular space between the large and small diameters of the piston.

Lubrication System

Lubrication of air compressor cylinders is generally accomplished by means of a mechanical force-feed lubricator that is driven by the compressor.

Cooling System

Most compressors are cooled by seawater supplied from the ship's fire and flushing system. The cooling water is usually available to each unit through at least two sources. Compressors located outside the larger machinery spaces are generally equipped with an attached circulating water pump as a standby source of cooling water. Some small low pressure compressors are air-cooled by a fan mounted on or driven by a compressor shaft. Removal of heat is necessary for economy of compression. During compression the temperature of the air is increased, the effect being to expand the air under compression to a larger volume, necessitating a corresponding increase of work to compress it. Therefore, multistaging with cooling of the air between stages reduces the power requirement for a given capacity. Intercooling reduces the maximum temperature in each cylinder, thereby reducing the amount of heat to be removed by the water jacket; also the lower temperature in the cylinder ensures good lubrication of the piston and the valves.

As previously noted, cooling of the air is required for most economical compression. Another reason for cooling the air between stages and after the last stage is to condense any moisture that may be present. The resulting condensate is then drained off. If the moisture is not removed from the air, it will be carried into the accumulator or into the air lines, where it can cause serious trouble.

The intercoolers used between stages and the aftercooler used after the last stage are of the same general construction except that the aftercooler is designed to withstand a higher working pressure than the intercoolers can withstand.

Accumulators

The accumulator is found in all compressed air plants, although the size of the unit varies according to the needs of the system. The accumulator (also called a receiver) helps to eliminate pulsations in the discharge line of the air compressor, acts as a storage tank during intervals when the demand for air exceeds the capacity of the compressor, and allows the compressor to shut down during periods of light load. Overall, the accumulator functions to retard increases and decreases in the pressure of the system, thereby lengthening the start–stop–start cycle of the compressor.

Control System

The control system of a reciprocating air compressor may include one or more devices such as automatic temperature shutdown devices, start–stop controls, constant-speed controls, and speed–pressure governors.

Automatic temperature shutdown devices are fitted to all recent designs of high pressure air compressors. Such a device stops the compressor automatically (and does not allow it to restart automatically) when the cooling water temperature rises above a safe limit. Some compressors are fitted with a device that shuts down the compressor if the temperature of the air leaving any stage exceeds a preset value.

Control or regulating systems for naval air compressors are mainly of the start–stop type. With this type of control, the compressor starts and stops automatically as the accumulator pressure rises or falls to predetermined limits. On electrically driven compressors, the system is very simple: the accumulator pressure operates against a pressure switch that opens when the pressure upon it reaches a given limit and closes when the pressure drops a predetermined amount.

Unloading System

Air compressor unloading systems are installed for the removal of all but the friction loads on the compressors. An unloading system automatically removes the compression load from the compressor while the unit is starting and automatically applies the load after the unit is up to operating speed. For units that have the start–stop type of control, the unloading system is separate from the control system. For compressors equipped with the constant-speed type of control, the unloading system is an integral part of the control system.

A number of different unloading methods are used, including closing or throttling the compressor intake, holding intake valves open, relieving intercoolers to the atmosphere, opening a bypass from the discharge to the intake, and using various combinations of these methods.

Rotary-Centrifugal Air Compressors

The one nonreciprocating type of air compressor that is found on board ship is variously referred to as a rotary compressor, a centrifugal compressor, or a "liquid piston" compressor. Actually, the unit is something of a mixture, operating partly on rotary principles and partly on centrifugal principles; most accurately, perhaps, it might be called a rotary-centrifugal compressor.

The rotary-centrifugal compressor is used to supply low-pressure compressed air. Because this compressor can supply air that is completely free of oil, it is often used as the compressor for pneumatic control systems and for other applications where oil-free air is required.

The rotary-centrifugal compressor, shown in figure 15–2, consists of a round, multibladed rotor that revolves freely in an elliptical casing. The elliptical casing is partially filled with high-purity water. The curved rotor blades project radially from the hub. The blades, together with the side shrouds, form a series of pockets or buckets around the periphery. The rotor, which is keyed to the shaft of an electric motor, revolves at a speed high enough to throw the liquid out from the center by centrifugal force, resulting in a solid ring of liquid revolving in the casing at the same speed as the rotor but following the elliptical shape of the casing. This action alternately forces the liquid to enter and to recede from the buckets in the rotor at high velocity.

To follow through a complete cycle of operation, let us start at point A in figure 15–2. The chamber (1) is full of liquid. The liquid, because of centrifugal force, follows the casing, withdraws from the rotor, and pulls air in through the inlet port. At (2) the liquid has been thrown outward from the chamber in the rotor and has been replaced with atmospheric air. As the rotation continues, the converging wall (3) of the casing forces the liquid back into the rotor chamber, compressing the trapped air and forcing it out through the discharge port. The rotor chamber (4) is now full of liquid and ready to repeat the cycle, which takes place twice in each revolution.

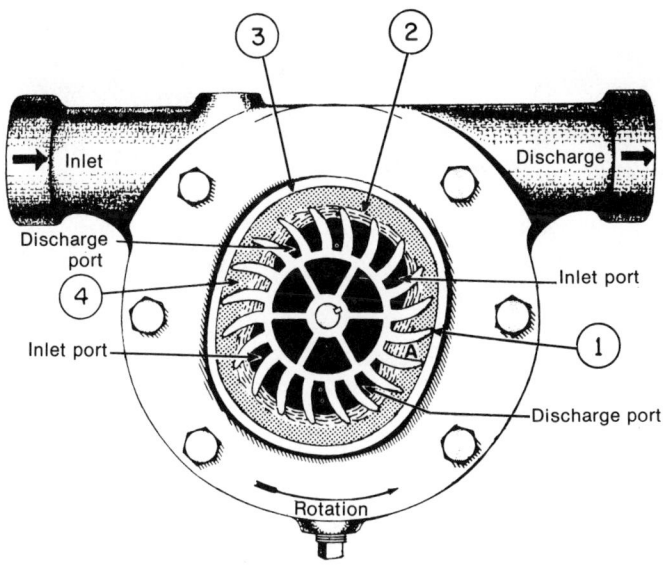

Figure 15–2. Cutaway of a rotary-centrifugal compressor.

A small amount of seawater must be constantly supplied to the compressor to make up for that which is carried over with the compressed air. The water that is carried over with the compressed air is removed in a refrigeration-type dehydrator.

Safety Precautions

There are many hazards associated with the process of air compression. Serious explosions have occurred in high-pressure air systems because of a diesel effect. Explosions may be caused by dust-laden air, presence of oil vapor in the compressor receiver, and by leaky or dirty valves resulting in abnormally high temperatures. Every effort must be made to have only clean, dry air at the compressor intake. Never use benzine, kerosene, or other light oils to clean compressor intake filters, cylinders, or air passages. These oils vaporize easily and will form a highly explosive mixture with the air under compression.

Air compressor accidents have also been caused by improper maintenance procedures such as disconnecting parts while they are under pressure, replacing parts with units designed for lower pressures, and installing stop valves or check valves in improper locations. Improper operating procedures have also caused air compressor accidents, with resulting serious injury to personnel and damage to equipment.

16

Refrigeration and Air Conditioning Plants

Refrigeration equipment is used on board ship for a number of purposes, including the refrigeration of ship's stores, the refrigeration of cargo, the cooling of water, and the conditioning of air for certain spaces. The distinction between refrigeration and air conditioning should be noted. Refrigeration is only a cooling process; air conditioning is a process of treating air so as to simultaneously control its temperature, humidity, cleanliness, and distribution to meet the requirements of the conditioned spaces.

Refrigeration

The purpose of refrigeration is to cool spaces, objects, or materials and to maintain them at temperatures below the temperature of the surrounding atmosphere. In order to produce a refrigeration effect, it is merely necessary to expose the material to be cooled to a colder object or environment and allow heat to flow in its "natural" direction—that is, from the warmer material to the colder one. For example, a pan of hot water placed on a cake of ice will be cooled by the flow of heat from the hot water to the ice. We can maintain this refrigeration effect as long as the ice lasts. But no matter how much ice we have, we cannot produce a refrigeration effect any greater than the cooling of the water to 32°F. We cannot, for example, cause the water to freeze by this method, since freezing would require the removal of the latent heat of fusion from water after it had been cooled to 32°F, and for this process we would need a temperature difference that does not exist when both the water and the ice are at 32°F. When the purpose of refrigeration is the production of ice or the maintenance of temperatures lower than 32°F at atmospheric pressure, it is obvious that ice is not a suitable refrigerant.

Refrigeration is a process involving the flow of heat; thus it is a thermodynamic process. From previous discussion in this text, we may surmise that a closed cycle would be most practicable for a large-scale refrigeration system. When we try to visualize such a cycle, however, it may appear at first glance that the cycle will have to run backward. In our previous discussion, we have been primarily concerned with a closed cycle in which thermal energy (in the form of heat) is converted into mechanical energy (in the form of work). Now, instead of wanting to convert heat into work, we want to remove heat from a body, and we want to continue to remove heat from this body even after its temperature has been lowered below that of its surroundings, in order to maintain the body at its lowered temperature. In other words, we want to extract heat from a cold body and discharge it to a warm area.

The question is: How can this be done, since we know from the second law of thermodynamics that heat cannot, of itself, flow from a colder body or region to a warmer one? It is entirely possible to extract heat from a body at a low temperature and discharge it to a body or region at a higher temperature, provided that a suitable expenditure of energy is made. The energy supplied to the refrigeration cycle for this purpose is in the form of work (mechanical energy) done on the working fluid (refrigerant) by a compressor.* In a refrigeration cycle, the refrigerant must alternate between low and high temperatures. When the refrigerant is at a low temperature, heat flows from the space or object to be cooled to the refrigerant. When the refrigerant is at a high temperature, heat flows from the refrigerant to a condenser. The energy supplied as work is used to raise the temperature of the refrigerant to a high enough value that the refrigerant will be able to reject heat to the condenser. This point is discussed in more detail later in this chapter, but should be noted now, since it is basic to the understanding of a mechanical refrigeration cycle.

Because the energy transformations in a refrigeration cycle occur in an order that is precisely the reverse of the sequence in a power cycle, the refrigeration cycle is sometimes said to be one in which heat is pumped "uphill." This view of a refrigeration cycle is entirely legitimate, provided that the "reverse order" of energy transformations does not imply actual thermodynamic reversibility. True thermodynamic reversibility is here, as elsewhere in the observable world, considered to be an impossibility. A refrigeration cycle does not give us something for nothing. Instead, we must put energy into the cycle in order to extract heat at a low temperature and discharge it at a higher temperature.

---

*A compressor provides the required energy in a vapor-compression refrigeration cycle, which is the cycle most commonly used in naval refrigeration plants. Other kinds of refrigeration cycles use other forms of energy to accomplish the same purpose—namely, to raise the temperature of the refrigerant after it has absorbed heat from the space or object to be cooled.

Definition of Terms

Some of the standard terms used in the discussion of refrigeration are defined in this section. A few of these terms were defined in Chapter 1, but are briefly noted here because of their importance in the study of refrigeration.

Unit of Heat. The British thermal unit (Btu) is the standard unit of heat measurement used in refrigeration, as in most other engineering applications. By definition, 1 Btu is equal to 778.26 foot-pounds.

Specific Heat. The specific heat of a substance is the quantity of heat required to raise the temperature of a unit mass of the substance 1 degree. In British systems of measurement, specific heat is expressed in Btu per pound per degree Fahrenheit.

Sensible Heat. Sensible heat is the term used to identify heat that is reflected in a change of temperature.

Latent Heat of Vaporization. The heat required to change a liquid to a gas (or, on the other hand, the heat that must be removed from a gas in order to condense it to a liquid) without any change in temperature is called the latent heat of vaporization.

Latent Heat of Fusion. The heat that must be removed from a liquid in order to change it into a solid (or, on the other hand, the amount of heat that must be added to a solid to change it to a liquid) without any change in temperature is called the latent heat of fusion.

Refrigerating Effect. Since the heat removed from an object that is being refrigerated is absorbed by the refrigerant, the refrigerating effect of a refrigeration cycle is defined as the heat gain per pound of refrigerant.

Refrigeration Ton. The unit that measures the amount of heat removal and thereby indicates the capacity of a refrigeration system is known as the refrigeration ton. The refrigeration ton is based on the cooling effect of 1 ton (2000 pounds) of ice at 32°F melting in 24 hours. The latent heat of fusion of ice (or water) is approximately 144 Btu. Therefore, the number of Btu required to melt one ton of ice is 144 × 2000, or 288,000 Btu. The standard refrigeration ton is defined as the transfer of 288,000 Btu in 24 hours. On an hourly basis, the refrigeration ton is 12,000 Btu per hour (288,000 divided by 24 equals 12,000).

It should be noted that the refrigeration ton is not necessarily a measure of the ice-making capacity of a machine, since the amount of ice that can be made depends upon the initial temperature of the water and other factors.

The R-12 Plant

The refrigeration system most commonly used in the Navy utilizes R-12 as the refrigerant.* Chemically, R-12 is dichlorodifluoromethane (CCl_2F_2). The boiling point of R-12 is so low that the substance cannot exist as a liquid unless it is confined and put under pressure; for example, R-12 boils at $-21°F$ at atmospheric pressure, at $0°F$ at 9.17 psig, at $50°F$ at 46.69 psig, and at $100°F$ at 116.9 psig. Because of its low boiling point, R-12 is well suited for use in refrigeration systems designed for only moderate pressures. It also has the advantage of being practically nontoxic, nonflammable, nonexplosive, and noncorrosive; and it does not poison or contaminate foods.

The R-12 refrigeration system is classified as a mechanical system of the vapor-compression type. It is a mechanical system because the energy input is in the form of mechanical energy (work). It is a vapor-compression system because compression of the vaporized refrigerant is the process that allows the refrigerant to discharge heat at a relatively high temperature.

The R-12 Cycle

The basic cycle of an R-12 refrigeration cycle is shown schematically in figure 16–1. For an introduction to the system, it will be helpful to trace the refrigerant through the entire cycle, noting especially the points at which the refrigerant changes from liquid to vapor and from vapor to liquid, and noting also the concomitant flow of heat in one direction or another.

As shown in figure 16–1, the cycle has two pressure sides: the low-pressure side, extending from the orifice of the thermostatic expansion valve up to and including the intake side of the compressor cylinders; and the high-pressure side, extending from the discharge side of the compressor to the thermostatic expansion valve. The condensing and evaporating pressures and temperatures indicated in figure 16–1 are not standard for all refrigeration plants, since pressures and temperatures are established as part of the design of any refrigeration system. It should be noted, also, that the pressures and temperatures shown in figure 16–1 are theoretical rather than actual values, even for this particular system. If the system were in actual operation, the pressures and temperatures would vary slightly because they are dependent upon the temperature of the cooling water

---

*In accordance with recent policy, refrigerants used in the Navy are no longer identified by trade names. Instead, they are identified by the letter R followed by the appropriate number, or else they are identified simply as "refrigerants." For example, the refrigerant formerly known as "Freon 12" is now identified either as R-12 or simply as a refrigerant.

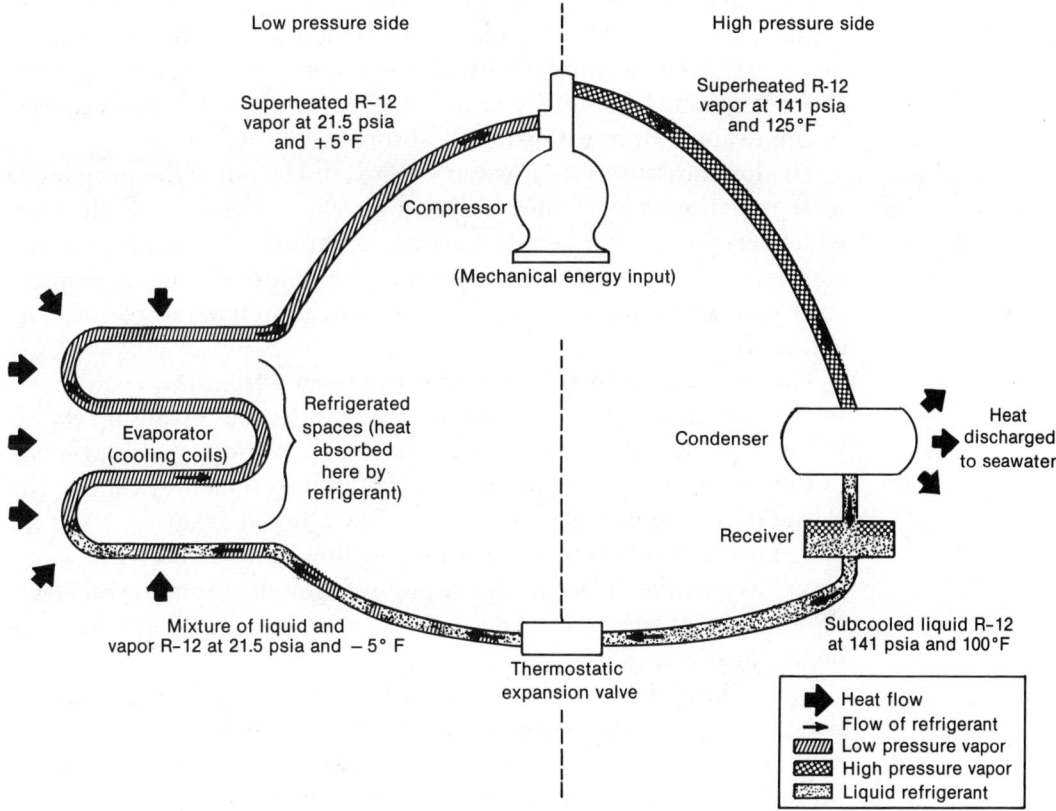

Figure 16–1. Schematic representation of an R–12 refrigeration cycle.

entering the condenser, the amount of heat absorbed by the refrigerant in the evaporator, and other factors.

Liquid R-12 enters the thermostatic expansion valve at high pressure, from the high-pressure side of the system. The refrigerant leaves the outlet of the expansion valve at a much lower pressure and enters the low-pressure side of the system. Because of the relatively low pressure, the liquid refrigerant begins to boil and to flash into vapor.

From the thermostatic expansion valve, the refrigerant passes into the cooling coil (evaporator). The boiling point of the refrigerant under the low pressure in the evaporator is extremely low—much lower than the temperature of the spaces in which the cooling coil is installed. As the liquid boils and vaporizes, it picks up its latent heat of vaporization from the surroundings, thereby cooling the space. The refrigerant continues to absorb heat until all the liquid has been vaporized, and the vapor has become slightly superheated. As a rule, the amount of superheat is about 10°F.

Refrigeration and Air Conditioning Plants

The refrigerant leaves the evaporator as a low-pressure superheated vapor, having absorbed heat and thus cooled the space. The remainder of the cycle is concerned with disposing of this heat and getting the refrigerant back into a liquid state so that it can again vaporize in the evaporator and thus again absorb heat.

The low-pressure superheated vapor is drawn out of the evaporator to the suction side of the compressor. The compressor is the unit which keeps the refrigerant circulating through the system; in the compressor cylinders, the refrigerant is compressed from a low-pressure vapor to a high-pressure vapor, and its temperature rises accordingly.

The high-pressure R-12 vapor is discharged from the compressor to the condenser. Here the refrigerant condenses, giving up its superheat, its latent heat of vaporization, and its heat of compression to the cooling seawater that flows through the condenser tubes. The refrigerant, still at high pressure, is now a liquid again.

From the condenser, the refrigerant flows into a receiver, which serves as a storage place for the liquid refrigerant. From the receiver, the refrigerant goes to the thermostatic expansion valve, and the cycle begins again.

From this brief summary of an R-12 vapor-compression refrigeration system, it may be seen that the cycle is indeed one in which heat is "pumped uphill" as a result of the arrangements that cause the refrigerant to go through successive phases of expansion, evaporation, compression, and condensation.

Major Components

The major components of a shipboard R-12 refrigeration plant are shown diagrammatically in figure 16–2. The primary parts of the system are the thermostatic expansion valve, the evaporator, the compressor, the condenser, and the receiver. Additional equipment required to complete the plant includes piping, pressure gages, thermometers, various types of control switches and control valves, strainers, relief valves, sight flow indicators, dehydrators, and charging connections. Figure 16–3 shows most of the components on the high-pressure side of an R-12 system, as actually installed on board ship.

In the following discussion of the major components of an R-12 system, we will treat the system as though it had only one evaporator, one compressor, and one condenser. As may be seen from figure 16–2, however, a shipboard refrigeration system may (and, indeed, usually does) include more than one evaporator and may include additional compressor and condenser units to provide operational flexibility and to protect against loss of refrigerating capacity.

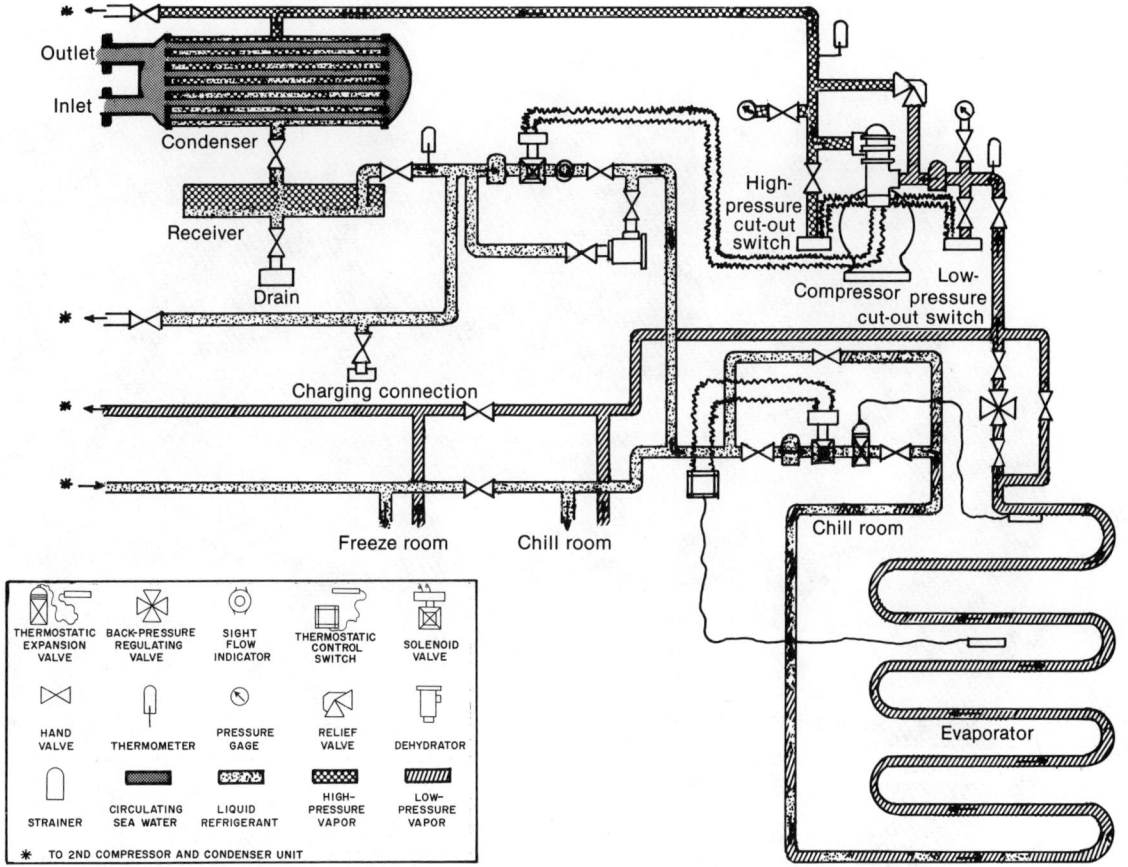

Figure 16–2. Diagram of an R–12 refrigeration system.

Thermostatic Expansion Valve. The thermostatic expansion valve, shown in figure 16–4, is essentially a reducing valve between the high-pressure side and the low-pressure side of the system. The valve is designed to proportion the rate at which the refrigerant enters the cooling coil to the rate of evaporation of the liquid refrigerant in the coil; the amount of refrigerant needed depends, of course, on the amount of heat being removed from the refrigerated space.

A thermal bulb for the thermostatic expansion valve is clamped to the cooling coil, near the outlet. The bulb contains R-12. Control tubing connects the bulb with the area above the diaphragm in the thermostatic expansion valve. When the temperature at the bulb rises, the R-12 expands and transmits a pressure to the diaphragm; this causes the diaphragm to be moved downward, thus opening the valve and allowing more refrigerant to enter the cooling coil. When the temperature at the bulb falls, the pressure above the diaphragm

Refrigeration and Air Conditioning Plants

Figure 16–3. High-pressure side of an R–12 installation on board ship.

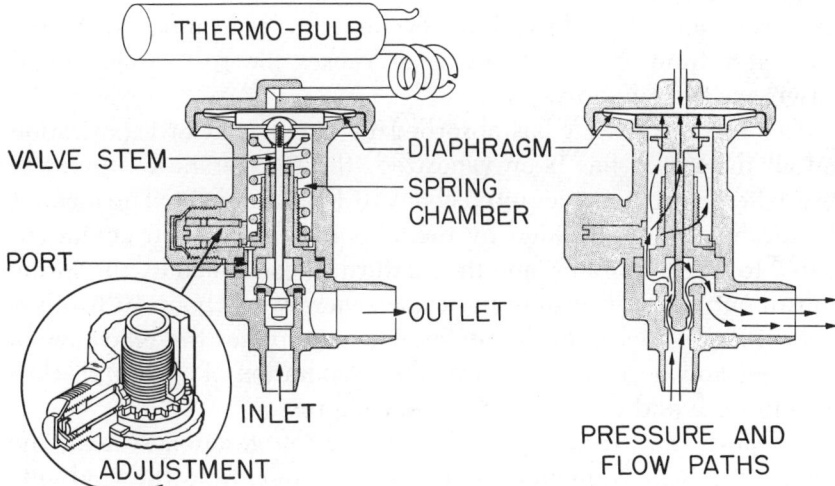

Figure 16–4. Diagram of a thermostatic expansion valve.

Figure 16–5. Typical arrangement of evaporator tubing.

is decreased, and the valve tends to close. Thus the temperature near the evaporator outlet controls the operation of the thermostatic expansion valve.

Evaporator. The evaporator consists of a coil of copper tubing installed in the space to be refrigerated. Figure 16–5 shows some of this tubing. The liquid R-12 enters the tubing at a very much reduced pressure, and its boiling point is therefore very much lowered. As it passes through the expansion valve, going from the high-pressure side of the system to the low-pressure side, some of the refrigerant boils and vaporizes because of the reduced pressure, and some of the remaining liquid refrigerant is cooled to its boiling point. Then, as the refrigerant passes through the evaporator, the heat flowing to the evaporator from the surrounding air causes the rest of the liquid refrigerant to boil and vaporize.

After the refrigerant has absorbed its latent heat of vaporization and all the liquid has been vaporized, the refrigerant continues to absorb heat until it has acquired about 10°F of superheat. The amount of superheat is determined by the amount of liquid refrigerant admitted to the evaporator; and this, in turn, is controlled by the spring adjustment of the thermostatic expansion valve. About 10°F of superheat is considered desirable because it increases the efficiency of the plant and because it ensures the evaporation of all liquid, thus preventing liquid carryover into the compressor.

Compressor. In a vapor-compression refrigeration system, the compressor is the unit that pumps heat "uphill" from the cold side to the hot side of the system.

The heat absorbed by the refrigerant in the evaporator must be removed before the refrigerant can again absorb latent heat in the evaporator. The only way in which the vaporized refrigerant can be made to give up the latent heat of vaporization that it absorbed in the evaporator is by condensation. In view of the relatively high temperature of the available cooling medium (seawater), the only way to make the vapor condense is by first compressing it.

The vapor drawn into the compressor is at very low pressure and very low temperature. In the compressor, both the pressure and the temperature are raised. Since an increase in pressure causes a proportional rise in temperature, and since the condensation point of a vapor is determined by the pressure, raising the pressure of the vaporized refrigerant provides a condensation temperature high enough to permit the use of seawater as a cooling and condensing medium. In other words, the compressor raises the pressure of the vaporized refrigerant sufficiently high to permit heat transfer and condensation to take place in the condenser.

In addition to this primary function, the compressor also serves to keep the refrigerant circulating and to maintain the required pressure differential between the high-pressure side and the low-pressure side of the system.

Condenser. The compressor discharges the high-pressure, high-temperature refrigerant vapor to the condenser, where it flows around the tubes through which seawater is being pumped. As the vapor gives up its superheat to the circulating seawater, the temperature of the vapor drops to the condensation point. As soon as the temperature of the vapor drops to its condensing point at the existing pressure, the vapor condenses, and in the process gives up the latent heat of vaporization that it picked up in the evaporator. The refrigerant, now in liquid form, is subcooled slightly below its boiling point at this pressure to ensure that it will not flash into vapor.

Circulating water is obtained through a pressure reducer from a branch connection from the firemain or by means of an individual pump taking suction from the sea. A water-regulating valve (not shown in figure 16–2) is usually installed to control the flow of cooling water through the condenser. The purge connection is on the refrigerant side; it is used to remove air and other noncondensable gases that are lighter than the R-12 vapor.

Most condensers used in naval refrigeration plants are water-cooled. However, some small units have air-cooled condensers, which consist of tubing with external fins to increase the heat-transfer surface. Most air-cooled condensers have fans to ensure positive circulation of air around the condenser tubes.

Receiver. The receiver acts as a temporary storage space and surge tank for the liquid refrigerant that flows from the condenser. The receiver also serves as a vapor seal to prevent the entrance of vapor into the liquid line to the thermostatic expansion valve.

Types of Air Conditioners

In general, three types of mechanical cooling equipment (air-conditioning units) are used on board ship: refrigerant ciruclating systems, chilled-water circulating systems, and self-contained air conditioners.

A refrigerant circulating system is shown in figure 16–6. As may be seen, this system is essentially a refrigeration system consisting of a compressor, a condenser, cooling coils, a fan, an air filter, and the necessary controls.

Hot moist air from the space to be cooled is drawn through a duct, where it mixes with fresh air drawn from outside. The fan blows the air over the cooling coil, and the refrigerant inside the coil cools the surface of the coil. The moisture drips off into a pan below the coil and is carried off by drain piping. The cool dry air leaving the coil is blown into the compartment to be cooled, where it absorbs the excess heat and moisture from the air already in the space. The air is then returned to the cooling coil, and the cycle is repeated. Air is exhausted from the space being cooled in order to allow fresh air to be drawn into the space. The cooling coils are installed in the ventilation ducts leading to the spaces to be cooled. The refrigerant used in this system is usually R-12.

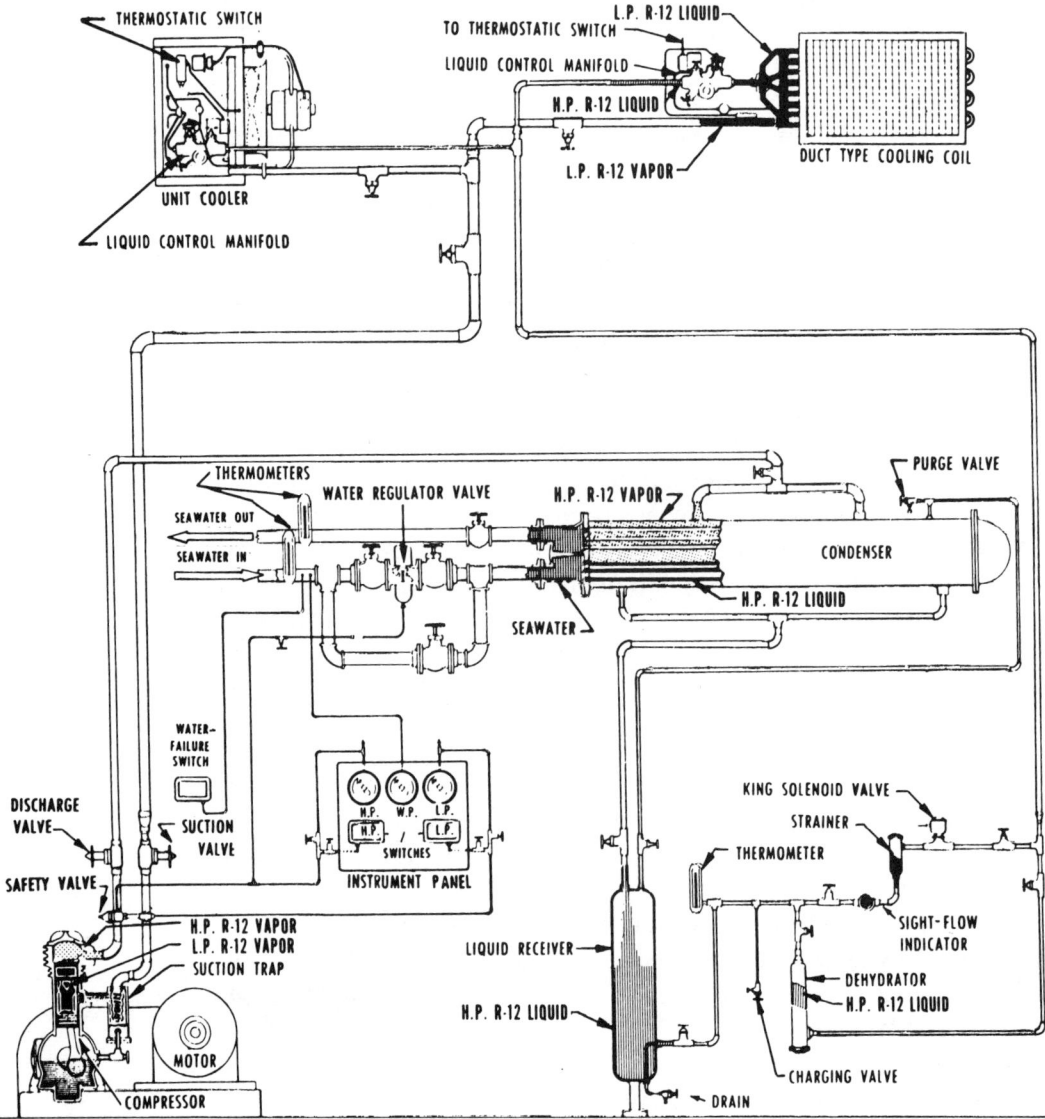

Figure 16–6. Refrigerant circulating type of mechanical cooling system.

Two types of chilled-water circulating systems are used for mechanical cooling on board ship. Both systems utilize chilled water as the secondary refrigerant, but one type uses R-12 as the primary refrigerant, and the other uses R-11 or R-14. R-12 systems use reciprocating compressors; R-11 or R-14 systems use centrifugal compressors.

Both types of chilled-water circulating systems operate on the same general principle. The secondary refrigerant (chilled water) is circulated to the various cooling coils. Heat from the spaces being cooled is absorbed by the chilled water and is removed from the water by the primary refrigerant in a water cooler.

The primary refrigerant vapor goes from the evaporator (water chiller) to the compressor, where it is compressed. It is then discharged to the condenser. In the condenser, the primary refrigerant vapor condenses, giving up its superheat, its latent heat of vaporization, and its heat of compression to the cooling water that flows through the condenser tubes. The liquid primary refrigerant then passes through an expansion device and back to the water chiller.

The secondary refrigerant picks up heat in the coils of the air-conditioned space and carries this heat to the chiller. The function of the chiller is to transfer the heat from the secondary refrigerant to the primary refrigerant, which surrounds the tubes of the chiller. As this heat is transferred, the liquid primary refrigerant absorbs its latent heat of vaporization, boils, and vaporizes. The quantity of liquid refrigerant thus evaporated varies directly with the amount of heat picked up by the secondary refrigerant. The vaporized primary refrigerant goes to the compressor, and the same sequence of events is repeated in a cyclical manner.

Self-contained air conditioners are installed in some ships that were originally built without mechanical cooling systems. A self-contained air conditioner is built with the entire unit in one metal cabinet. The compressing element in the unit is usually of the hermetically sealed type, with the motor and the compressor contained in a welded steel shell.

Safety Precautions

Never allow R-12 to come in contact with a flame or red-hot metal! When exposed to excessively high temperatures, R-12 breaks down into phosgene gas, an extremely poisonous substance. No smoking is allowed while handling R-12 or if R-12 is present in the atmosphere.

Because R-12 is such a powerful freezing agent that even a very small amount can freeze the delicate tissues of the eyes, causing permanent damage, it is essential that goggles be worn by all personnel who may be exposed to a refrigerant, particularly in its liquid form. If refrigerant does get in the eyes, the person suffering the injury should receive medical treatment immediately in order to avoid permanent damage to the eyes. In the meantime, put drops of clean olive oil, mineral oil, or other nonirritating oil in the eyes, and make sure that the person does not rub his eyes. *Caution:* Do not use anything except clean, nonirritating oil for this type of eye injury.

(Note: If large leaks are indicated, the soap method should be used to detect leaks; for minute leaks, the halide torch should be employed.)

If R-12 comes in contact with the skin, it may cause frostbite. This injury should be treated like any other case of frostbite. Immerse the affected part in a warm bath for about 10 minutes; then dry it carefully. *Do not* rub or massage the affected area.

R-12 is considered a fluid of low toxicity. However, in closed spaces, high concentrations displace the oxygen in the air and thus do not sustain life. If a person should be overcome by R-12, remove him *immediately* to a well-ventilated place and get medical attention at the earliest opportunity. Watch his breathing. If the person is not breathing, give artificial respiration.

17

Basic Hydraulics and Applications

The word hydraulics is based on the Greek word for water. It originally meant a study of the physical behavior of water at rest and in motion. Use has broadened its meaning to cover the physical behavior of all liquids, including the oils used in the vast majority of naval hydraulic systems. Of interest to us is the study of particular fluids under various conditions of flow, and ways of directing this flow to useful ends.

Hydraulic systems operate on the principle that, since liquids are for all practical purposes noncompressible, force exerted at any point on an enclosed liquid is transmitted equally in all directions. Hence a hydraulic system permits the accomplishment of a great amount of work with relatively little effort on the part of shipboard personnel.

Physical Properties of Liquids

Shapelessness. While solids always have a definite shape, liquids have no outer form of their own. They quickly conform in shape to their containers. Because of their own shapelessness, we can lead liquids almost anywhere in a pipe or a hose by means of gravity or an applied force. The force can be applied by several sources—a pump, high-pressure air, or manual devices, to name just a few.

Incompressibility. As mentioned above, it is the incompressibility of a liquid that makes a hydraulic system so valuable. When a force is applied to a confined liquid, the liquid exhibits substantially the same effect of rigidity as a solid. If an appropriate exit is provided, this effect can be combined with fluidity to transmit a force. For example, a force of 15 pounds on 1 cubic inch of water will decrease its volume by 1/20,000! It would take a force of 32 tons to reduce it by a mere 10 percent. This is typical of other liquids, including the various grades of oil used in naval hydraulic systems. Additionally, when pressure on a liquid is removed, the liquid immediately returns to its former volume.

Transmission of Forces through Liquids. When we strike the end of a bar, the main force of the flow is carried longitudinally from one end of the bar to the other. This happens because the bar is rigid. The direction of the blow almost entirely determines the direction of the transmitted force. When we apply a force to the end of a column of confined liquid, however, it is transmitted not only straight through to the other end, but also equally and undiminished in every direction throughout the column—forward, backward, and sideways—so that the containing vessel is literally filled with pressure. For this reason a flat fire hose takes on a circular cross section when it is filled with water under pressure. The outward push of the water is equal in every direction. Should there be leaks, water would leave the hose at the same velocity through the leaks, no matter what side of the hose they might happen to be on.

Figure 17–1 is an example of a simple hydraulic system. For the system of figure 17–1, assume that the surface area of piston A in contact with the hydraulic fluid is 1 square foot, while the area of piston B is 4 square feet. Now a force of 100 pounds is applied to piston A. Since the fluid is incompressible, this force is transmitted to the surface of piston B. The force acting on piston A is 100 pounds per square foot, since it was applied to a surface of 1 square foot. On piston B, however, the force is acting on an area of 4 square feet. Hence the resultant output is a force of 400 pounds. Although the example above is oversimplified, the principles it illustrates are the same as those employed in more complex shipboard hydraulic systems. With the same effort it takes to turn the steering wheel on an automobile, tons of mass such as rudders, gun mounts, aircraft ele-

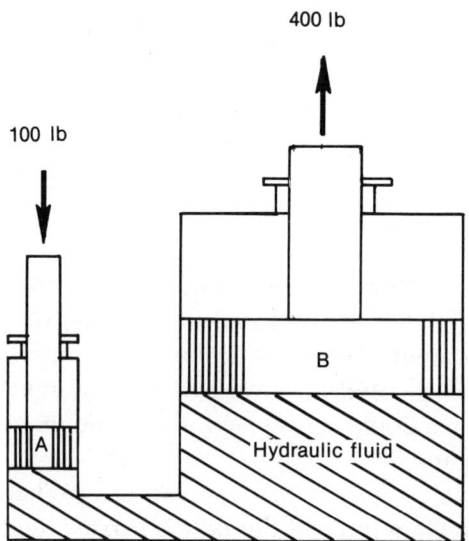

Figure 17–1. *Hydraulic principle*.

vators, and submarine control surfaces and antennas are precisely positioned.

Problems with Hydraulic Systems. The extreme flexibility of hydraulic elements gives rise to a number of system characteristics. Since fluids have no shape of their own, they must be positively confined throughout the entire system, and prevented from going anywhere except where we want them to go. Special thought must be given to structural organization and the relationship of parts of a hydraulic system; we must provide strong piping and containers and prevent leaks. This characteristic is acute with the high pressures obtained in many hydraulic installations; in fact, the thousands of pounds per square inch encountered in shipboard hydraulic systems are often the highest pressures encountered anywhere in the ship. The problems associated with a severe leak of hydraulic oil under a pressure of 3000 pounds per square inch could easily endanger the entire ship and crew.

The pressures set up in hydraulic systems must be controlled, as must the movement of the component fluids. This movement causes friction, within the liquid itself and against containing surfaces, which if excessive can lead to serious losses in efficiency. Dirt must be filtered from the system, or it will accumulate and clog small passages to score tightly fitted parts.

Electrohydraulic Speed Gear

The units of auxiliary machinery, such as the anchor windlass, are required to operate at variable speeds over a considerable range. In addition, there must be close control of speed between maximum and minimum limits. A common requirement of some auxiliary machinery units is high starting torque and ability to accelerate to maximum speed quickly. To meet these requirements in most naval ships, the electrohydraulic drive has been adopted. Electrohydraulic drives are used for driving or controlling steering gears, windlasses, winches, capstans, airplane cranes, ammunition hoists, airplane elevators, and distant control valves. This chapter contains information on electrohydraulic drives and some machinery with which they are used.

In an electrohydraulic drive, rotary motion is transmitted by a combination of an electrically driven hydraulic pump (A-end) and a hydraulic motor (B-end). (See figure 17–2.) The B-end is made to rotate by the hydraulic force of oil acting on pistons. Movement of the pistons' A-end is controlled by a tilt box in which a socket ring is mounted, as shown in figure 17–3.

Moving the tilting box one way or the other will control the di-

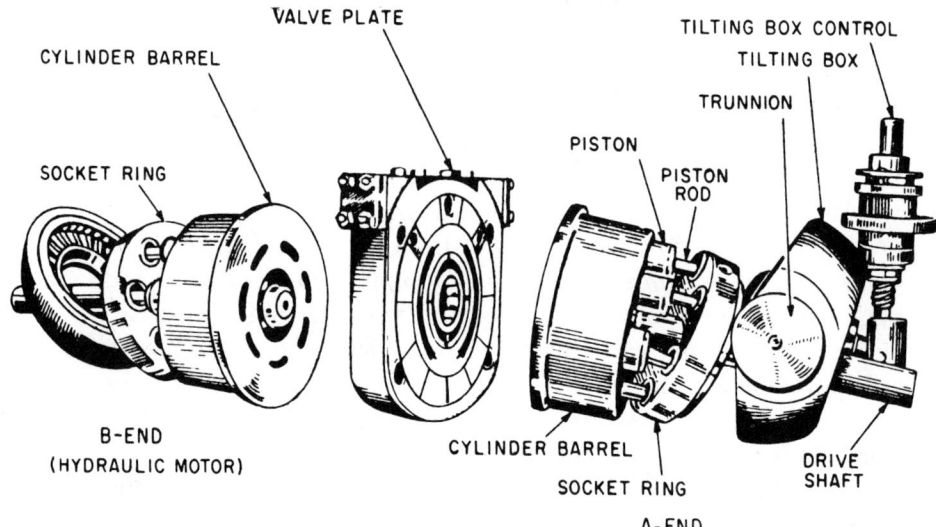

Figure 17–2. *Waterbury pump and motor.*

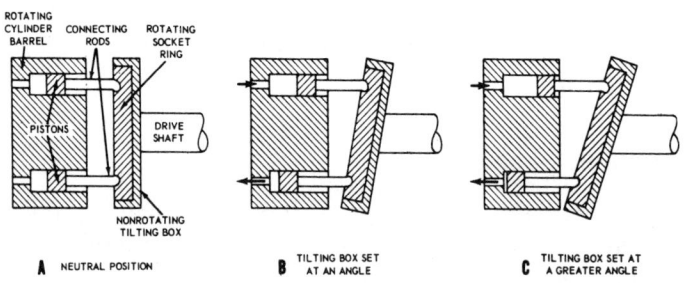

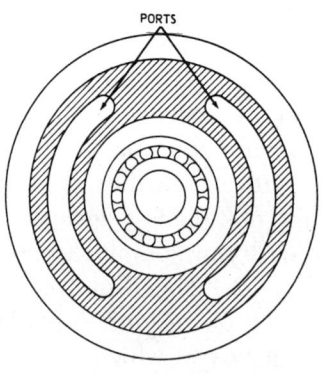

Figure 17–3. *Tilt box position diagram.*

rection of flow; and the amount of angle at which the tilting box is placed in either direction controls the length of piston movement, which in turn controls the amount of fluid flow. The socket ring in the A-end is always in motion, but with the tilt box in a neutral or vertical position no oil is pumped to the B-end. Any movement of the tilt box, regardless of how slight, causes pumping action to start, and therefore causes immediate action in the B-end. Advantages of this arrangement are that it:

1. Incorporates tubing that requires very little space and readily transmits fluids around corners.
2. Allows flexibility in location of components.
3. Operates at variable speeds, when required to.
4. Permits close control of speeds from minimum to maximum limits.
5. Permits rapid shifting from no-load to full load without damage to machinery.
6. Accelerates quickly.
7. Has a high rate of efficiency.
8. Has a favorable power-to-weight ratio.

Hydraulic Accumulators

An accumulator is a pressure storage reservoir in which hydraulic fluid is stored under pressure from an external source, usually a pump. Accumulators are used for:

1. Hydraulic shock suppression.
2. Fluid makeup in a closed system.
3. Leakage compensation.
4. Source of emergency power in case of loss of normal power.
5. Holding high pressure without keeping the pump in continuous operation.

Accumulators can be of three basic types: piston type, bag or bladder type, and direct-contact gas-over-hydraulic-fluid type. The most widely used type of naval hydraulic accumulator is the piston type, employed in the hydraulic systems of virtually all submarines and several surface ships. The piston-type accumulator consists of a cylindrical body with a floating or rod-guided piston. The hydraulic fluid (oil) is pumped into one end of the cylinder, and the piston is forced toward the opposite end of the cylinder against a captive charge of gas, usually air.

As in an internal combustion engine, allowing air and oil to mix under pressure could result in combustion. Pressure surges cause heat generation which might ignite an air/oil mixture. To prevent this situation from occurring, piston rings and seals are used to keep the air and oil separated. In case the oil or air should leak past the seal, a telltale drain system is used in some accumulators to vent the oil or air to the top of the piston through a cavity drilled in the piston for this purpose. The leakage is then easily detected, and the accumulator can be taken off service until repaired. The drain allows operation of the accumulator with minor leakage, since mixing of oil and air is prevented.

Figure 17–4 is a simplified hydraulic system of the kind used to raise and lower antennas and masts. In the system of figure 17–4, the main functions of the accumulator are to maintain a high pressure on the system without keeping a pump in continuous operation and to act as a source of limited emergency power for a short time in case of failure of normal power to the operating pump. Hydraulic shock suppression is not of concern in submarine use, since the hydraulic pumps are of the screw type which are virtually pulse-free—an important concern on board a sound-conscious submarine. Note the accumulator level indicator in the system. When the oil level in the

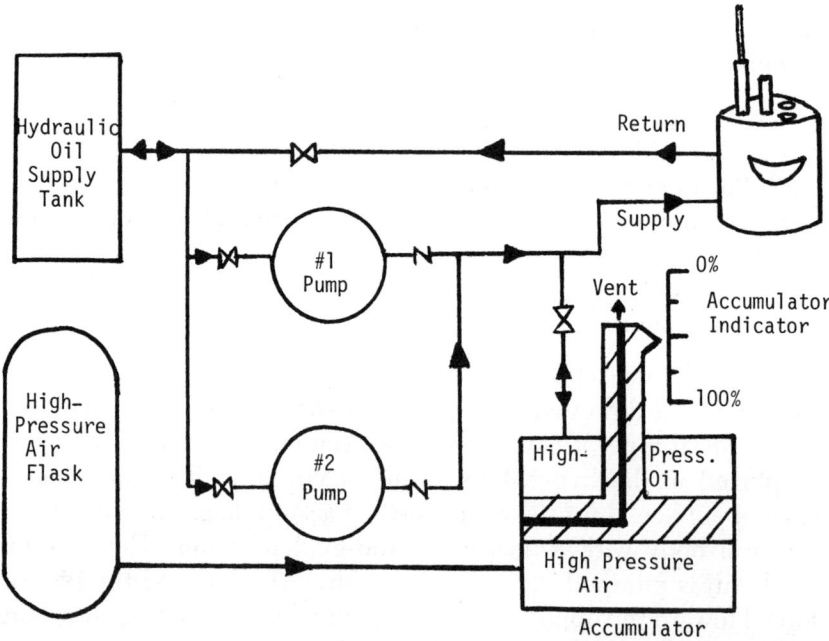

Figure 17–4. Schematic diagram of a simplified submarine-type hydraulic system.

accumulator reaches 25 percent of capacity, the indicator shuts a limit switch, which starts a remotely selected pump. The pump discharges oil to the system, charging the accumulator. When the accumulator has been charged to 90 percent of capacity, the indicator shuts another limit switch, this one de-energizing the operating pump. The system continues to cycle in this manner, the frequency depending on the load on the system (i.e., the number of masts being operated).

A system similar to the one previously described is used in submarines for the major hydraulic loads, including steering, stern and fair-water planes, anchor windlass, torpedo tubes, and a host of seawater valves and miscellaneous equipment. This main hydraulic system, being much larger, operates at a higher pressure and has three or more hydraulic pumps. As in the example of figure 17–4, each accumulator level indicator controls its associated pump or pumps.

The bladder or bag type accumulator is used in the gun mount and/or steering hydraulic systems in some surface ships. This type of accumulator consists of a shell or case with a flexible bladder inside. The bladder is precharged with air or inert gas to a specified value. Fluid is then forced into the area around the bladder, further compressing the gas in it. This type of accumulator has the advantage that as long as the bladder is intact, there is no exposure of fluid to the gas charge, and therefore less danger of an explosion.

18

Steering Systems

Electrohydraulic Steering Gear

The steering gear is the mechanism that transmits power from the steering engine to the rudder. The term steering gear is, however, applied frequently to the entire assembly of the driving engine and the transmitting mechanism. For shipboard use a good steering gear must have the following characteristics:

1. It must be rugged.
2. It must be so designed that the rudder cannot overcome the controlling wheel.
3. It must have high mechanical advantage.

Both electrohydraulic and electromechanical steering gear arrangements provide these desired characteristics.

Most steering installations on board naval ships are of the electrohydraulic type. The development of this type was prompted primarily by the large momentary electrical power requirements for electromechanical steering gears. Also, the elimination of DC power from ships made switching and speed control more difficult with the electromechanical type. Additional advantages of the electrohydraulic steering gear are low power consumption, sensitive response to wheel movements, small space requirements, low weight, and flexibility of management.

To modify the electrohydraulic system described in Chapter 17 for the reciprocating motion required in steering gears, the B-end of the electrohydraulic drive system is replaced by a piston or a ram. The force of the hydraulic fluid causes movement of the piston or ram, which subsequently turns the rudder.

Steering Gear Arrangements

There are various types of electrohydraulic steering gear layouts in use, but their operating principles are about the same. However, axial piston variable displacement pumps are used in most installations. These pumps are connected by piping to the ram cylinders of the steering gear (see figure 18–1). Two pipes from each pump are united at a main transfer valve, which is a multiported valve that permits the ram cylinders to be connected to either pump while the pipes from the other pump are connected for bypassing.

Various methods are used for connecting the hydraulic rams to the tiller, the arrangements depending on the design and on the space available for the installation. The gear illustrated in figure 18–1 is typical of those installed in destroyers. The installation includes a single ram set athwartship. The ram operates the rudder through a single yoke tiller fitted with a sliding block.

Principles of Operation

Regardless of the type of equipment (double ram or single ram, axial pump or radial pump) included in electrohydraulic steering installations, the principles of operation are basically the same. The discharge volume and direction of flow from the variable displacement pumps are controlled by the operation of the tilting block in the pump. This control is accomplished mechanically by means of trick wheels in the steering gear room, and by remote control from one or more steering stations.

Any movement, right or left, of the control from any of the various steering stations places the hydraulic pump on stroke and causes the pump to supply liquid under pressure to the hydraulic ram, resulting in a corresponding right or left movement of the rudder. This rudder movement actuates the follow-up gear, which in turn immediately acts to return the pump control to neutral but does not accomplish this until the assigned rudder position has been attained. The rudder is held in the assigned position by the hydraulic pump until another movement is originated at the steering station of movement of the rudder. If the rudder is moved by sea state action, the feedback system will cause the pump to reposition the rudder to the zero or neutral position.

Emergency Steering Systems

All naval combatant and auxiliary ships equipped with electrohydraulic steering gears are also equipped with an auxiliary steering gear. This emergency steering system generally consists of a relief and shuttle valve, hand-operated hydraulic pump, and the piping, valves, and fittings necessary to complete the system. The emergency equipment is installed in or near the steering gear compartment.

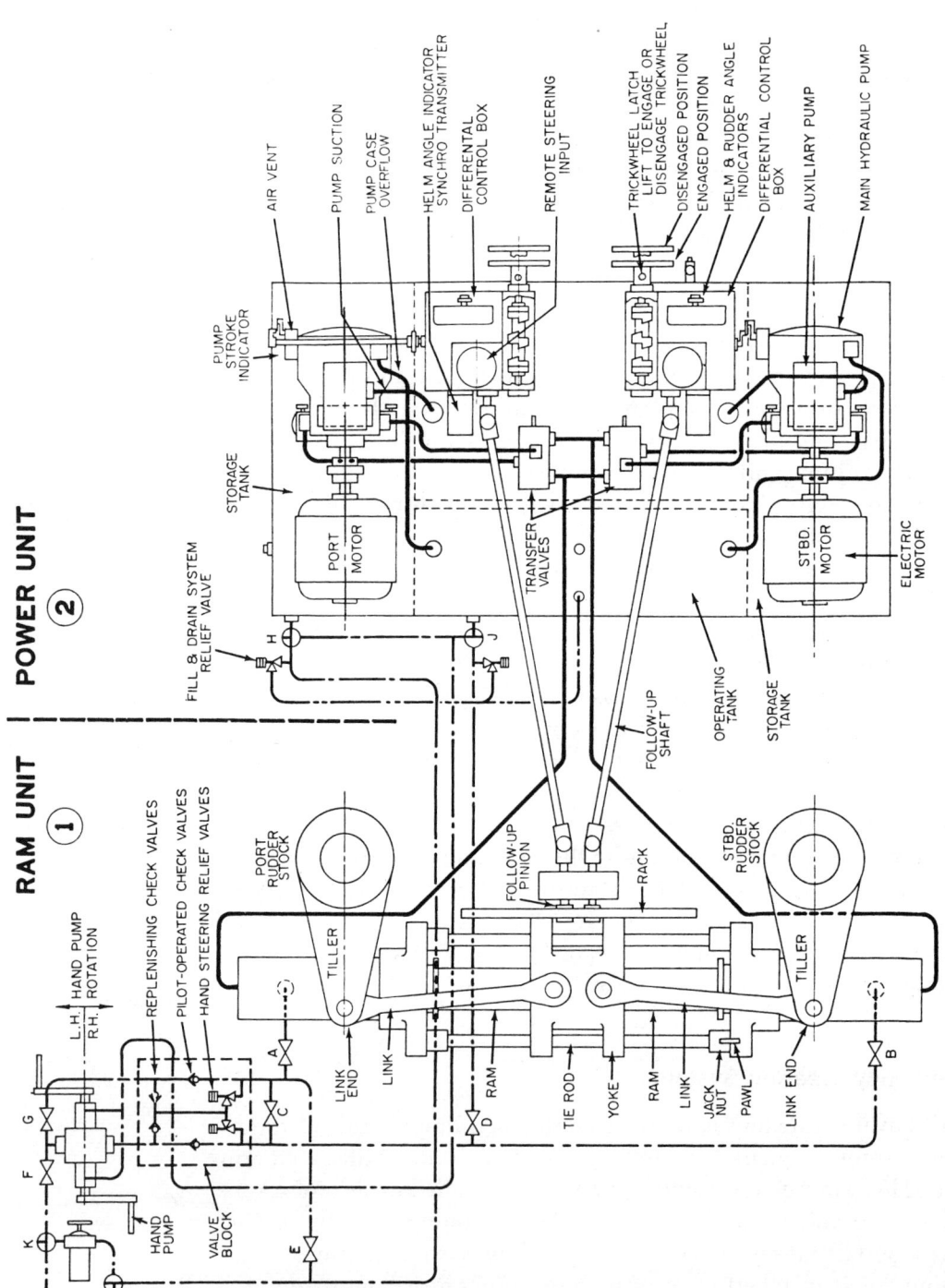

Figure 18-1. Electrohydraulic steering gear.

To prevent the pressure developed by the hand pump from causing motoring or leakage through the main hydraulic units, the piping from the emergency pump to the main hydraulic system is so arranged that the high-pressure stop valves may be closed. The emergency pump is usually connected to the main hydraulic system in a manner whereby all ram cylinders will be in use. Since it is necessary to block off the emergency system under normal steering gear operation, the emergency lines are usually connected to the drain valves to eliminate the necessity of additional high-pressure valves.

Remote Control

The electric remote control systems, used with electrohydraulic steering gears, are divided into two general types: the direct current, pilot motor type; and the alternating current, synchronous motor type.

The direct current, pilot motor type of distant control, used with early electrohydraulic steering gears, consists of a small reversible direct current motor that is connected through a differential gear to the control shaft of a variable-stroke, rotary pump. The control of the pilot motor is effected by means of a magnetic contractor control panel, located adjacent to the motor, and through master controllers located at the distant control stations. The motor is equipped with a magnetic brake which promptly stops and holds the motor when the master controller is returned to the neutral position.

The alternating current, synchronous transmission type of distant control is used in modern vessels. It consists of self-synchronous-type transmitters located in the several steering stations and controlled by the motion of the wheel, suitable electric leads, and a self-synchronous-type receiver connected through a differential-type follow-up mechanism to the control shaft of the variable-stroke hydraulic pumps. Self-synchronous transmitters and receivers are alternating current electric generator-motors so designed that the rotor of the receiver follows exactly, in speed and amount of angular displacement, the motion of the transmitter rotor. Motion of the steering wheel, which is mounted on an extension of the shaft of the transmitter rotor, is, therefore, transmitted directly to the steering engine control mechanism, which acts to cause the steering engine to produce the desired rudder angle. The follow-up system, which has the function of maintaining the movement of the rudder proportional to that of the steering wheel, so governs the action of the control that the engine is stopped as soon as the rudder matches each motion of the wheel. The electrical systems of remote rudder control are simple and reliable; there are no great friction losses, and all transmission lines may pierce bulkheads and decks at the most desirable locations.

It should be noted that the newer classes of ships have differing remote steering systems. Currently up-to-date information on the general types is not available; therefore, newly assigned officers should obtain the steering gear technical manual for their class ship when assigned, for an understanding of the exact system on board.

19

Fundamental Electrical Theory

Shipboard electrical systems include a great variety of equipment that provides numerous services indispensable to the operation of a modern naval ship. These systems distribute power throughout the ship for offensive and defensive weapons, the ship's movement, and shipboard habitability. Since the systems and equipment utilizing electric power are often under the cognizance of a division other than the electrical division, a joint responsibility frequently exists for the operation, maintenance, and repair of electrical systems and equipment.

The various techniques and equipment developed from applying the principles of electricity require some degree of theoretical knowledge, as well as practical experience and skill, for safe operation.

Like steam, electricity transports energy. Steam carries the heat energy produced in the ship's boilers to the turbines where it is expended in doing work. Electric current carries the electric energy produced by the generators to the electric motors where it is expended in performing work.

Terms

Conductors. All materials will conduct electricity but some of them offer more resistance than others. Metals such as silver, copper, aluminum, and iron offer little resistance and are called conductors.

Insulators. In contrast to good conductors, some substances such as wood, paper, procelain, rubber, mica, and plastics offer a high resistance to an electric current and are known as insulators. Electric circuits throughout the ship are made up of copper wires covered with rubber or some other insulator. The wire offers little resistance to the current, while the insulation keeps the current from passing to the steel structure of the ship.

Current. Current is the rate at which electricity flows through a conductor or circuit. The practical unit, called the ampere, specifies the rate at which the electric current is flowing. The term ampere (I) is similar to the term knot, which means nautical miles per hour. In other words, the amperage of a current is the number of particles passing a point in a circuit each second. (A current of electricity can be compared to the rate of flow of water through a pipe.)

Electromotive Force. Before an electric current can flow through a wire, there must be a source of electric "pressure," just as there must be a pump to build up water pressure before water will flow through a pipe. This electric pressure (E) is known as electromotive force (emf), potential difference, or voltage (V); a generator or a battery is the most common source. If you increase the pressure or voltage in the conductor, a greater current will flow, just as an increased pressure on water in a pipe will increase the flow.

Resistance. Electrical resistance is that quality of an electric circuit that opposes the flow of current through it. The unit of resistance is known as the ohm.

Power. Power is the rate at which work is done. In an electrical circuit, the unit is the watt.

Direct Current. When emf is unidirectional and of more-or-less constant magnitude, it is called DC (direct current). Batteries, for example, produce only direct current.

Alternating Current. In an AC (alternating current) circuit, the magnitude and direction of current flow are constantly changing. If these changes were plotted, they would describe a sine curve (figure 19-1). Starting from zero, the emf builds up to a maximum in one direction, falls back to zero, builds up to a maximum in the other

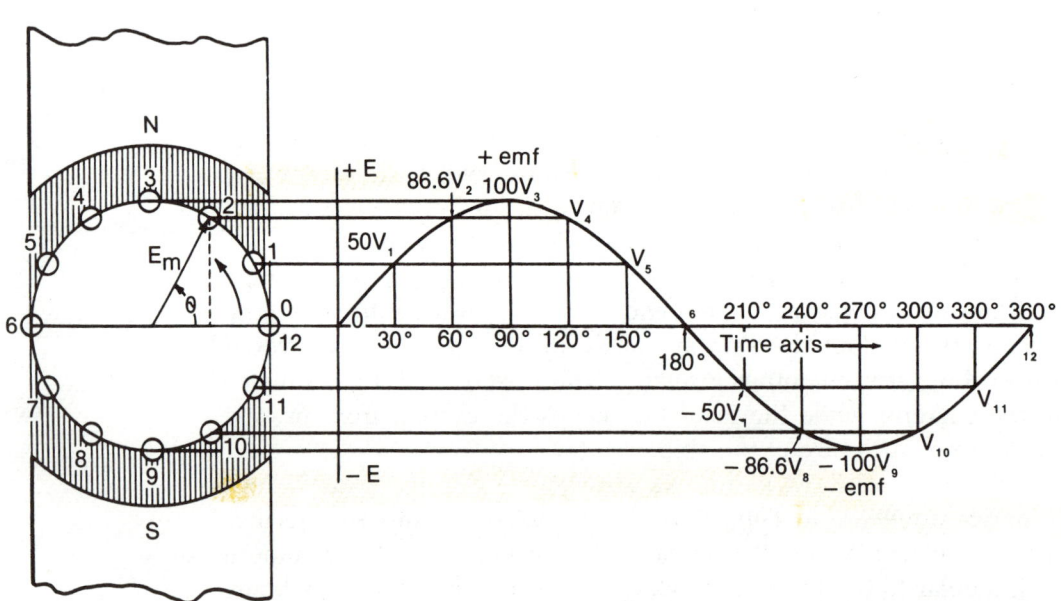

Figure 19–1. *Generation of sine-wave voltage.*

direction, and then returns to zero ready to start again. Each time the emf does this, it is called a cycle. The frequency of an AC circuit is the number of such cycles per second. The unit of frequency measurement is the hertz (Hz), where 1 Hz = 1 cycle/sec.

Electromagnetic Induction. One of the most useful and widely employed applications of magnets is in the production of vast quantities of electric power from mechanical sources. The mechanical power may be provided by a number of different devices, including gasoline engines, diesel engines, water turbines, steam turbines, and gas turbines. The final conversion of the energy to electricity is done by generators employing the principle of electromagnetic induction.

Three conditions must exist before a voltage can be produced by electromagnetic induction. First, we must have a magnetic field; second, a conductor; and third, relative motion between the field and the conductor. In accordance with these conditions, when a conductor is moved across a magnetic field so as to cut the lines of force, a voltage is produced. If the ends of a conductor are connected to a low-reading voltmeter or galvanometer and the conductor is moved rapidly down through a magnetic field, there is a momentary reading on the meter. When the conductor is moved up through the field, the meter deflects in the opposite direction. If the conductor is held stationary and the magnet is moved so that the field cuts across the conductor, the meter is deflected in the same manner as when the conductor was moved and the field was stationary.

The voltage developed across the conductor terminals by electromagnetic induction is known as an induced emf, and the resulting current that flows is called induced current. The induced emf exists only as long as relative motion occurs between the conductor and the field.

There is a definite relationship between the direction of flux, the direction of motion of the conductor, and the direction of the induced emf. When two of these directions are known, the third can be found.

Direct Current Generators

A DC generator is a rotating machine that converts mechanical energy into electrical energy. This conversion is accomplished by rotating an armature, which carries conductors, in a magnetic field, thus inducing an emf in the conductors.

A DC generator consists essentially of a steel frame or yoke containing the pole pieces and field windings; an armature consisting of a group of copper conductors mounted in a slotted cylindrical core; a commutator for maintaining the current in one direction through

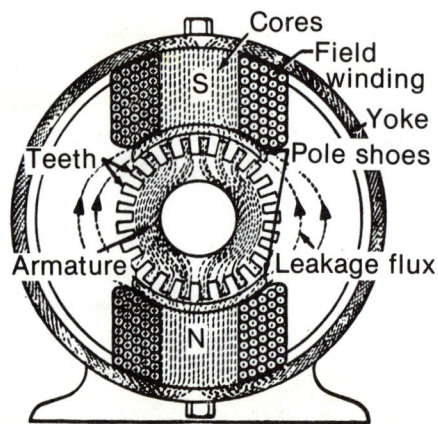

Figure 19–2. Magnetic circuit of a two-pole generator.

the external circuit; and brushes with brush holders to carry the current from the commutator to the external load circuit.

The frame, in addition to providing mechanical support for the pole pieces, serves as a portion of the magnetic circuit in that it provides a path for the magnetic flux between the poles.

The field windings of a DC generator receive current either from an external DC source or directly across the armature, thus becoming electromagnets. They are connected so that they produce alternate north and south poles, and, when energized, they establish magnetic flux in the field yoke, pole pieces, air gap, and armature core, as shown in figure 19–2.

The armature is mounted on a shaft and is rotated through the field by an outside energy source (prime mover). Thus we have a magnetic field, a conductor, and relative motion between the two—which, it will be remembered, are the three essentials for producing a voltage by magnetism. If the output of the armature is connected across the field windings, the voltage and the field current at start will be small because of the small residual flux in the field poles. However, as the generator continues to run, the small voltage across the armature will circulate a small current through the field coils, and the field will become stronger. In a self-excited generator, this action causes the generator voltage to rise quickly to the proper value and the machine is said to "build up" its voltage.

The simplest generator armature winding is a loop or single coil. Rotating this loop in a magnetic field will induce an emf whose strength is dependent upon the strength of the magnetic field and the speed of rotation of the conductor.

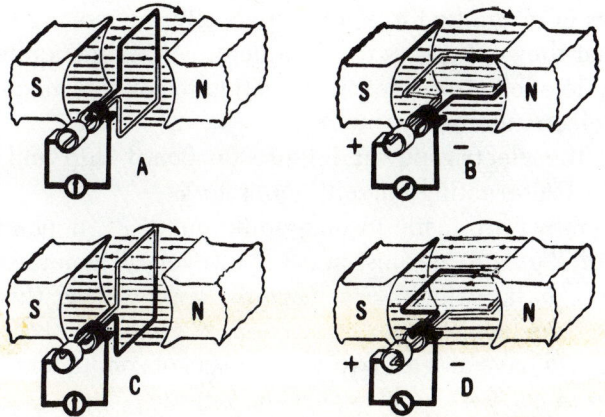

Figure 19–3. Single-coil generator with commutator.

A single-coil generator with each coil terminal connected to a bar of a two-segment metal ring is shown in figure 19–3. The two segments of the split ring are insulated from each other and the shaft, thus forming a single commutator which mechanically reverses the armature coil connections to the external circuit at the same instant that the direction of generated voltage reverses in the armature coil.

The emf developed across the brushes is pulsating and unidirectional. A pulsating direct voltage of this characteristic (called ripple) is unsuitable for most applications. In practical generators, more coils and more commutator bars are used to produce an output voltage waveform with less ripple.

Voltage control is either manual or automatic. In most low power cases, the process involves changing the resistance of the field circuit, thus controlling the field current which permits control of the terminal voltage. This means that the field current is the control variable. The major difference between the various voltage regulator systems is merely the method by which the field circuit resistance is controlled.

The construction of a DC motor is essentially the same as that of a DC generator. The DC generator converts mechanical energy into electrical energy, and the DC motor converts the electrical energy into mechanical energy. A DC generator may be made to function as a motor by applying a suitable source of direct voltage across the normal output electrical terminals.

Alternating Current Generators

Just as a current flowing in a conductor produces a magnetic field around the conductor, the reverse of this process is true. A voltage can be generated in a circuit by moving a conductor so that it cuts across lines of magnetic force, or, conversely, by moving the lines of force so that they cut across the conductor. An AC generator utilizes this principle of electromagentic induction to convert mechanical energy into electrical energy.

Most of the electric power for use on board ship and ashore is generated by alternating current generators.

AC generators are made in many different sizes, depending upon their intended use. Regardless of size, however, all generators operate on the same basic principle: a magnetic field cutting through conductors, or conductors passing through a magnetic field. Thus all generators will have at least two distinct sets of conductors. They are: (1) a group of conductors in which the output voltage is generated, and (2) a group of conductors through which direct current is passed to obtain an electromagnetic field of fixed direction. The conductors in which the output voltage is generated are always referred to as the armature windings. The conductors in which the electromagnetic field originates are always referred to as the field windings.

In addition to the armature and field, there must also be relative motion between the two. To provide this relative motion, AC generators are built in two major assemblies—the stator and the rotor. The rotor rotates inside the stator, and the rotor may be driven by any one of a number of commonly used prime movers, including steam turbines, gas turbines, and internal combustion engines.

In the revolving-armature AC generator, the stator provides a stationary electromagnetic field. The rotor, acting as the armature, revolves in the field, cutting the lines of force and producing the desired output voltage. In this generator, the armature output is taken through sliprings and thus retains its alternating characteristics.

For a number of reasons, the revolving-armature AC generator is seldom used. Its primary limitation is the fact that its output power is conducted through sliding contacts (sliprings and brushes). These contacts are subject to frictional wear and sparking. In addition, they are exposed, and thus liable to arc-over at high voltages. Consequently, revolving-armature generators are limited to applications of low power and low voltage.

The revolving-field AC generator (figure 19–4) is by far the most commonly used type. In this type of generator, direct current from a separate source is passed through windings on the rotor by means of sliprings and brushes. This maintains a rotating electromagnetic field of fixed polarity (similar to a rotating bar magnet). The rotating magnetic field, following the rotor, extends outward and cuts through the armature windings embedded in the surrounding stator. As the

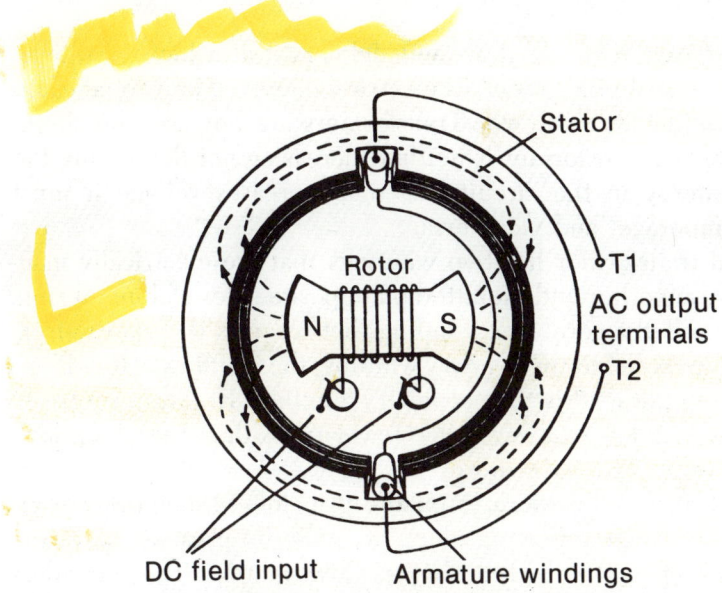

Figure 19-4. Essential parts of a rotating-field AC generator.

rotor turns, alternating voltages are induced in the windings, since magnetic fields of first one polarity and then the other cut through them. Since the output power is taken from stationary windings, the output may be connected through fixed terminals directly to the external loads, as through terminals T1 and T2 in figure 19-4. This is advantageous because there are no sliding contacts in the output circuit, and the whole output circuit is continuously insulated, thus minimizing the danger of arc-over.

Sliprings and brushes are still used on the rotor to supply direct current to the field; they are adequate for this purpose because the power level in the field is much lower than in the armature circuit.

The only practicable way to regulate the voltage output of an AC generator is to control the strength of the rotating magnetic field. The strength of the electromagnetic field may be varied by changing the amount of current flowing through the coil. Thus, voltage regulation in an AC generator is accomplished by varying the field current. This allows a relatively large AC voltage to be controlled by a much smaller DC voltage and current.

Transformers

A transformer is a device that has no moving parts and that transfers energy from one circuit to another by electromagnetic induction. The energy is always transferred without a change in frequency but usually with changes in voltage and current. A step-up transformer receives electrical energy at one voltage and delivers it at a higher voltage; a step-down transformer receives electrical energy at one voltage and delivers it at a lower voltage. Transformers are not used on direct current. Since a transformer is neither increasing nor decreasing the electrical energy in the circuit, when it increases voltage it must decrease amperage, and vice versa.

A typical transformer has two windings that are electrically insulated from each other and mounted on opposite sides of a metal ring. The ring is called a core. The winding that receives the energy from the AC source is called the primary winding; the winding that delivers the energy to the load is the secondary winding. In a step-up transformer, the primary windings have fewer turns than the secondary; in a step-down, they have more.

Since a transformer operates on the principle of electromagnetic induction, all of the following elements must be present: (1) a conductor, (2) a magnetic field, and (3) relative motion. The secondary windings satisfy the first condition, and the current through the primary windings satisfies the second. But, if a transformer has no moving parts, how can we obtain relative motion? The answer: Use alternating current. As the current constantly fluctuates from zero to maximum to zero to maximum, the magnetic flux will be constantly growing and then shrinking. The alternate expansion and collapse of the lines of force provide the relative motion between field and conductor.

Control and Safety Devices

The distribution of electric power requires the use of many devices to control the current and to protect the circuits and equipment.

Control devices are those electrical accessories that govern, in some predetermined way, the power delivered to any electrical load. In its simplest form, the control applies voltage to (or removes it from) a single load. In more complex control systems, the initial switch may set into action other control devices that govern the motor speeds, the compartment temperatures, the depth of liquid in a tank, the aiming and firing of guns, or the direction of guided missiles.

Switchboards make use of hand-operated (manual) switches as well as electrically operated controls. Manually operated switches are those familiar electrical items that can be operated by motions of the hand, as with a pushing, pulling, or twisting motion. The type of action required to operate the manually operated switch is indicated by the names of the controls—push-button switch, pull-chain switch, or rotary switch.

Automatic switches are devices that perform their function of control through the repeated closing and opening of their contacts, without requiring a human operator. Limit switches and float switches are representative automatic switches.

The simplest protective device is a *fuse*, consisting of a metal alloy strip or wire and terminals for electrically connecting the fuse into the circuit. Normally, when a circuit is overloaded or when a fault develops, the fuse element melts and opens the circuit that it is protecting. However, all fuse openings are not the result of current overload or circuit faults. Abnormal production of heat, aging of the fuse element, poor contact due to loose connection, oxides or other corrosion products forming within the fuse holder, and unusually high ambient temperatures will alter the heating conditions and the time required for the element to melt.

A more complex type of protective device is the *circuit breaker*. In addition to acting as protective devices, circuit breakers perform the function of normal switching and are used to isolate a defective circuit while repairs are being made.

Circuit breakers are available in many types; some may be operated both manually and electrically, while others are restricted to one mode of operation. Electrically operated circuit breakers employ an electromagnet to trip a release mechanism that causes the breaker contacts to open.

Voltage Produced by Chemical Action

Chemical energy is transformed into electrical energy within the cells of a battery. Shipboard uses of electricity from this source include power supply for emergency lighting (with dry cell batteries) and the starting of small engines (with wet cell batteries).

The most common dry cell battery consists of a cylindrical zinc container, a carbon electrode, and an electrolyte of ammonium chloride and water in paste form. The zinc container is the negative electrode of the cell; it is lined with a nonconducting material to insulate it from the electrolyte. When a circuit is formed, the current flows from the negative zinc electrode to the positive carbon electrode.

In a common wet cell storage battery, the electrodes and the electrolyte are altered by the chemical action that takes place when the cell delivers current. Such a battery may be restored to its original condition by forcing an electric current through it in the opposite direction to that of discharge.

The most common wet cell storage battery in use is the lead-acid battery, having an emf of 2.2 volts per cell. In the fully charged state, the positive plates are pure lead peroxide, and the negative plates are pure lead immersed in a dilute sulfuric acid electrolyte.

When a circuit is formed, the chemical action between the ionized electrolyte and dissimilar metal plates converts chemical energy to electrical energy. As the storage battery discharges, the sulfuric acid is depleted by being gradually converted to water, while both positive and negative plates are converted to lead sulfate. This chemical reaction is represented by the following equation, the reversibility of which is dependent upon electrical energy being added during the charging cycle:

$$Pb + PbO_2 + 2H_2SO_4 \underset{\text{CHARGING}}{\overset{\text{DISCHARGING}}{\rightleftarrows}} 2PbSO_4 + 2H_2O$$

The capacity of a battery is measured in ampere-hours. The capacity is equal to the product of the current (in amperes) and the time (in hours) during which the battery is supplying this current to a given load. The capacity depends upon many factors, the most important of which are: (1) the area of the plates in contact with the electrolyte, (2) the quantity and specific gravity of the electrolyte, (3) the general condition of the battery, and (4) the final limiting voltage.

20

Shipboard Electrical Distribution

Ships of the U. S. Navy use three-phase, 450-volt, 60-cycle electrical power as their primary supply to all electrical loads. The purpose of this section is to use the generator model developed in the previous discussion to demonstrate how this voltage, frequency, and phase relationship is produced.

House current is adequately described as single-phase, 120-volt, 60-cycle, alternating electrical current, as is illustrated in figure 20–1. This current is generated by a generator of the rotating field type, whose armature is wound with a continuous wiring arrangement such that power is continuously being produced in this single conductor. Three-phase electrical power, on the other hand, uses three separate windings wound on the single stator core to produce three voltages which are a function of the position of the rotating electromagnetic

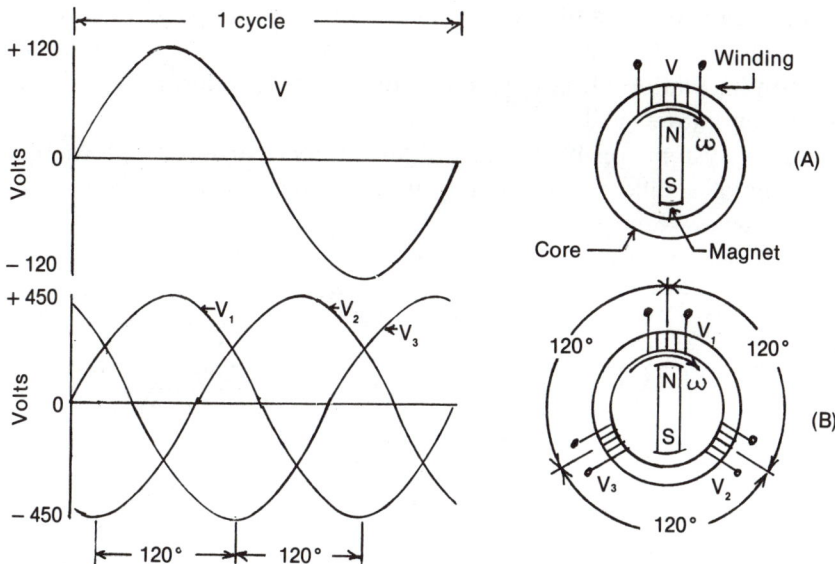

Figure 20–1. (A) Single-phase 120-volt current generated by a rotating field AC generator, utilizing a single conductor armature. (B) Three-phase 450-volt AC current generated by a rotating field AC generator utilizing three separate windings on the armature.

301

field produced on the rotor. This arrangement allows a single generator to produce the output illustrated in figure 20–1.

Why is such a system used? By using three separate conductors on a single core, the ability to generate a greater amount of electrical power is increased without significantly increasing the size and weight of the electrical generator. Also, this system allows us to use an ungrounded electrical system, which provides the ability to continue to generate electrical power even if one conductor is damaged (figure 20–2). It is important to note that three-phase systems are not used to enhance personnel safety, but are used explicitly to increase equipment reliability.

Electrical Distribution Systems

In the previous section, the generation of electrical power was discussed. The electrical distribution system is the means of transferring this electrical power from the source (generators) to the various loads in the ship (motors, lighting, weapon systems, etc.). A typical ship's AC power system consists of the ship's service distribution system (SSDS), which provides normal and alternate sources of power; the auxiliary/emergency power system; and the casualty power system. The casualty power system is not found in all ships, being used primarily in surface combatants. The several different AC power distribution systems utilize transformers to provide power at various voltages, with the primary electrical source being 450-volt, 60-cycle, three-phase alternating current produced by the ship's service turbine generators (SSTG).

Many systems on board ship require direct current or alternating current of a different frequency from the normal 60-cycle source.

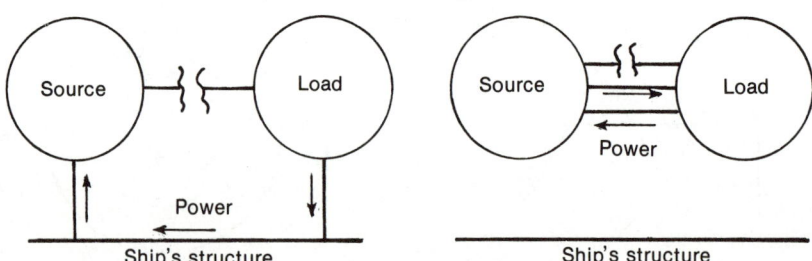

Figure 20–2. (A) Single-phase, grounded power system (utilizes ship's structure as a part of distribution system). (B) Three-phase, ungrounded system (Note: In single phase system, a break in the conductor interrupts all power to the load, whereas loss of one phase in the three-phase system only reduces power-carrying capability.)

Primary loads using this type of power are the electronic, solid state fire control and weapon systems equipment installed on board.

Direct current can be produced from alternating current by either of two methods. The first method is known as rectifying and is accomplished by electrically inverting the negative half of the AC current's sinusoidal pulse so that the time average value of the current approximates a unidirectional, steady state direct current value. (See figure 20–3.) The second method is by using a motor generator set, the operation of which is beyond the scope of this text.

Frequency different from 60 cycles may be generated by using a generator of the same basic construction as the turbine generator, but which is smaller and contains a greater number of electrical windings on the armature and/or is operated at a higher speed than the typical SSTG. Most commonly, motor generators sets are used to convert 60 Hz power to the desired frequency.

Ship's Service Distribution System (SSDS)

The SSDS provides normal and alternate power to all ship's loads. The distinction between these two types of power is important and will be developed as the SSDS is presented. During all reading, refer to figure 20–4.

Power for the SSDS is provided by the SSTGs. Depending on the type of ship, these generators may be driven by either steam turbines, gas turbines, or diesel engines. The number, type, and capacity of the generators are determined by the load that they must supply. The SSTGs are located in the ship and supply switchboards located nearby. Figure 20–4 is a schematic of how a distribution system looks but does not show their exact location. Number one and two SSTGs supply the forward switchboards which are located on the port side of the ship above the waterline, while numbers three and four SSTGs supply the after switchboards which are located on the starboard side above the waterline. The forward switchboards are designated as the controlling group, and it is from these panels that an operator controls the entire electric plant.

Switchboards are metal enclosures that house various instruments, indicating devices, and protective and regulating apparatus required for controlling the operation of the generators and the distribution of electrical power. They also contain buswork—large, metal, high current carrying conductors, which distribute the electrical power from the generator output to the distribution breakers also located in the switchboard enclosure. The various SSTGs and distribution switchboards are interconnected by bus ties, or electrical cabling, so

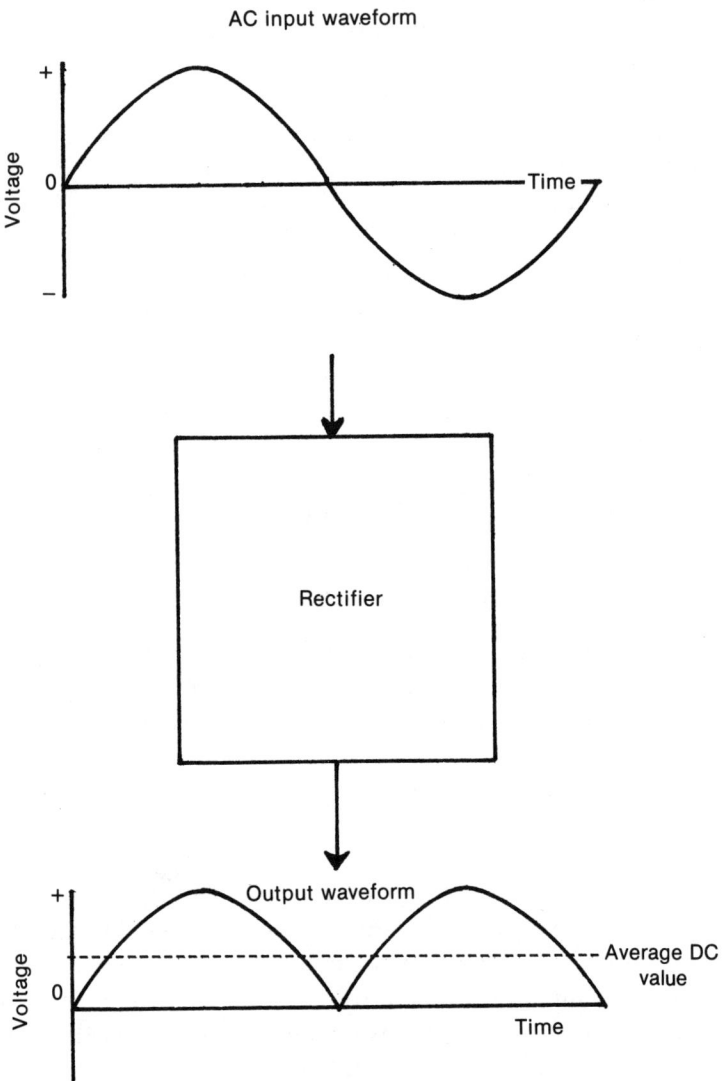

Figure 20–3. Input and output waveforms of a full wave rectifier (Note: Inversion of negative half of sine wave yields a time-averaged value that approximates a direct current signal.)

that any switchboard can be connected to any power source by closing the right circuit breaker. This allows a remote generator to be operated in parallel with a local generator to increase the power delivery capability to any particular switchboard, or allows the after (forward) generators to supply the forward (after) switchboards if a switchboard's normal power supply is lost.

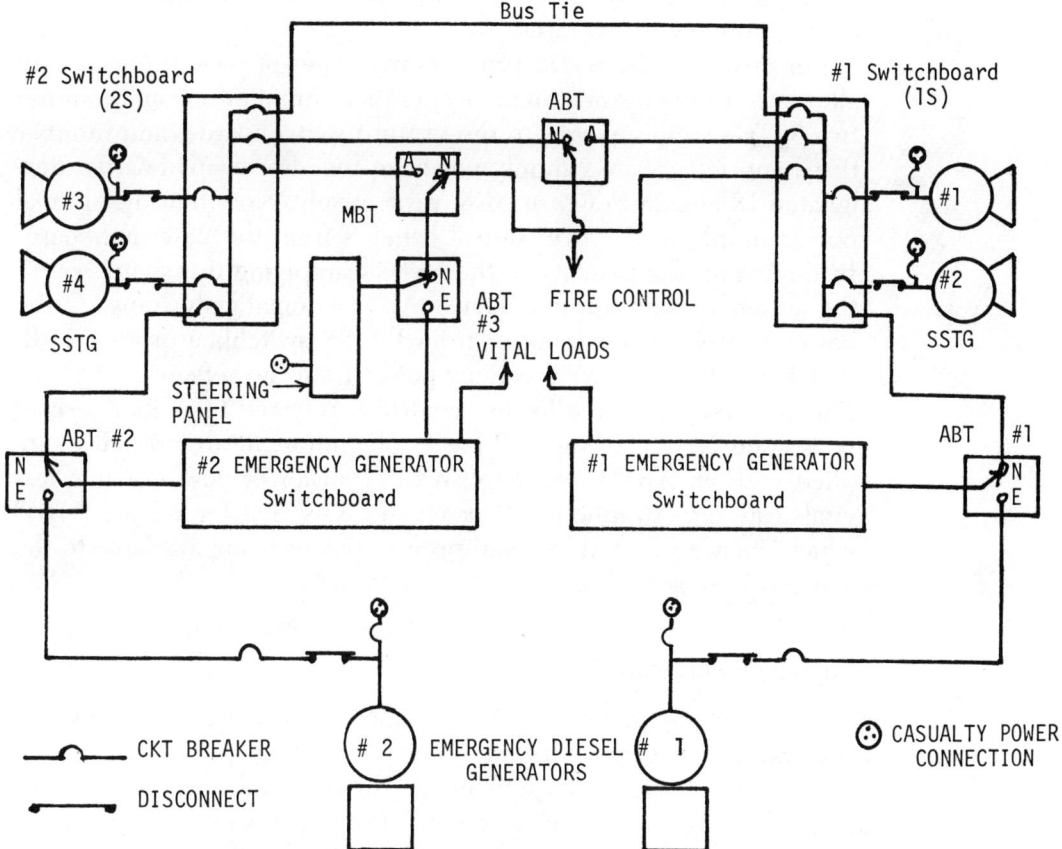

Figure 20–4. Schematic diagram of a ship's electrical distribution system.

In large installations, distribution of electrical power is from the generator and distribution switchboards to load centers, to distribution panels, and finally to the loads. This concept allows protection of power sources by providing a means of selectively eliminating loads locally rather than securing the power source each time it is desired to secure power to a load.

In practice, distribution of power is accomplished remotely by opening or closing circuit breakers. Circuit breakers are electromechanical devices using solenoids and springs to open or close contacts allowing the interruption or conduction of power in an electrical circuit. These breakers also can be actuated by automatic trip devices if an unsafe situation is sensed. Some trips used are overcurrent trips (too much current flowing in the circuit), reverse power trips (power sensed flowing in the wrong direction in the circuit), undervoltage trips (voltage too low in the circuit), and under frequency trips (frequency too low in the circuit).

Functionally, the SSDS provides two types of power: normal and alternate. In the normal mode of operation, number one and number two SSTGs supply power to the forward switchboards, and number three and four SSTGs supply power to the after switchboards, designated 1S and 2S in figure 20–4, respectively. Note that the normal power supply to the fire control panel is from the 2S switchboard. In the event of a casualty to the SSTGs supplying this switchboard, the automatic bus transfer device (ABT) automatically transfers the power source for fire control from the 2S switchboards to the 1S switchboards which are receiving power from the remaining SSTGs. This new power source for fire control is referred to as its alternate power source. Note that vital pieces of equipment are normally supplied with an ABT to accomplish the transfer of power, while less vital loads are supplied with a manual bus transfer device (MBT) which must be shifted by hand prior to power being available to the affected load.

Auxiliary/Emergency Power

Auxiliary/emergency power is supplied by two or more diesel generators. These generators and their associated emergency switchboards are physically located in the forward and after ends of the ship. They are normally located low in the ship as well, thus providing vertical and horizontal separation between the SSDS and the auxiliary/emergency distribution system. This arrangement minimizes the possibility of a total loss of electrical power generating capability due to battle damage. The number and capacity of the emergency generators is determined by the vital loads that must be supplied under the worst casualty conditions. The operation of the auxiliary/emergency power system is automatic and best explained by way of example.

The steering panel shown in figure 20–4 is normally powered by number one or two SSTG via the SSDS normal power supply. If the SSTG supply is lost, and the after switchboards are available to supply power, number three ABT remains in the "N" position, and the manual bus transfer device must be manually shifted to its alternate power supply for the steering panel. However, if power were lost to both the forward and after switchboards, this would be sensed by the forward and after emergency diesel, which would automatically start and supply power to the forward and after emergency switchboards. Numbers one, two, and three ABTs would shift to the "E" position, allowing the diesel generators to supply electrical power to the steering gear, as well as other vital loads. In this case the steering panel would be receiving power via the auxiliary/emergency power distribution system.

Casualty Power Distribution System

Damage to the SSDS and the auxiliary/emergency distribution system in wartime led to the development of the casualty power distribution system (CPDS). This system provides for making temporary connections to vital circuits and equipment, and is limited to the facilities necessary to keep the ship afloat and permit her to get out of the danger area. The system also provides a limited amount of electrical power to armament, such as antiaircraft guns and their directors to protect the ship when she is in a damaged condition.

The CPDS includes suitable lengths of portable electrical cable stowed in racks throughout the ship. Permanently installed casualty power bulkhead terminals form an important part of the CPDS. They are used for connecting portable cables on opposite sides of bulkheads, so that power may be transmitted through compartments without loss of watertight integrity; also included are permanently installed riser terminals between decks. The vital equipment selected to receive power will have a terminal box mounted on or near the equipment or panel concerned and connected in parallel with the normal feeder for the equipment.

Sources of supply for the CPDS are provided at each ship's service and emergency generator switchboard. A casualty power riser terminal is installed on the rear side of the switchboard or switchgear group and connected to the buses through a 225- or 250-ampere circuit breaker. This circuit breaker is connected between the generator circuit breaker and the generator disconnect links so that if the disconnect links are pulled, the generator may be isolated from the switchboard and used exclusively for casualty power purposes, if desired.

Electrical Safety

After an accident has happened, investigation almost invariably reveals that it could have been prevented by the exercise of simple safety precautions, which are then posted for future guidance, but which never undo the consequences of the accident that has gone before. Always observe safety precautions, and keep accidents from happening. Always remember that: (1) electrocution strikes without warning; (2) hurrying reduces caution and invites accidents; (3) taking time to be careful saves time in the end; (4) taking chances is an invitation to trouble; (5) if you do not know the safe way, it pays to find it out before exposing yourself to danger; and (6) every electrical circuit is a potential source of danger and must be treated as such.

Need for Safety Precautions

Safety precautions must always be observed by persons working around electric circuits and equipment to avoid injury from electric shock or short circuits caused by accidentally placing or dropping a metal tool, ruler, flashlight case, or other conducting article across an energized line. The arc and fire that can be produced from even a relatively low voltage circuit may cause extensive damage to equipment and serious injury to personnel.

That the danger from shock from the 450-volt AC ship's service system is reasonably well recognized by operating personnel is shown by the relatively few reports of serious shock received from this voltage despite its widespread use. On the other hand, a number of shipboard fatalities have been reported due to contact with 115-volt circuits. Despite a fairly widespread, but totally unfounded, popular belief to the contrary, low-voltage (115 volts and below) circuits are very dangerous and can cause death when the resistance of the body is lowered by moisture, and especially when current passes through the chest. Shipboard conditions are particularly conducive to severity of shock because the body is likely to be in contact with the ship's metal structure, and the body resistance may be lowered because of perspiration or damp clothing. Extra care is therefore needed.

Grounding

Safety is using plastic-cased in lieu of metal-cased portable electric tools when a choice exists. Safety is using rubber or plastic-covered, in lieu of metal-type, portable receptacles when a choice exists. Metal enclosing cases, bases, frames, and structural parts of electrical equipment that are not intended or expected to operate at potentials above ground, should be grounded. Normally, on board steel-hulled vessels such grounds are inherently provided because the metal enclosure, cases, or frames are in contact with one another and the metal structure of the vessel. Where such inherent grounding is not provided by the mounting arrangements, for instance equipments supported on shock mounts that insulate it from the hull, and on board wooden-hulled vessels, ground connections should be provided to the ground. If enclosing cases and similar parts are not grounded, a breakdown of insulation may raise them to line voltage and create a hazard that is more serious because it is unexpected. Such breakdowns from a circuit or a machine to an ungrounded metal object will not be shown by ground tests of measurements of insulation resistance.

Safety Rules for Working On Electrical Circuits

All electrical circuits are to be considered alive until it is positively determined that they are dead. To check a circuit, test the live side with a voltmeter or voltage tester, then test the dead side with the same device, and then retest the live side. This is done to ensure that the testing device is in good condition. Opening the switch to kill the power circuit does not always kill associated or remote control circuits. Always test such circuits in addition to testing the power circuit. Make sure all alternate power sources are secured.

Whenever it is necessary to check a circuit to see if it is alive, a voltmeter, voltage tester, or other suitable means shall be used. Never trust the insulation on live circuits when considering personnel safety. It may look perfect, yet not prevent a shock. Sufficient current leakage may be present to cause a fatal shock. Except in cases of emergency, never work on an energized circuit. It must be considered that the circuit is energized until a personal check has been made to see that the switch is open and tagged, and the circuit has been tested with a voltmeter or voltage tester.

Energized switchboards are a great source of danger. No work shall be undertaken on switchboards (energized or de-energized) without first obtaining the approval of the electrical and engineering officer. The commanding officer's approval shall be obtained prior to the commencement of work on any energized switchboard, and appropriate additional safety precautions shall be observed.

Circuits to be overhauled or repaired should be de-energized by opening all switches through which the power could be supplied, and the circuit should be tested with a voltmeter or a voltage tester. These switches should be tagged: "DANGER SHOCK HAZARD. Do not change position of switch except by direction of NAME . . . RATE/RANK." Warning tags (see figure 20–5) may be used for this purpose. In case more than one repairman is engaged in repair work on an electrical circuit, a tag for each party should be placed on the supply switches. After the work has been completed, each party should remove his own tags but no other.

The covers of fuse boxes and junction boxes should be kept securely closed except when work is being done. Safety devices such as interlocks, overload relays, and fuses should never be altered or disconnected except for replacement. Safety or protective devices must never be changed or modified in any way without specific authori-

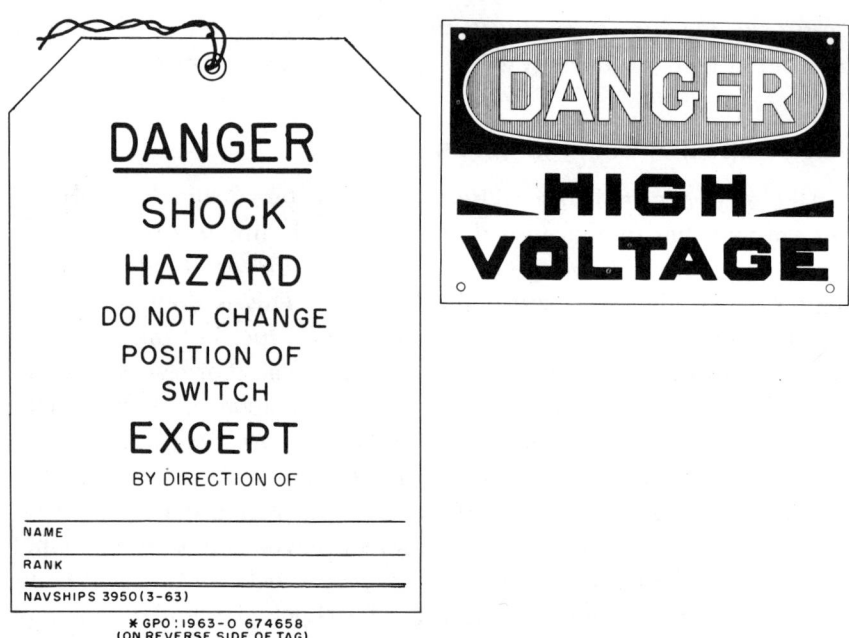

Figure 20–5. Electrical warning tags used in the Navy.

zation. Fuses should be removed and replaced after the circuit has been de-energized. When a fuse blows, it should be replaced only with a fuse of the correct current and voltage rating. When possible, a circuit should be carefully checked before the replacement is made, since a burned-out fuse is often the result of a circuit fault.

Safety Rules for Portable Tools and Equipment

Naval specifications for metal-cased portable tools require the electric cord for the tool to be provided with a distinctively marked grounding conductor in addition to the conductors for supplying power to the tool. Specifications require that green be used for the grounding conductor in cables for all new metal-cased portable tools and equipment. The end of the grounding conductor that is within the tool should be connected to the metal housing; the other end should be grounded—that is, connected to the ship's metal structure. In order to provide a convenient means of connecting the grounding conductor to ground, the Navy has standardized the use of grounded-type plugs and receptacles that automatically make this connection when the plug is inserted in the receptacle.

Safety is wearing rubber gloves when plugging in and operating any portable electric equipment. (Portable electrical equipment means such items as metal-cased power tools, receptables, extension cords, lights, electrode holders, meters, or other metal-cased devices held in the hand or carried while connected to an electric power outlet.)

Some of the portable tools now in use on board naval vessels may not yet be provided with the grounded-type plug. In addition, there is a wide range of miscellaneous portable electric equipment that may be issued without being provided with a cord that has a grounding conductor and a grounded plug. This equipment includes galley equipment, office equipment, shop equipment, medical equipment, barber shop equipment, and laundry equipment. All electrically operated equipment now on board ship without a grounding conductor and grounding plug, should be provided with one by ship's force (those members attached to the command).

In connecting the cord and plug, the grounding conductor of the cord should be connected to the ground contact of the plug at one end, and to the metal equipment casing at the other end. Extreme care must be exercised to see that the ground connection is properly made. If the grounding conductor that is connected to the metallic equipment casing is inadvertently connected to a line contact of the plug, a dangerous potential will be placed on the equipment casing. This will almost certainly cause a fatal shock to the man handling the portable equipment when it is plugged into a power receptacle, since line voltage will be on the exposed parts of the portable metal-cased equipment. To guard against this danger, the connections should be tested after they have been made. Tests of portable equipment should be made in a workshop equipped with a nonconducting-surface workbench and diamond-tread rubber deck covering. Electricians making the tests should wear rubber gloves during the tests.

If grounded receptacles have not yet been installed in the spaces where the tool or equipment is to be used, other types of plugs and receptacles may be used if the grounding conductor in the tool or equipment cord is connected to the ship's metal structure by other methods; for example, by means of a spring clip or by securing the grounding conductor to a convenient screw or bolt. In those cases where the tool or equipment does not include an extra conductor for grounding, an additional conductor should be obtained and connected between the metal casing of the tool or equipment and the ship's structure. The additional conductor used to ground the case of the equipment must be of ample cross-sectional area and current-carrying capacity to be effective. It should have a cross-sectional area equal at least to the cross-sectional area of the power conductors in the tool, and preferably a greater one. If the tool or equipment housing has two or more metallic parts that are not electrically connected, each should be connected to the grounding conductor. Care should be taken to secure a good contact between the grounding conductor and the metal to which it is connected, by scraping any paint away and scratching a clean surface. The ground connection should be made before inserting the power supply. Frequent inspection and check of the connections within portable electric tools and equipment should be made to ensure that the supply cord and its connections are suitably insulated, and that the ground connection is intact.

Because a metal ship is a hazardous location, personnel who must use portable electric tools connected to extension cords should take care to plug the device into the extension cord prior to inserting the extension cord into a bulkhead receptacle. Likewise, the extension cord should be unplugged from the bulkhead receptacle before the device is unplugged from the extension cord.

Reinforcement of flexible cable at junctions with portable equipment usually consists of a molded rubber sleeve. Examine to ensure that the sleeve is sound and free of cracks. Reenforcement consisting of coiled metal spring is dangerous, since it can conceal broken cable insulation and exposed conductors. Replace spring coil with rubber, and examine the attached plug for damage. Before issue of any portable electrical equipment, the item, with associated extension cords connected, should be plugged into a dummy receptacle and tested for resistance from equipment housing to ship's structure using a voltmeter. Move or work cable with a twisting motion. A change in resistance will indicate broken strands in the grounding conductor. If this is found, the cable must be replaced. It is further suggested that, at the discretion of the commanding officer, a list be established of portable equipment requiring testing, and that the following items be considered:

1. Portable hand electric tools that are permanently checked out or are on loan to other shipboard departments or divisions should be tested once a week.

2. Electrical equipment that is frequently touched, such as hotplates, coffee makers, toasters, portable vent sets, and movie projectors, should be tested once a week.

3. Electric equipment that is infrequently touched, such as bulkhead mounted vent fans energized from receptacles, should be tested once a year.

Personal Equipment

Personally owned or non-naval-standard electric lights, fans, and tools are not justified. They generally present a shock hazard because of inferior insulation, leakage currents, and flimsy structure. Periodic inspection should be made to eliminate them from the ship. Adequate numbers and types of naval lights, fans, and tools are available to meet all needs. Personally owned hobby equipment such as hand-held motor-driven carving tools are frequently found to be of flimsy construction and unsafe for use on board ship. Such equipment may be retained on board ship as an exception, subject to the following safety precautions:

1. Before use of portable, electrically operated hobby devices is permitted on board ship, they should be inspected and tested by the ship's electricians. Equipment that passes this inspection should be tagged as safe, giving the date of inspection. This equipment should be frequently reinspected, at least once every six months.

2. At any time that the hobby tool is damaged or is obviously defective, that is, if molded housings, cords, or plugs contain cracks or breaks, or if the cord insulation breaks when sharply bent, the tool must not be used until repaired and reinspected by the electrical gang.

No personal electrical equipment, such as radios, television sets, record players, wire or tape recorders, or other personal appliances as listed above, shall be used on board ship without the engineer officer's approval. Periodic inspection should be made to enforce this vital safety regulation.

21

Ship Construction

Principles of Ship Design

Major factors to be considered in the construction of any warship are mission, armament, protection, seaworthiness, strength, stability, maneuverability, speed, and cruising range.

The *mission* of a warship is a prime determinant in the construction planning for that ship. As an example, for an ASW destroyer, the sonar gear, associated detection gear, weapons, and required personnel are the basic units about which a hull is constructed. The need for ships of different sizes and shapes within a type, that is, the variance between two such ships as a guided missile destroyer and an ASW destroyer, is an indication of the required mission. An even greater differences is evident when one compares the differences of ships used in amphibious operations or in auxiliary missions.

Armament is the gage by which the offensive power of a ship is measured. Normally we think of armament as meaning guns, rockets, missiles, and so on. Depending on the ship's mission, however, the term also includes aircraft used for offensive purposes and landing craft (such as LCVPs and LCMs) suitable for amphibious operations.

Protection comprises those features that are provided to thwart or minimize the effects of enemy attack. Included in this category is internal subdivision by longitudinal and transverse bulkheads for limiting the spread of fire and flooding caused by damage. Torpedo defense systems are also found in large ships.

Seaworthiness is the term used to describe a ship's ability to operate in all kinds of wind, weather, and seas. Again, the ability of a ship to accomplish her assigned mission must be considered in the proper evaluation of seaworthiness. Stability, size, and freeboard are controlling factors.

Stability concerns the ability of the ship to return to an upright position when heeled over by an external force. This is a partial measure of the ship's ability to absorb punishment involving underwater damage and flooding. In addition, stability has an important influence on the period of roll, which, to some extent, determines a vessel's utility as a gun or aircraft platform.

Maneuverability is the characteristic that permits rapid changes of course and speed and includes the ability to turn in a small diameter. The need for maneuverability varies considerably for the various types of ships, from the highly maneuverable FFs and DDs to the slower-moving auxiliary and amphibious-type ships.

Speed is determined by weight of the ship, its underwater shape, and the power and efficiency of the propulsion plant.

Cruising range, also called endurance, refers to the ability to remain at sea for long periods of time and traverse long distances. It is determined by fuel capacity, fresh water capacity, efficiency of the propulsion plant with respect to fuel consumption, and provision capacity (dependent on storage space and refrigeration). The ability to replenish at sea has a direct relationship to cruising range, especially today when units usually operate in coordinated maneuvers and operations, utilizing replenishment capabilities to multiply at-sea effectiveness. This factor must be considered in any final requirement analysis.

Obviously these qualities are not independent of each other. For example, a change in speed requirements affects the cruising range considerably. Excessive plating reduces the proportion of weight that can be used for machinery, and tends to reduce potential speed.

The designer of a ship tries to incorporate as many favorable features as possible in keeping with the general use to which the ship will be put. All ships represent a compromise in which some factors must dominate others. Destroyers, for example, sacrifice plating for speed. A very important consideration in all types of ships is the habitability features that must be incorporated for the comfort of the personnel manning the warships.

Basic Ship Structure

In considering the structure of a ship, it is common practice to liken the ship to a box girder. Like a box girder, a ship may be subjected to tremendous stresses. The magnitude of stress is usually expressed in pounds per square inch (psi).

When a pull is exerted on each end of a bar, as in part A of figure 21–1, the bar is under the type of stress called *tension*. When a pressure is exerted on each end of a bar, as in part B of figure 21–1, the bar is under the type of stress called *compression*. If equal but opposite forces are exerted on the upper and lower bars, as shown in part C of figure 21–1, the pins connecting these bars are subjected to a stress at right angles to their length. This stress is called *shear*. When a shaft, bar, or other material is subjected to a twisting motion,

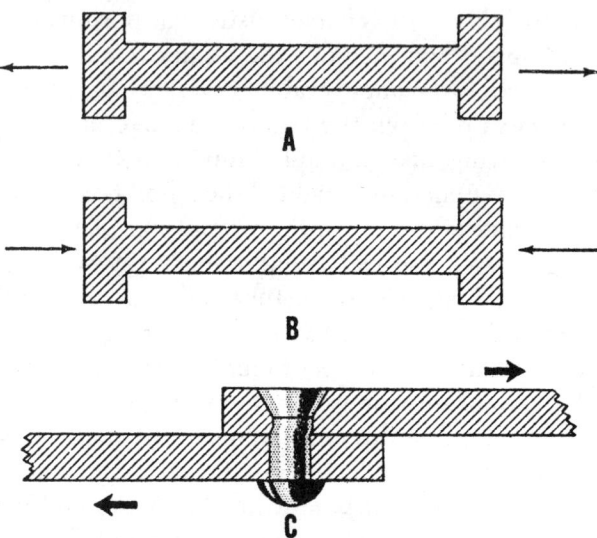

Figure 21–1. Stresses in metal (A) tension; (B) compression; (C) shear.

the resulting stress is known as *torsional stress*. Torsional stress is not illustrated in figure 21–1.

When a material is compressed, it is shortened. When it is subjected to tension, it is lengthened. This change in shape is called *strain*. The change of shape (strain) may be regarded as an effect of stress.

If a simple beam is supported at its two ends and various vertical loads are applied over the center of the span, the beam will bend (figure 21–2). As the beam bends, the upper section of the beam compresses, and the lower part stretches. Somewhere between the top and bottom of the beam, there is a section that is neither in compression nor in tension; this is known as the *neutral axis*. The greatest stresses in tension and compression occur near the middle of the length of the beam, where the loads are applied.

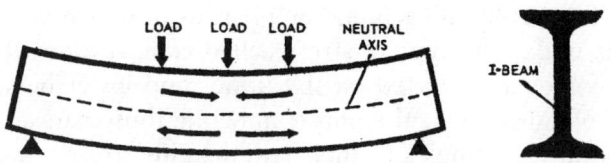

Figure 21–2. I-beam with load placed over center.

316   Introduction to Naval Engineering

Longitudinal Bending and Stresses

In an I-beam, the greater mass of structural material is placed in the upper and lower flanges to resist compression and tension. Relatively little material is placed in the web that holds the two flanges so that they can work together; the web, being near the neutral axis, is less subject to tension and compression stresses than are the flanges. The web does take care of shearing stresses, which are sizable near the supports.

A ship in a seaway can be considered similar to this I-beam (or more correctly, it can be likened to a box girder) with supports and distributed loads. The supports are the buoyant forces of the waves; the loads are the weight of the ship's structure and the weight of everything contained within the ship.

The ship shown in figure 21–3 is supported by waves, with the bow and stern each riding a crest and the midship region in the trough. This ship will bend with compression at the top and tension at the bottom. A ship in this condition is said to be *sagging*. In a sagging ship, the weather deck tends to compress under compressive stress, and the bottom plating tends to stretch under tensile stress. A sagging ship is undergoing longitudinal bending—that is, it is bending in a fore-and-aft direction.

When the ship advances half a wavelength, so that the crest is amidships and the bow and stern are over troughs, as shown in figure 21–4, the stresses are reversed. The weather deck is now in tension, and the bottom plating is in compression. A ship in this condition is said to be *hogging*. Hogging, like sagging, is a form of longitudinal bending. The effects of longitudinal bending must be considered in the design of the ship, with particular reference to the overall strength that the ship must have.

In structural design, the terms hull girder and ship girder are used to designate the structural parts of the hull. The structural parts of

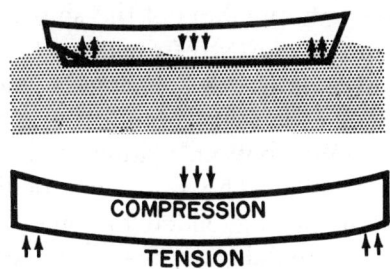

Figure 21–3. Sagging.

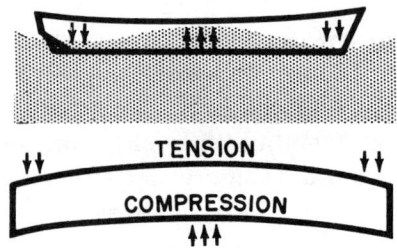

Figure 21–4. Hogging.

the hull are those parts that contribute to its strength as a girder and provide what is known as longitudinal strength. Structural parts include the framing (transverse and longitudinal), the shellplating, the decks, and the longitudinal bulkheads. These major strength members enable the ship girder to resist the various stresses to which it is subjected.

The ship girder is subjected to rapid reversal of stresses when the ship is in a seaway and is changing from a hogging condition to a sagging condition (and vice versa), since these changes occur in the short time required for the wave to advance half a wavelength. Other dynamic stresses are caused by pressure loads forward due to the ship's motion ahead, by panting of forward plating due to variations of pressure, by the thrust of the propeller, and by the rolling and pitching of the ship. (Note: Panting is a small in-and-out working of the plating at the bow.)

Transverse and longitudinal stresses result from the pressure of the water on the ship's sides, which subjects the transverse and longitudinal framing, deck beams, and shellplating below water to a hydrostatic load. Local stresses occur in the vicinity of masts, windlasses, winches, and heavy weights. These areas are strengthened by thicker deck plating or by deeper or reinforced deck beams.

Hull Members

The principal strength members of the ship girder are at the top and bottom where the greatest stresses occur. The top flange includes the strength deck plating, the deck stringers, and the sheer strakes of the side plating. The bottom flange includes the keel, the outer bottom plating, the inner bottom plating, and any continuous longitudinal framing installed. The side webs of the ship girder are composed of the side plating, aided to some extent by any long, continuous fore-and-aft bulkheads. Some of the strength members of a destroyer hull girder are indicated in figure 21–5.

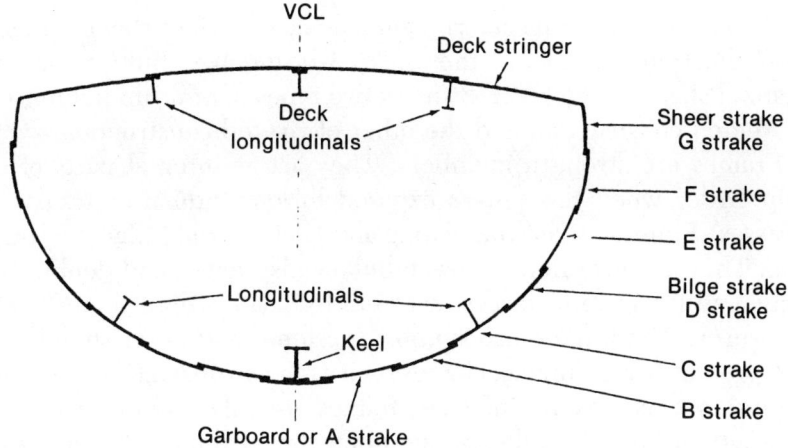

Figure 21–5. Strength member of a destroyer hull girder.

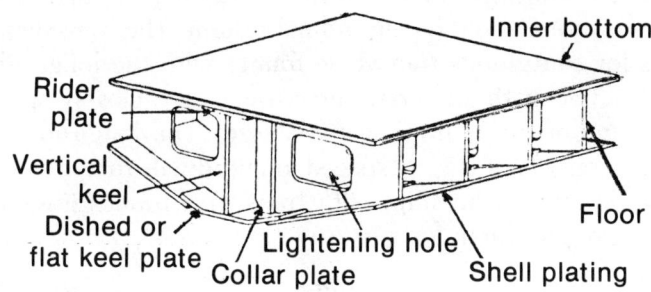

Figure 21–6. Type of keel structure.

Keel

The most important structural member of a ship is the keel, which runs the length of the ship's bottom from the stem to the stern post. It acts as a backbone, performing a function similar to that of the human spine. The keel of a metal ship does not project below the bottom as does the fin keel of a sailboard, but lies entirely within the ship. It is built up of plates and angles into an I-beam shape. The lower flange of the I-beam structure is a flat plate keel that forms the center strake of the bottom plating. The web of the I-beam is the center vertical keel. The height of the center vertical keel varies from about 2 feet in small ships to nearly 7 feet in large ships. The upper flange of the I-beam is called the rider plate. If the vessel is fitted with an inner bottom, the rider plate forms the center strake of the inner bottom plating. At the ends of the vessel, the keel is joined to two heavy castings (the stem and stern posts), which complete the backbone. One type of keel structure is shown in figure 21–6.

Framing

Frames used in ship construction may be of various shapes. Figure 21–7 illustrates frames of the angle, I-beam, tee, bulb angle, and channel shapes. Figure 21–8 shows two types of built-up frames, one of welded construction and the other of riveted construction.

Frames are strength members. They act as integral parts of the ship girder when the ship is exposed to longitudinal or transverse stresses. Frames stiffen the plating and keep it from bulging or buckling. They act as girders between bulkheads, decks, and double bottoms, and transmit forces exerted by load weights and water pressures. The frames also support the inner and outer shell locally and protect against unusual forces such as those caused by underwater explosions. As may be inferred, frames are called upon to perform a variety of functions, depending upon the location of the frames in the ship.

There are two important systems of framing in current use: the transverse system and the longitudinal system. The transverse system provides for continuous transverse frames with the longitudinals intercostal between them. Transverse frames are closely spaced, and a small number of longitudinals are used. The longitudinal system of framing consists of closely spaced longitudinals that are continuous along the length of the ship, with transverse frames intercostal between the longitudinals.

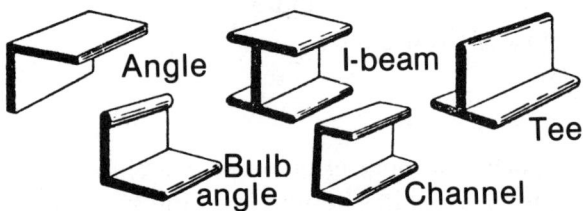

Figure 21–7. Types of frames used on board ship.

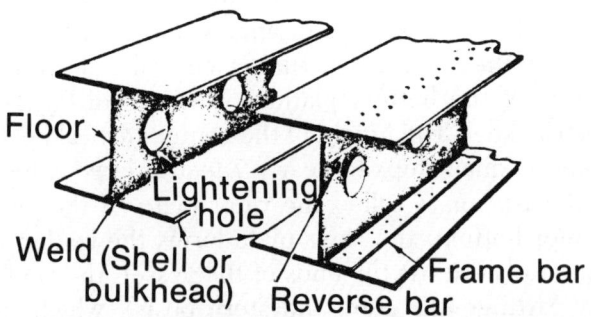

Figure 21–8. Two types of built-up frames.

Transverse frames are attached to the keel and extend from the keel outward around the turn of the bilge and up to the edge of the main deck. Closely spaced along the length of the ship, they define the form of the ship.

Longitudinals run parallel to the keel along the bottom, bilge, and side plating. The longitudinals provide longitudinal strength, stiffen the shellplating, and tie the transverse frames and the bulkheads together. The longitudinals in the bottom are of the built-up type.

Where two sets of frames intersect, one set must be cut to allow for the other set. The frames that are cut, and thereby weakened, are known as intercostal frames; those that continue through are called continuous frames. Both intercostal and continuous frames are shown in figure 21–9.

A cellular form of framing results from a combination of longitudinal and transverse framing systems utilizing closely spaced deep framing. Cellular framing is used in most naval ships. In the bottom framing, which is probably the strongest part of a ship's structure, the floors are integrated into a rigid cellular construction. Heavy loads such as the ship's propulsion machinery are bolted to foundations that are built directly on top of the bottom framing.

Plating

The outer bottom and side plating forms a strong, watertight shell. Shellplating consists of approximately rectangular steel plates arranged longitudinally in rows or courses called *strakes*. The strakes are lettered, beginning with the A strake (also called the garboard strake) which is just outboard of the keel and working up to the uppermost side strake (called the sheer strake).

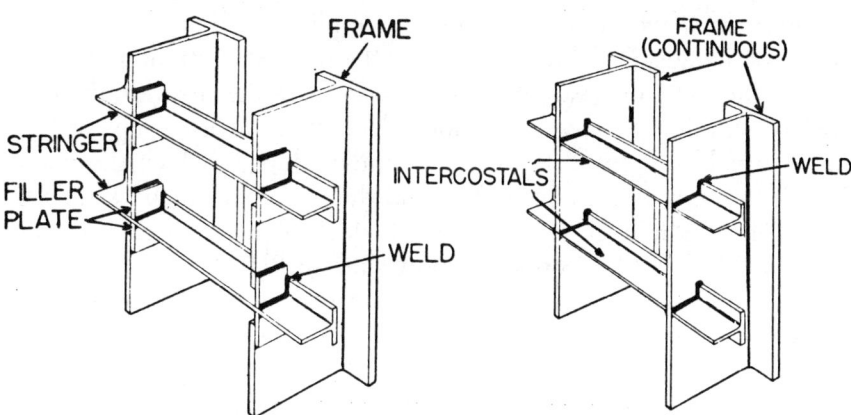

Figure 21–9. Examples of intercostal and continuous frames.

The end joint formed by adjoining plates in a strake is called a *butt*. The joint between the edges of adjoining strakes is called a *seam*.

Since the hull structure is composed of a great many individual pieces, the strength and tightness of the ship as a whole depends very much upon the strength and tightness of the connections between the individual pieces. In modern naval ships, welded joints are used to a very great extent. However, riveted joints are still used for some applications.

Decks

The ship is divided by a series of decks and platforms into tiers of compartments, the decks forming the floors and ceilings of the compartments (the bulkheads forming the walls). The floor of a ship's compartment always is called the deck, and the ceiling is always called the overhead. (The words floor and ceiling have other meanings on board ship. As in figure 21–6, a floor is a transverse partition in the double bottoms. The ceiling is a term applied to the planing with which the side of a ship is sheathed.)

The deck normally is composed of rectangular steel plates joined into strakes similar to the shellplating. The plates in the outermost strake of deck plating are stringer plates; they are connected to the shellplating and are important structural members of the ship. Deck plating is strengthened by transverse and longitudinal deck and beams and deck girders on the underside of the deck. The beams and girders are usually composed of I or T beams fastened to the shell frames by means of triangular steel brackets. Decks above the waterline usually are arched (cambered) so that they are higher at the centerline. The camber provides greater strength and aids in the drainage of water.

The term *strength deck* is generally applied to the deck that acts as the top flange of the hull girder. It is the highest continuous deck—usually the main or weather deck. However, the term strength deck may be applied to any continuous deck that carries some of the longitudinal load. In destroyers and similar ships in which the main deck is the only continuous high deck, the main deck is the strength deck. The flight deck is the strength deck in recent large aircraft carriers (CVs) and helicopter support ships (LPH).

The main deck is supported by deck beams and deck longitudinals. Deck beams are the transverse members of the framing structure. The beams are attached to and supported by the frames at the sides, as shown in figure 21–10. In most naval construction, light deck beams are interspaced at regular intervals with deep deck beams. Deck

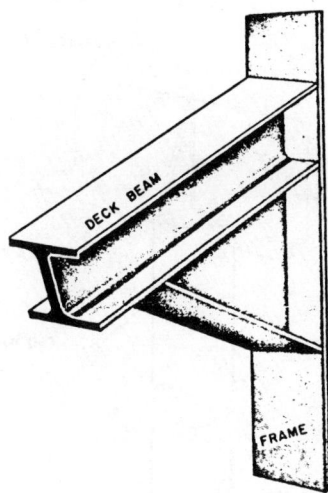

Figure 21–10. Example of how beams are attached to and supported by the frames at the sides.

longitudinals are used to provide longitudinal strength. When possible, the heaviest longitudinals are located at the center and near the outboard edges.

The outboard strake of deck plating which connects with the shell-plating is called the deck stringer. The deck stringer, which is heavier than the other deck strakes, serves as a continuous longitudinal stringer, providing longitudinal strength to the ship's structure.

A deck is named in two ways: by its position in the ship and by its use or function. Decks extending from side to side and from stem to stern are complete decks; decks occurring only in certain portions of the vessel are partial decks. The uppermost complete deck is the main deck. The complete decks below this are the second deck, third deck, and so on, normally numbered downward. Partial decks that do not extend continuously from bow to stern have special names such as:

1. 'Tween deck, which is below the main deck but above the lowest complete deck.

2. Upper deck, which is above the main deck from the bow to abaft amidships in merchant ships. It is referred to in naval ships as the 01 level. Succeeding levels above are named 02 level, 03 level, and so on.

3. Platform deck, which is below the lowest complete deck. Platforms are numbered downward as first platform, second platform, and so forth.

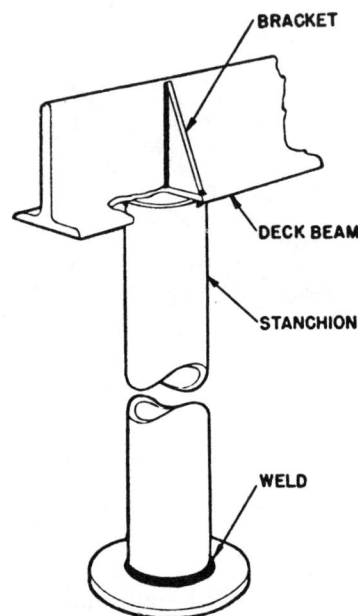

Figure 21–11. Stanchion.

Miscellaneous working platforms or flats consisting of gratings are located in the machinery spaces to aid in the operation of the ship's engines.

Upper Decks and Superstructure

The decks above the main decks are not strength decks in most ships other than CVs. The upper decks are usually interrupted at intervals by expansion joints. The expansion joints keep the upper decks from acting as strength decks (which they are not designed to be) and thus prevent cracking and buckling of deck houses and superstructure.

Stanchions

In order to reinforce the deck beams and to keep the deck beam brackets and side frames from carrying the total load, vertical stanchions or columns are fitted between decks. Stanchions are constructed in various ways of various materials. Some are made of pipe or rods; others are built up of various plates and shapes, welded or riveted together. The stanchion shown in figure 21–11 is in fairly common use; this pipe stanchion consists of a steel tube fitted with special pieces for securing it at the upper end (head) and at the lower end (heel).

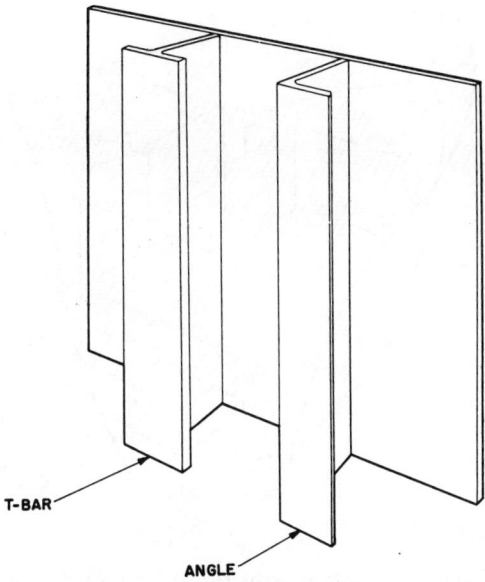

Figure 21–12. Two types of bulkhead stiffeners.

Bulkheads

Bulkheads are the vertical partitions that, extending athwartships and fore and aft, provide compartmentation to the interior of the ship. Bulkheads may be either structural or nonstructural. Structural bulkheads, which tie the shellplating, framing, and decks together, are capable of withstanding fluid pressure; these bulkheads usually provide watertight compartmentation. Nonstructural bulkheads are lighter; they are used chiefly for separating activities on board ship.

Bulkheads consist of plating and reinforcing beams. The reinforcing beams are known as bulkhead stiffeners. Two types of bulkhead stiffeners are shown in figure 21–12. Bulkhead stiffeners are usually placed in the vertical plane and aligned with deck longitudinals; the stiffeners are secured at top and bottom to any intermediate deck by brackets attached to deck plating. The size of the stiffeners depends upon their spacing, the height of the bulkhead, and the hydrostatic pressure that the bulkhead is designed to withstand.

Bulkheads and bulkhead stiffeners must be strong enough and have adequate stiffeners to resist excessive bending or buckling in case of flooding in the compartments that they bound. If too much deflection takes place, some of the seams might fail.

In order to form watertight boundaries, structural bulkheads must be joined to all decks, shellplating, bulkheads, and other structural

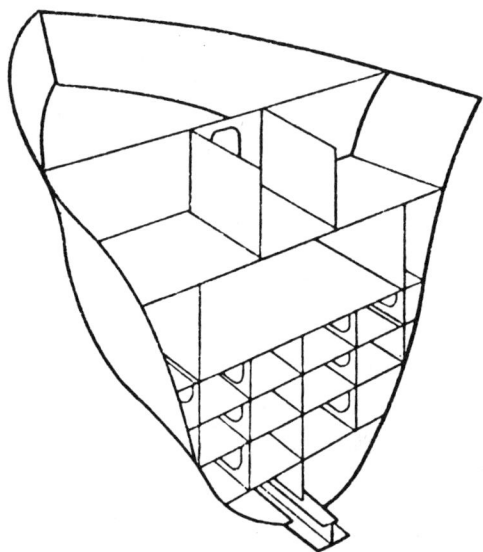

Figure 21–13. Compartmentation provided by transverse and longitudinal bulkheads.

members with which they come in contact. Main transverse bulkheads extend continuously through the watertight volume of the ship, from the keel to the main deck, and serve as flooding boundaries in the event of damage below the waterline.

In general, naval ships are divided into as many watertight compartments, both above and below the waterline, as are compatible with the missions and functions of the ships. The compartmentation provided by transverse and longitudinal bulkheads is illustrated in the bow section shown in figure 21–13.

Ship Compartmentation

Every space in a naval ship (except for minor spaces such as peacoat lockers, linen lockers, cleaning gear lockers, etc.) is considered a compartment and is assigned an identifying letter–number symbol. This symbol is marked on a label plate secured to the door, hatch, or bulkhead of the compartment.

For ships constructed after March 1949, the compartment numbers consists of a deck number, frame number, relation-to-centerline-of-ship number, and letter showing use of the compartment. These designations are separated by dashes.

The main deck is always numbered 1. The first deck or horizontal division below the main deck is numbered 2, the second below is numbered 3, and so on, consecutively, for subsequent lower division boundaries. When a compartment extends down to the bottom of the ship, the number assigned the bottom compartments is used. The first horizontal division above the main deck is numbered 01, the second above is numbered 02, and so forth, consecutively, for subsequent upper divisions. The deck number becomes the first part of the compartment number and indicates the vertical positions within the ship.

The frame number at the foremost bulkhead of the enclosing boundary of a compartment is its frame location number. Where these forward boundaries are between frames, the frame number forward is used. Fractional numbers are not used. The frame number becomes the second part of the compartment number.

Compartments located so that the centerline of the ship passes through them carry the number 0. Compartments located completely to starboard of the centerline are given odd numbers, and those completely to port of the centerline are given even numbers. When two or more compartments have the same deck and frame number and are entirely to port or entirely to starboard of the centerline, they have consecutively higher odd or even numbers, as the case may be, numbering from the centerline outboard. In this case, the first compartment outboard of the centerline to starboard is 1, the second is 3, and so on. Similarly, the first compartment outboard of the centerline to port is 2, the second 4, and so on. When the centerline of the ship passes through more than one compartment, the compartment having that portion of the forward bulkhead through which the centerline of the ship passes carries the number 01, 02, 03, and so on, in any sequence found desirable. These numbers indicate the relation to the centerline, and are the third part of the compartment number.

The fourth and last part of the compartment number is the capital letter that identifies the assigned primary usage of the compartment. A single capital letter is used, except that in dry and liquid cargo ships a double letter designation is used to identify compartments assigned to cargo carrying. The compartment letters for ships are shown in figure 21–14. An example of a compartment symbol in a ship is given in figure 21–15.

Ship Material Considerations

Corrosion

When metals are exposed to the combined action of air and moisture, corrosion ensues. In general, corrosion consists of oxidation of

```
A—Supply and storage
C—Control
E—Machinery
F—Fuel
L—Living quarters
M—Ammunition
T—Trunks and passages
V—Voids
W—Water
```

Figure 21-14. *Example of compartment letters.*

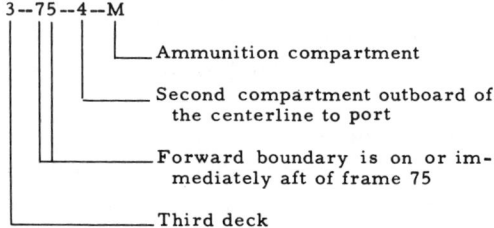

Figure 21-15. *Example of a compartment symbol.*

the parent metal, although complex carbonates and hydrates may also be formed. Some metals are more susceptible to corrosion than others. Certain metals and alloys resist corrosion as, for example, stainless steel and Monel metal. Some corrosion-resistant metals such as tin, lead, and zinc have insufficient physical strength to permit their use as structural members. However, they are suitable in many cases for application as a protective coating; for example, tin is often used as an internal lining for copper piping, and zinc as a surface coating for light steel plating (this process is called galvanizing).

Many metals resist corrosion because the surface layer of oxide initially formed protects the undelying parent metal from further oxidation. This is not true, however, of most ferrous alloys, such as steel. In their case, the superficial corroded layer, called rust, actually accelerates corrosion of the metal underneath.

Rusting is also aggravated by other factors, among the most important of which are heat and acid moisture. The presence of salt from seawater furnishes the necessary acid medium. It is apparent, therefore, that conditions on board ship are highly favorable to rapid rusting of steel structures. Besides the loss of structural strength following corrosion of strength members, the danger to watertight integrity is serious. Even if the metal is not completely rusted through, weak spots will appear, which may fail when the boundary concerned is subjected to a head of water as a result of damage or flooding. Therefore, the problem of preventing rust is of vital importance. The success of the widespread use of steel in warship construction is dependent upon provision of satisfactory protective rust-inhibiting coatings accompanied by the use of corrosion-resistant material in construction.

Welds

Ship construction methods of today utilize two systems of making structural connections, riveting and electric arc welding. Each of these types of connection has its own frailties, or susceptibility to loosening.

Connections effected by use of continuous welding are inherently tight, inasmuch as the weld itself forms a continuous, impermeable connection between the two members joined.

In a typical watertight compartment, the joints and connections susceptible to loosening may be considered in three categories;

1. Bulkhead and deck boundaries
2. Seams in bulkhead and deck plating
3. Doublers around openings, such as doors, hatches, and manholes

With regard to through fittings, the tendency in modern construction is to use the welded type, in view of its simplicity and inherent tightness.

From the maintenance point of view, welded connections, although initially tight, can become loose. The most common causes for such failure are working of the ship by violent high-speed maneuvering (in light ships) and machine vibration, which tend to cause relative motion between adjacent structural members in joints that may result in welds cracking.

If welds crack, the faulty weld should be "V'd" out well beyond the limits of the crack and rewelded with due attention to welding technique in order to avoid inducing further failure in adjacent connections; the ends of the crack should be drilled when possible. Welds are sometimes porous, owing to improper welding techniques. Such defects show up under test and should be corrected similarly.

Welds are also subject to cracking due to the combined action of locked-up stresses, present as a result of improper welding during construction, and stresses induced by the ship's motion in a seaway. Inattention to a crack in welding usually results in growth of the crack. Failure to inspect with sufficient frequency or care may well result in minor defects growing into major ones before they are discovered. Defects in the boundaries of liquid-filled tanks manifest themselves; elsewhere, defects must be the object of search.

From the viewpoint of maintenance, every ship must be as nearly self-supporting as possible in view of the infrequency of Navy-yard overhauls and upkeep periods in time of war. The necessity for shipboard personnel trained in skilled welding is obvious.

Accesses

The closures and fittings that pierce watertight bulkheads and decks are each a source of possible leakage and therefore deserve careful consideration with regard to points of weakness.

Examination shows the possible avenues for leakage in both doors and hatches include cracked welds in way of frames and deteriorated or faulty gaskets and knife edges.

Rubber gaskets are installed to provide a tight all-around fit. Exposure to heat, sun, oils and greases, or paint coatings will cause rapid deterioration of rubber gaskets. Even under the best conditions they will lose their elasticity and "life" with age. Therefore, extreme care must be taken to keep oils, greases, and paints off rubber gaskets and to replace hardened or cracked gaskets without delay. If gaskets or knife edges are found to have paint on them, it should not be removed with abrasives (which will destroy the "fit"); they should be cleaned with a block of soft wood.

The other half of the joint, the knife edge, also requires upkeep. Knife edges may be distorted by impact of heavy objects, as when ammunition is carelessly handled through a door or hatch, or may become uneven because of corrosion. They must be kept clean and bright to assure freedom from rust. The crew should be indoctrinated in the necessity of handling heavy weight in such a manner that knife edges will not be struck or bent. A badly damaged knife edge usually requires Navy-yard work for satisfactory repair; hence, the best remedy is to prevent the occurrence of the damage.

22

Stability and Buoyancy

This chapter deals with the principles of stability, stability curves, and the inclining experiment.

Principles of Stability

A floating body is acted upon by forces of gravity and forces of buoyancy. The algebraic sum of these forces must equal zero if equilibrium is to exist.

Any object exists in one of three states of stability: stable, neutral, or unstable. We may illustrate these three states by placing three cones on a table top, as shown in figure 22–1. When cone A is tipped so that its base is off the horizontal plane, it tends, up to a certain angle of inclination, to reassume its original position. Cone A is thus an example of a stable body—that is, one that tries to attain its original position through a specified range of angles of inclination.

Cone B is an example of neutral stability. When rotated, this cone may come to rest at any point, reaching equilibrium at some angle of inclination.

Cone C, balanced upon its apex, is an example of an unstable body. Following any slight inclination by an external force, the body will come to rest in a new position where it will be more stable.

From Archimedes' law, we know that an object floating on or submerged in a fluid is buoyed up by a force equal to the weight of the fluid it displaces. The weight (displacement) of a ship depends upon the weight of all parts, equipment, stores, and personnel. This total weight represents the effect of gravitational force. When a ship is

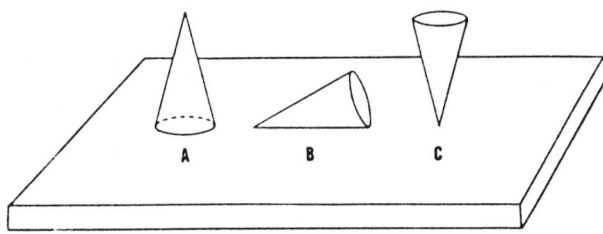

Figure 22–1. Three states of stability.

floated, she sinks into the water until the weight of the fluid displaced by her underwater volume is equal to the weight of the ship. At this point, the ship is in equilibrium—that is, the forces of gravity (G) and the forces of buoyancy (B) are equal, and the algebraic sum of all forces acting upon the ship is equal to zero. This condition is shown in part A of figure 22–2. If the underwater volume of the ship is not sufficient to displace an amount of fluid equal to the weight of the ship, the ship will sink (part B of figure 22–2) because the forces of gravity are greater than the forces of buoyancy.

The depth to which a ship will sink when floated in water depends upon the density of the water, since the density affects the weight per unit volume of a fluid. Thus we may expect a ship to have a deeper draft in fresh water than in salt water, since fresh water is less dense (and therefore less buoyant) than salt water.

Although gravitational forces act everywhere upon the ship, it is not necessary to attempt to consider these forces separately. Instead, we may regard the total force of gravity as a single resultant or composite force that acts vertically downward through the ship's center of gravity (G). Similarly, the force of buoyancy may be regarded as a single resultant force that acts vertically upward through the center of buoyancy (B) located at the geometric center of the ship's underwater body. When a ship is at rest in calm water, the center of gravity and the center of buoyancy lie on the same vertical line.

Displacement

Since weight (W) is equal to the displacement, it is possible to measure the volume of the underwater body (V) in cubic feet and multiply this volume by the weight of a cubic foot of seawater, in order to find what the ship weighs. This relationship may be written as:

$$W = V \times \frac{1}{35}$$

or:

$$V = 35W$$

where V is the volume of displaced seawater, in cubic feet; W is its weight, in tons; and 35 is the cubic feet of seawater per ton (when dealing with ships, it is customary to use the long ton of 2240 pounds).

It is also obvious, then, that displacement will vary with draft. As the draft increases, the displacement increases. This is indicated in figure 22–3 by a series of displacements shown for successive draft lines on the midship section of a cruiser.

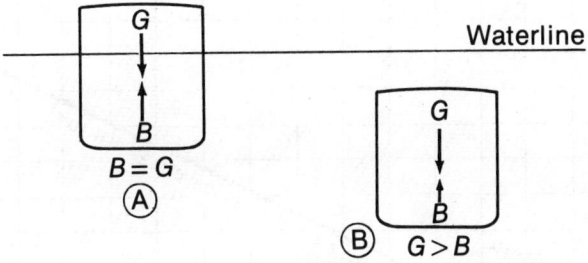

Figure 22–2. Interaction of force of gravity and force of buoyancy.

Figure 22–3. Displacement data.

The volume of an underwater body for a given draft line can be measured in the drafting room by using graphical or mathematical means. This is done for series of drafts throughout the probable range of displacements in which a ship is likely to operate. The values obtained are plotted on a grid on which feet of draft are measured vertically and tons of displacement horizontally. A smooth line is faired through the points plotted, providing a curve of displacement vs. draft, or a displacement curve as it is generally called. The result for a cruiser is shown in figure 22–4.

To use the curve shown in figure 22–4 for finding the displacement when the draft is given, locate the value of the mean draft on the draft scale at left and proceed horizontally across the diagram to the curve. Then drop vertically downward, and read the displacement from the scale. For example, if the mean draft is 24 feet, the displacement found from the curve is approximately 14,700 tons.

KB vs. Draft

As the draft increases, the center of buoyancy (*B*) rises with respect to the keel (*K*). Figure 22–5 shows how different drafts result in different values of *KB*, the height of the center of buoyancy from the keel (*K*). A series of values for *KB* is obtained, and these values are

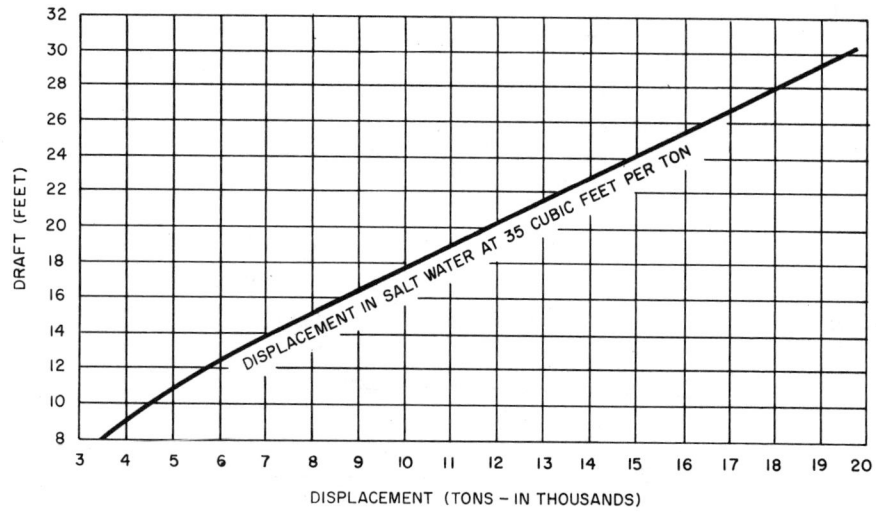

Figure 22–4. Displacement curve of a cruiser.

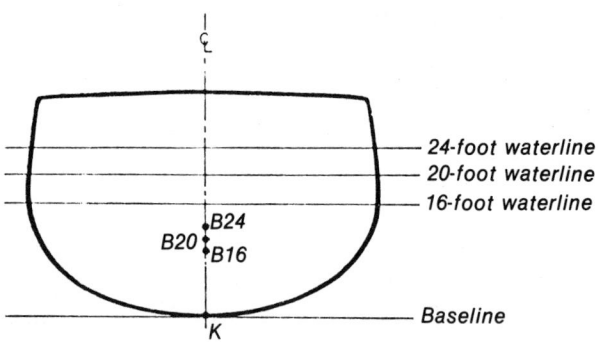

Figure 22–5. Successive centers of buoyancy (B) for different drafts.

plotted on a curve to show *KB* vs. draft. Figure 22–6 illustrates a typical *KB* curve.

To read *KB* when the draft is known, start at the proper value of draft on the scale at the left and proceed horizontally to the curve. Then drop vertically downward to the baseline (*KB*).

Thus, if a ship were floating at a mean draft of 19 feet, the *KB* found from the chart would be approximately 10.5 feet.

Reserve Buoyancy

The volume of the watertight portion of the ship above the waterline is known as the ship's reserve buoyancy. Freeboard, a rough measure of the reserve buoyancy, is the distance in feet from the waterline

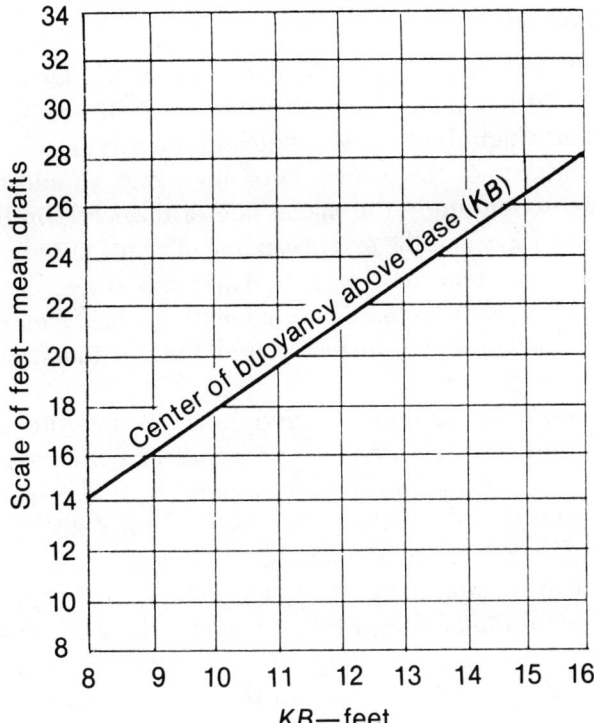

Figure 22–6. KB curve.

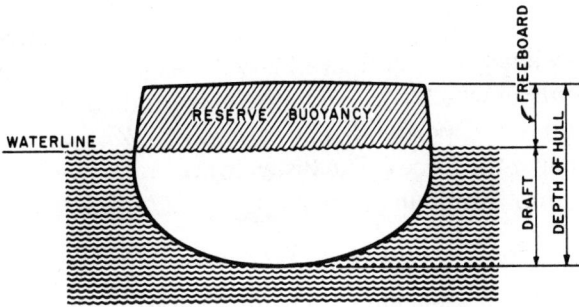

Figure 22–7. Reserve buoyancy, freeboard, draft, and depth of hull.

to the main deck along the ship's length, with some classification rules specifying a minimum freeboard at the midship section. As indicated in figure 22–7, freeboard plus draft is equal to the depth of the hull in feet.

When weight is added to a ship, draft and displacement increase in the same amount that freeboard and reserve buoyancy decrease. Reserve buoyancy is an important factor in a ship's ability to survive flooding due to damage. It also contributes to the seaworthiness of the ship in very rough weather.

Inclining Moments

The moment of a force is the tendency of the force to produce rotation or to move an object about an axis. The distance between the point at which the force is acting and the axis of rotation is called the moment arm or the lever arm of moment.* To find the value of a moment, we multiply the magnitude of the force by the distance between the force and the axis of rotation. The magnitude of the force is expressed in some unit of weight (pounds, tons, etc.), and the distance is expressed in some unit of length (inches, feet, etc.); hence the unit of the moment is the foot-pound, the foot-ton, or some similar unit.

When two forces of equal magnitude act in opposite and parallel directions and are separated by a perpendicular distance, they form a *couple*. The *moment of a couple* is found by multiplying the magnitude of one of the forces by the perpendicular distance between the lines of action of the two forces.

When a disturbing force exerts an inclining moment on a ship, causing the ship to heel over to some angle, there is a change in the shape of the ship's underwater body and a consequent relocation of the center of buoyancy. Because of this shift in the location of B, B and G no longer act in the same vertical line. Instead of acting as separate equal and opposite forces, B and G now form a couple.

The newly formed couple produces either a *righting moment* or an *upsetting moment*, depending upon the relative locations of B and G. The ship illustrated in figure 22–8 develops a righting moment, the magnitude of which is equal to the magnitude of one of the forces (B or G) times the perpendicular distance (GZ) that separates the lines of action of the forces. The distances GZ is known as the *righting arm* of the ship. Mathematically,

$$RM = W \times GZ$$

where RM is the righting moment (in foot-tons), W is the displacement (in tons), and GZ is the righting arm (in feet).

For example, a ship that displaces 10,000 tons and has a 2-foot righting arm at a certain angle of inclination has a righting moment of 10,000 tons times 2 feet, or 20,000 foot-tons. This 20,000 foot-tons represents the moment, which in this instance tends to return the ship to an upright position.

*The significance of the distance between the force and the axis of rotation may be seen if we consider a simple seesaw. If two persons of equal weight sit on opposite ends, equally distant from the center support, the seesaw balances. But if one person moves closer to or farther away from the center, the person farther away from the support moves downward because the *effect* of his weight is greater.

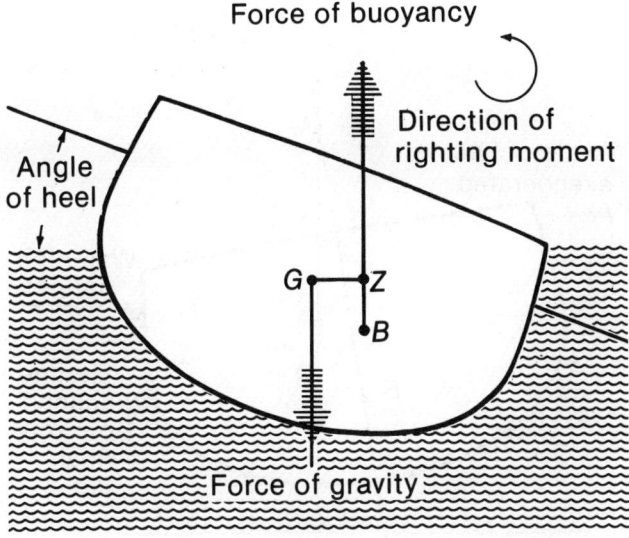

Figure 22–8. *Development of a righting moment when a stable ship inclines.*

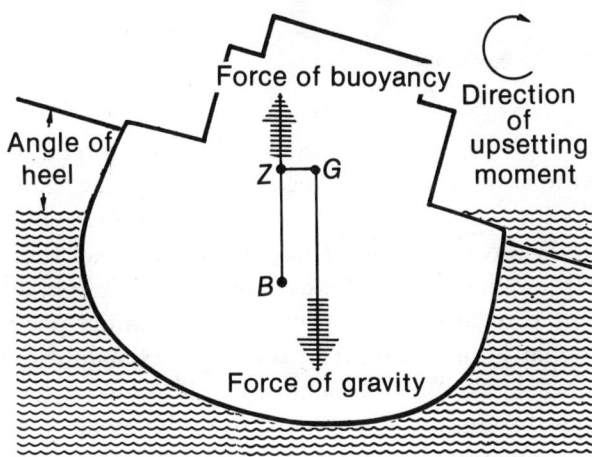

Figure 22–9. *Development of an upsetting moment when an unstable ship inclines.*

Figure 22–9 shows the development of an upsetting moment resulting from the inclination of an unstable ship. In this case, it is apparent that the high location of G and the new location of B contribute to the development of an upsetting moment rather than a righting moment.

The Metacenter (M)

A ship's metacenter is the intersection of two successive lines of action of the force of buoyancy as the ship heels through a very small

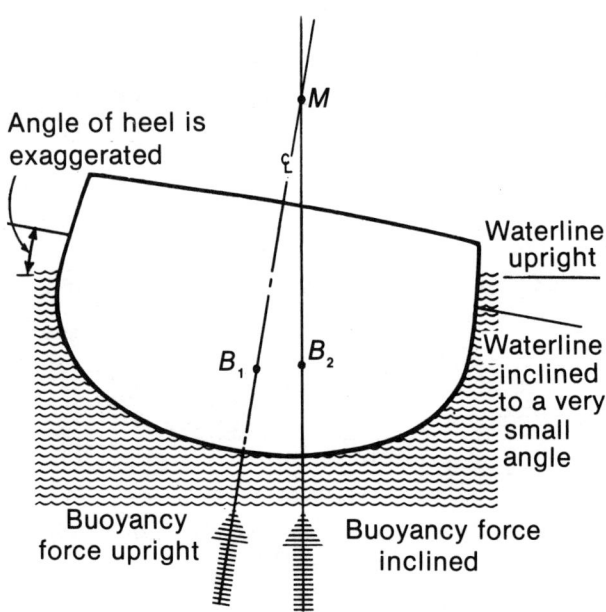

Figure 22–10. The metacenter.

angle. Figure 22–10 shows two lines of buoyant force. One of these represents the ship on an even keel; the other is for a small angle of heel. The point where they intersect is the initial position of the metacenter. When the angle of heel is greater than the angle used to compute the metacenter, M moves off the centerline, and the path of movement is a curve. However, it is the initial position of the metacenter that is most useful in the study of stability. In the discussion that follows, the initial position is referred to as M. The distance from the center of buoyancy (B) to the metacenter (M) when the ship is on even keel is the metacentric radius.

Metacentric Height (GM)

The distance from the center of gravity (G) to the metacenter is known as the ship's metacentric height (GM). Figure 22–11 shows a ship heeled through a small angle (the angle is exaggerated in the drawing), establishing a metacenter at M. The ship's righting arm GZ is one side of the triangle GZM. In this triangle GZM, the angle of heel is at M. The side GM is perpendicular to the waterline at even keel, and ZM is perpendicular to the waterline when the ship is inclined.

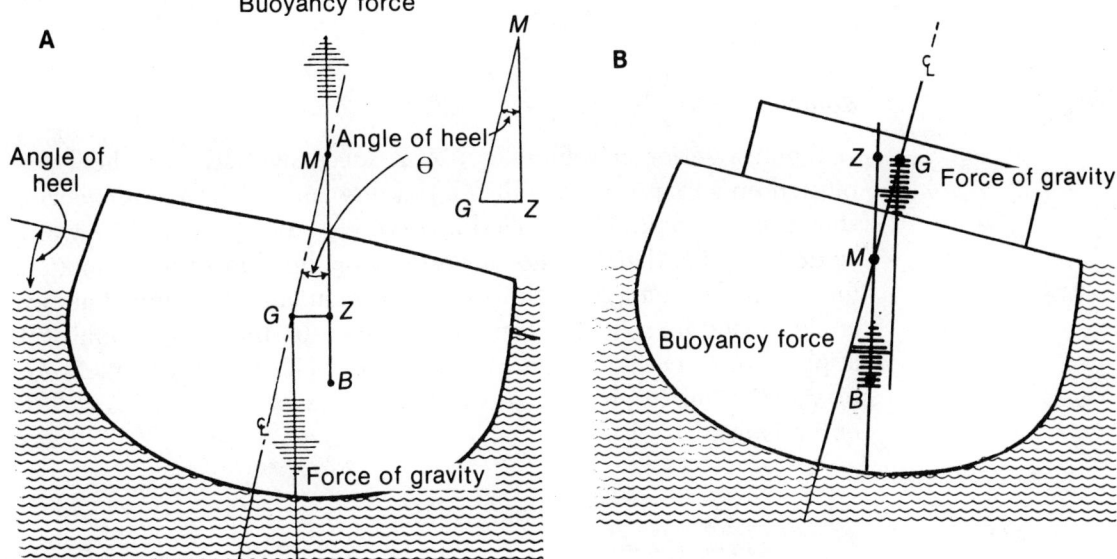

Figure 22–11. *Two conditions of a ship: (A) stable condition, G below M; (B) unstable condition, G above M.*

For any angle of heel up to 7°, there will be a definite relationship between *GM* and *GZ* because $GZ = GM \sin \theta$. Thus, *GM* acts as a measure of *GZ*, the righting arm.

GM is also an indication of whether the ship is stable or unstable at small angles of inclination. If *M* is above *G*, the metacentric height is positive, the moments that develop when the ship is inclined are righting moments, and the ship is stable (part A of figure 22–11). But if *M* is below *G*, the metacentric height is negative, the moments that develop are upsetting moments, and the ship is unstable (part B of figure 22–11).

Influence of Metacentric Height

When the metacentric height of a ship is large, the righting arms that develop at small angles of heel are also large. Such a ship resists roll and is said to be stiff. When the metacentric height is small, the righting arms are also small. Such a ship rolls slowly and is said to be "tender." Representative *GM* values for most naval ships of the following classes are: CGs, 4 to 6 feet; DDs, 3 to 4 feet; FFs, 3 to 5 feet; and KAs, 1 to 6 feet.

Large *GM* values and large righting arms are desirable for resistance to the flooding effects of damage. However, a smaller *GM* value is sometimes desirable for the slow, easy roll that makes for more accurate gunfire. Thus, the GM value for a naval ship is the result of compromise.

Stability Curves

When a series of values for GZ at successive angles of heel are plotted on a graph, the result is a stability curve. The stability curve shown in figure 22–12 is called a curve of static stability. The word static indicates that it is not necessary for the ship to be in motion for the curve to apply; if the ship were momentarily stopped at any angle during its roll, the value of GZ given by the curve would still apply. (Note: Design engineers usually use GM values as a measure of stability up to about 7° heel. For angles beyond 7°, a stability curve is used.)

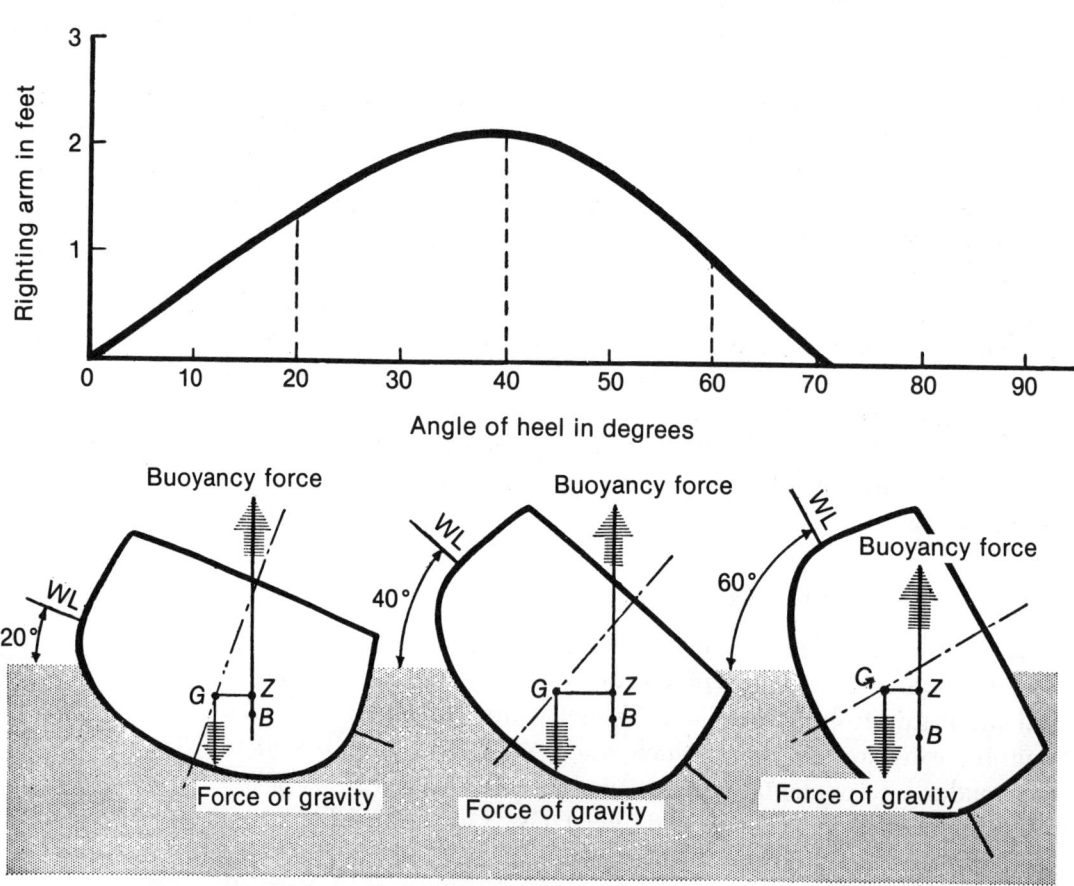

Figure 22–12. *Righting arms of a ship inclined at successively larger angles of heel.*

To understand the stability curve, it is necessary to consider the following: (1) The ship's center of gravity does not change position as the angle of heel is changed. (2) The ship's center of buoyancy is always at the center of the ship's underwater hull. (3) The shape of the ship's underwater hull changes as the angle of heel changes. Putting these facts together, we see that the position of G remains constant as the ship heels through various angles, but the position of B changes according to the angle of inclination. The righting arm increases with increasing angle of heel at a nearly constant rate up to about 30° for destroyer type ships. As the deck edge submerges, the GZ curve levels off and begins to decrease, eventually reaching a value of zero at large angles of heel. Certain hull types (generally those with vertical sides over large portions of the ship) have righting arm curves that slope upward to the point of deck edge submergence, and thus get "stiffer" at moderate angles of heel.

Effect of Draft on Righting Arm

A change in displacement will result in a change of draft and freeboard; and B will shift to the geometric center of the new underwater body. At any angle of inclination, a change in draft causes B to shift both horizontally and vertically with respect to the waterline. The horizontal shift in B changes the distance between B and G, and thereby changes the length of the righting arm, GZ. Thus, when draft is increased, the righting arms are reduced through the entire range of stability. Figure 22–13 shows how the righting arm is reduced when the draft is increased from 18 feet to 26 feet, when the ship is inclined at an angle of 20°.

A reduction in the size of the righting arm usually means a decrease in stability. When the reduction in GZ is caused by increased displacement, however, the total effect on stability is more difficult to

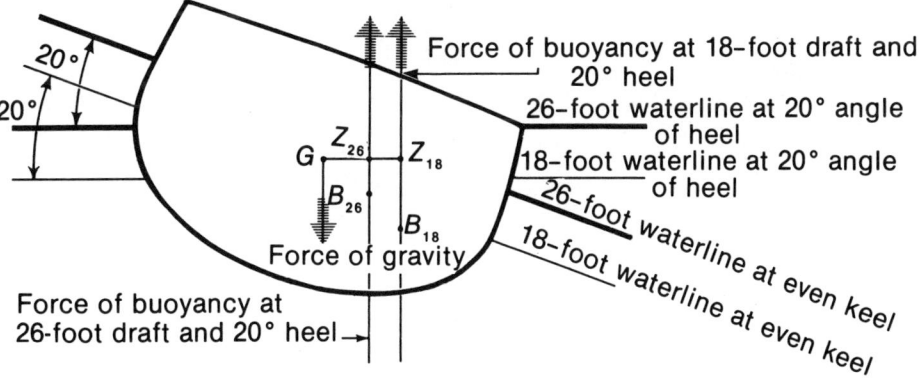

Figure 22–13. Effect of draft on the righting arm.

evaluate. Since the righting moment is equal to W times GZ, the righting moment will be increased by the gain in W at the same time that it is decreased by the reduction in GZ. The gain in the righting moment, caused by the gain in W, does not necessarily compensate for the reduction in GZ.

In brief, there are several ways in which an increase in displacement affects the stability of a ship. Although these effects occur at the same time, it is best to consider them separately. The effects of increased displacement are:

1. Righting arms (GZ) are decreased as a result of increased draft.

2. Righting moments may be increased as a result of the increased displacement (W), if $GZ \times W$ is increased.

23

Cross Curves of Stability

To facilitate stability calculations, the design activity inclines a line drawing of the ship at a given angle, and then lays off on it a series of waterlines. These waterlines are chosen at evenly spaced drafts throughout the probable range of displacements. For each waterline the value of the righting arm is calculated, using an assumed center of gravity rather than the true center of gravity. A series of such calculations is made for various angles of heel—usually 10°, 20°, 30°, 40°, 50°, 60°, 70°, 80°, and 90°—and the results are plotted on a grid to form a series of curves known as the cross curves of stability (figure 23–1). Note that, as draft and displacement increase, the curves all slope downward, indicating increasingly smaller righting arms.

The cross curves are used in the preparation of stability curves. To take a stability curve from the cross curves, a vertical line (such as line *MN* in figure 23–1) is drawn on the cross curve sheet at the displacement that corresponds to the mean draft of the ship. At the intersection of this vertical line with each cross curve, the corresponding value of the righting arm on the vertical scale at the left can be read. Then this value of the righting arm at the corresponding angle of heel is plotted on the grid for the stability curve. When a series of such values of the righting arms from 10° through 90° of heel have been plotted, a smooth line is drawn through them and the uncorrected stability curve for the ship at that particular displacement is obtained. The curve is not corrected for the actual height of the ship's center of gravity, since the cross curves are based on an assumed height of *G*. However, the stability curve does embody the effect of displacement on the righting arm curve for a given position of the center of gravity.

Effects of Weight Shifts

If one weight in a system of weights is moved, the center of gravity of the whole system moves along a path parallel to the path of the component weight. The distance that the center of gravity of the

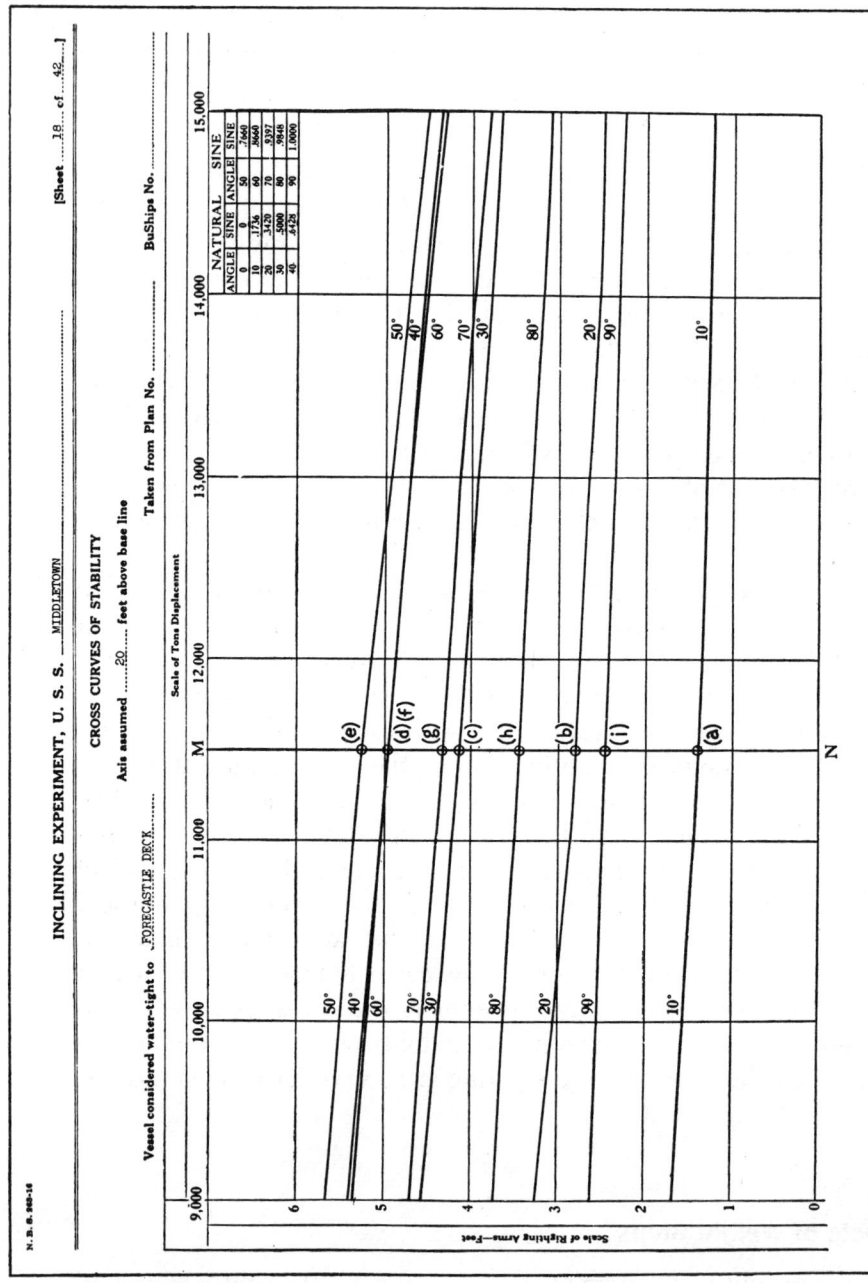

Figure 23-1. Cross curves of stability.

system moves may be calculated from the formula:

$$GG_1 = \frac{(w)(s)}{W}$$

where w is the component weight, in tons; s is the distance the component weight is moved, in feet; W is the weight of the entire system, in tons; and GG_1 is the shift in the center of gravity of the system, in feet.

Weight movements in a ship can take place in three possible directions—athwartships, fore and aft, and vertically (perpendicular to the decks). The most general type of movement is inclined with respect to all three of these. Such a diagonal movement can be divided into components in each of the three directions, and one component can be studied at a time without reference to the others. For example, if a weight is moved from the main deck, starboard side, aft, to a storeroom on the fourth deck, port side, forward, this movement may be regarded as taking place in three steps, as follows:

1. From main deck to fourth deck (down)
2. From starboard side to port side (across)
3. From stern to bow (forward)

Vertical Weight Shift

If a weight is moved straight up a vertical distance in a ship, the ship's center of gravity will move straight up on the centerline (figure 23–2). The vertical rise in G (explained later in the chapter) can be computed from the formula mentioned previously.

Example. A ship is operating with a displacement of 11,500 tons. Her ammunition, totaling 670 tons, is to be moved from the magazines to the main deck, a distance of 36 feet. Find the rise in G.

$$GG_1 = \frac{670 \times 36}{11,500} = 2.1 \text{ feet}$$

Since moving a weight that is already on board will cause no change in displacement, there can be no change in M, the metacenter. If M remains fixed, then the upward movement of the center of gravity results in a loss of metacentric height:

$$G_1M = GM - GG_1$$

where G_1M is the new metacentric height (after weight movement), in feet; GM is the old metacentric height (before weight movement), in feet; and GG_1 is the rise in center of gravity, in feet.

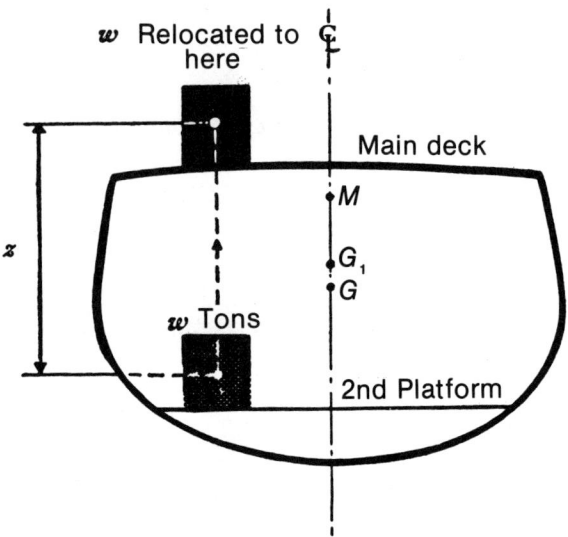

Figure 23–2. Shift in G due to a vertical weight shift.

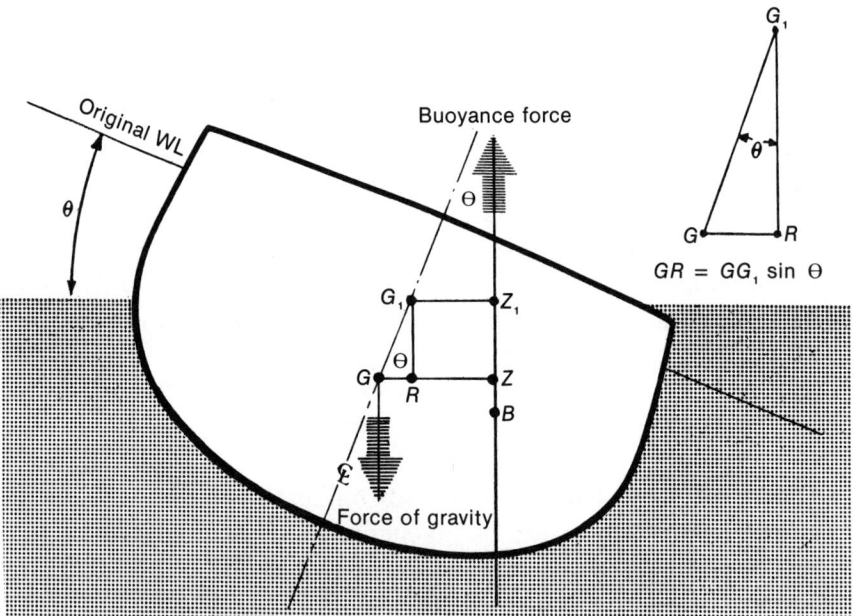

Figure 23–3. Loss of righting arm due to a rise in the center of gravity.

If the ammunition on the main deck is moved down to the sixth deck, the positions of G and G_1 will be reversed. The shift in G can be found from the same formula as before, the only difference being that GG_1 now becomes a gain in metacentric height instead of a loss.

If a weight is moved vertically downward, the ship's center of gravity, G, will move straight down on the centerline, and the correction is additive. In this case the sine curve is plotted below the abscissa. The final stability curve is that portion of the curve above the sine correction curve.

A vertical shift in the ship's center of gravity changes every righting arm throughout the entire range of stability. If the ship is at any angle of heel, such as in figure 23–3, the righting arm is GZ with the center of gravity at G. But if the center of gravity shifts to G_1 as a result of a vertical weight shift upward, the righting arm becomes G_1Z_1, which is smaller than GZ by the amount of GR. In the right triangle GRG_1, the angle of heel is at G_1; hence the loss of the righting arm may be found from

$$GR = GG_1 \times \sin \theta$$

This equation may be stated in words as: The loss of righting arm equals the rise in the center of gravity times the sine of the angle of heel. The sine of the angle of heel is a ratio that can be found by consulting a table of sines.

If the loss of GZ is found for 10°, 20°, 30°, and so forth, by multiplying GG_1 by the sine of the proper angle, a curve of loss of righting arms can be obtained by plotting values of $GG_1 \times \sin \theta$ vertically against angles of heel horizontally, which results in a sine curve. When plotted, the curve is as illustrated in figure 23–4.

The sine curve may be superimposed on the original stability curve to show the effect on stability characteristics of moving the weight up in a ship. Inasmuch as displacement is unchanged, the righting arms of the old curve need be corrected for the change of G only, and no other variation occurs. Consequently, if $GG_1 \times \sin \theta$ is de-

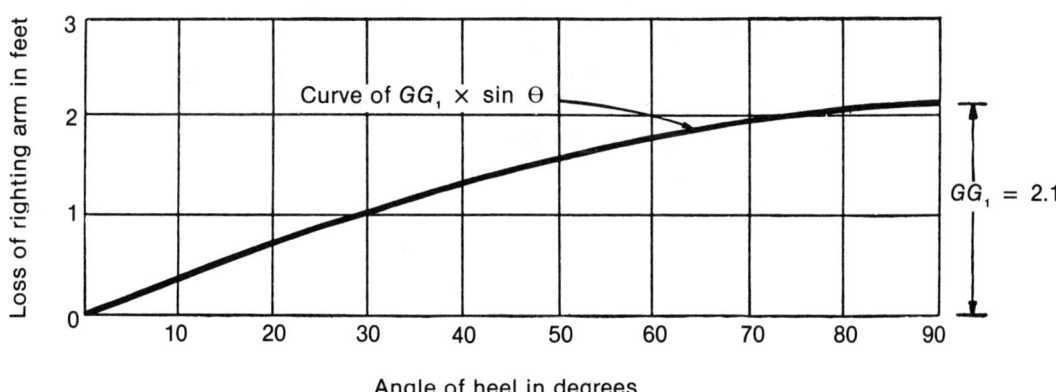

Figure 23–4. *Sine curve showing the loss of righting arm at various angles of heel.*

ducted from each *GZ* on the old stability curve, the result will be a correct righting arm curve for the ship after the weight movement.

In figure 23–5 a sine curve has been superimposed on an original stability curve. The dotted area is that portion of the curve that was lost due to moving the weight up, whereas the lined area is the remaining or residual portion of the curve. The residual maximum righting arm is *AB* and occurs at an angle of about 37°. The new range of stability is from 0° to 53°.

The reduced stability of the new curve becomes more evident if the intercepted distances between the old *GZ* curve and the sine curve are transferred down to the base, thus forming a new curve of static stability (figure 23–6). Whereas the old righting arm at 30° was *AB*, the new one has a value of *CB*, which is plotted up from the base to locate point *D* (*CB* = *AD*); thus a point is established at 30° on the new curve. A series of points thus obtained by transferring intercepted distances down to the base line delineates the new curve, which may be analyzed as follows:

1. *GM* is now the quantity represented by *EF*.
2. The maximum righting arm is now the quantity represented by *HI*.
3. The angle at which the maximum righting arm occurs is 37°.
4. The range of stability is from 0° to 53°.

Total dynamic stability is represented by the shaded area.

Horizontal Weight Shift

When the ship is upright, *G* lies in the fore-and-aft centerline, and all weights on board are balanced. Moving any weight horizontally will result in a shift in *G* in an athwartship direction, parallel to the weight movement. *B* and *G* are no longer in the same vertical line, and an upsetting moment exists at 0° inclination, which will cause the ship to heel until *B* moves under the new position of *G*. In calm water the ship will remain at this angle, and in a seaway it will roll about this angle of permanent list. This shift in *G* can be computed from the formula:

$$G_1G_2 = \frac{(w)(d)}{W}$$

where G_1G_2 is the athwartship shift of *G*, in feet; *w* is the weight moved over, in tons; *d* is the distance *w* moved, in feet; and *W* is the displacement of the ship, in tons.

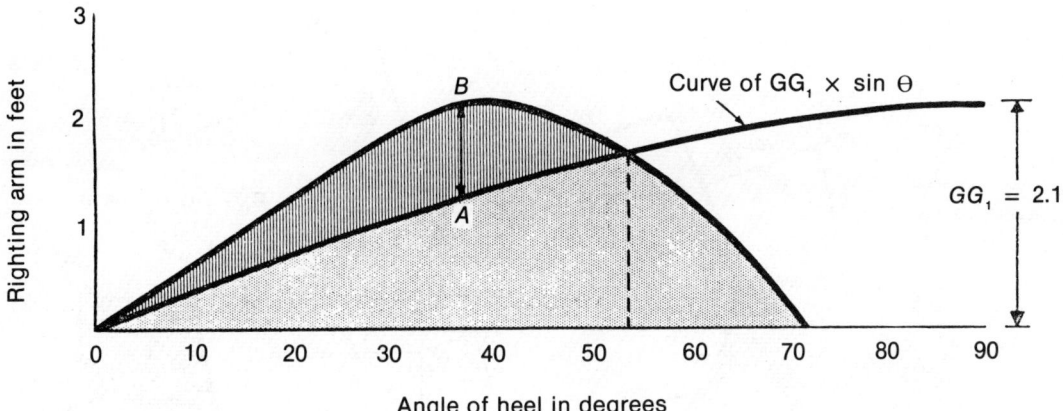

Figure 23–5. Sine curve superimposed on the original stability curve.

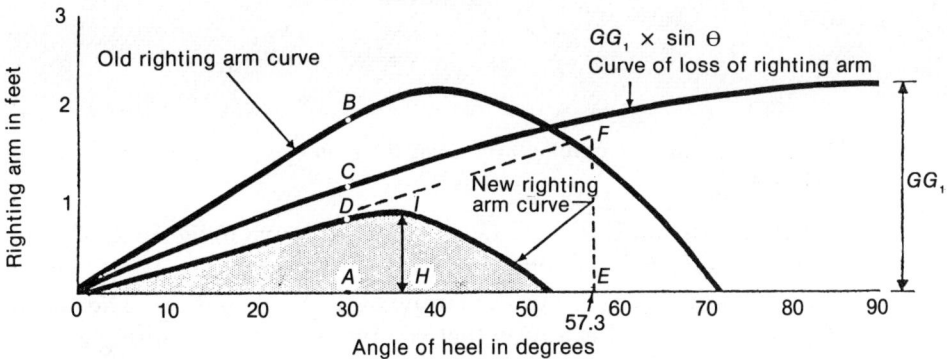

Figure 23–6. Curve of static stability as corrected for a loss of stability due to a vertical weight shift.

Going back to our original problem, let us further assume that ship's stores totaling 185 tons are shifted from port storerooms to starboard storerooms, a horizontal distance of 56 feet. Using the above formula:

$$G_1G_2 = \frac{185 \times 56}{11,500} = 0.90 \text{ foot}$$

In figure 23–7 the righting arm has been reduced from G_1Z_1 to G_2Z_2 by this weight shift. G_2Z_2 is smaller than G_1Z_1. However, the distance G_1T is equal to $G_1G_2 \times \cos \theta$. Thus, the loss of righting arm involved in an athwartship movement of G is equal at any angle of heel to $G_1G_2 \times \cos \theta$. This variable distance ($G_1G_2 \times \cos \theta$) is called the ship's inclining arm; when this value is multiplied by the displacement, W, the product is the ship's inclining moment.

Cross Curves of Stability   **349**

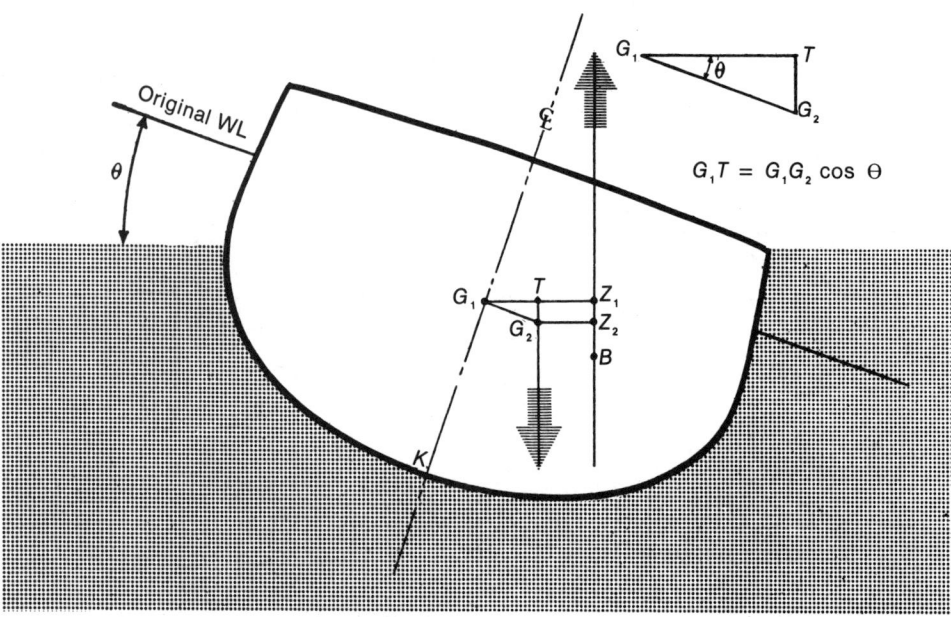

Figure 23–7. Loss of righting arm when the center of gravity is moved off the centerline.

The expression $G_1T = G_1G_2 \times \cos \theta$, as shown in figure 23–7, may be stated as: The inclining arm is equal to the athwartship shift in the center of gravity times the cosine of the angle of heel. The cosine of the angle of heel is a ratio that can be found by consulting a table of cosines. If the inclining arm is computed for 10°, 20°, 30°, and so on by multiplying G_1G_2 by the cosine of the proper angle, a curve of inclining arms can be obtained by plotting values of $G_1G_2 \times \cos \theta$ vertically against angles of heel horizontally, which results in a cosine curve. Note that the cosine curve (figure 23–8) is just the opposite of the sine curve (figure 23–5), but is otherwise identical in shape.

Just as the sine curve was superimposed on the *GZ* curve, so may the cosine curve be superimposed on the stability curve to show the effect on stability of moving a weight athwartship. The cosine curve has been placed on the original stability curve, corrected for the actual height of the center of gravity. The dotted area (figure 23–8) is the portion of the curve lost because of the weight shift, and the lined area is the remaining or residual portion of the curve. The residual maximum righting arm is *AB*, which develops at an angle of about 37°. The new range of stability is from 20° to 50°.

The new curve of static stability can be plotted on the base by transferring down the intercepted distances between the cosine curve

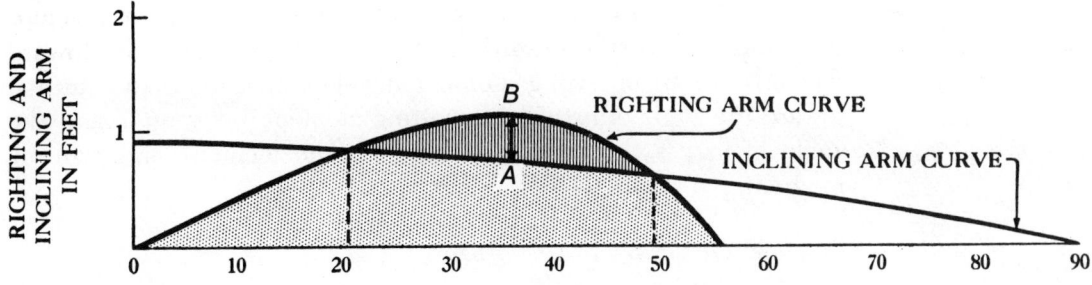

Figure 23–8. Cosine curve superimposed on the original stability curve.

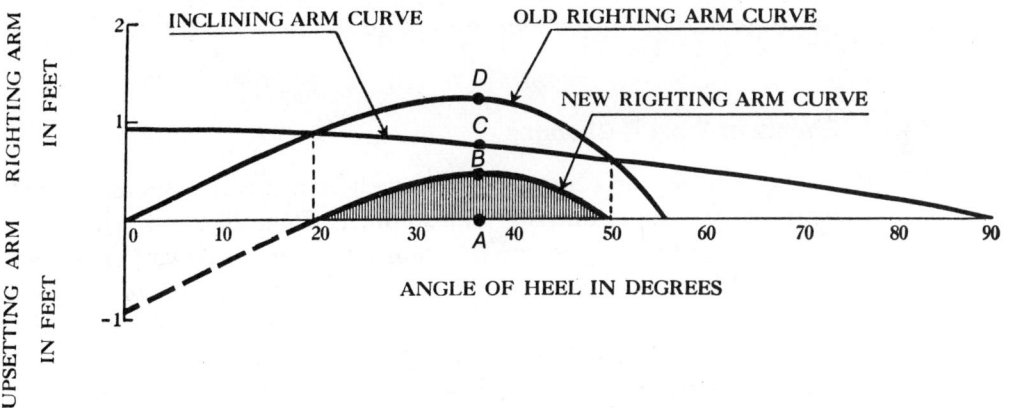

Figure 23–9. New curve of static stability after correction for a horizontal weight shift.

and the old GZ curve. For example, in figure 23–9, the old righting arm at 37° was AD; the loss of righting arm (inclining arm) at this angle is AC, leaving residual GZ of CD. This value has been plotted up from the base of AB to provide one point on the final curve. The residual stability may be analyzed on the new curve as follows:

1. The maximum righting arm is AB.
2. The angle of the maximum righting arm is at A.
3. The range of stability is 20° to 50°.
4. The total dynamic stability is represented by the lined area.

The ship will have a permanent list at 20°, which is the angle where B is under G; the inclining arm equals the original righting arm; the cosine curve crosses the original GZ curve; and the residual

Cross Curves of Stability 351

righting arm is zero. In a seaway the ship will roll about this angle of list. If it rolls farther to the listed side, a righting moment develops that tends to return it toward the angle of list. If it rolls back toward the upright, an upsetting moment develops that tends to return it toward the angle of list. The upsetting moment (between 0° and the angle of list) is the difference between the inclining and righting moments.

Diagonal Weight Shift

A weight may be shifted diagonally (a) so that it moves up or down and athwartship at the same time, or (b) by moving one weight up or down and another athwartship. A diagonal shift should be treated in two steps, first by finding the effect on GM and stability of the vertical shift, and second, by finding the effect of the horizontal movement. The corrections are applied as previously described.

Effects of Weight Changes

The addition or removal of any weight in a ship may affect list, trim, draft, displacement, and stability. Regardless of where the weight is added (or removed), when determining its various effects it should be considered first to be placed in the center of the ship, then moved up (or down) to its final height, next moved outboard to its final off-center location, and finally shifted to its fore or aft position.

Assume that a weight is added to a ship so that the list or trim is not changed, and G will not shift. The first thing to do is find the new displacement, which is the old displacement plus the added weight:

$$\text{New displacement} = W + w \text{ (tons)}$$

where W is the old displacement, in tons; and w is the added weight, in tons. With the new value of displacement, refer to the curves of form, and on the displacement curve find the corresponding draft, which is the new mean draft. Figure 23–10 shows typical displacement and other curves that are generally referred to as curves of form.

If the change in draft is not over 1 foot, the procedure can be reversed. Find the tons-per-inch immersion for the old mean draft from the curves of form, divide the added weight (in tons) by the tons-per-inch immersion in order to get the bodily sinkage in inches, and add this bodily sinkage to the old mean draft to get the new mean draft. Using the new mean draft, refer to the curves of form, and find the new displacement.

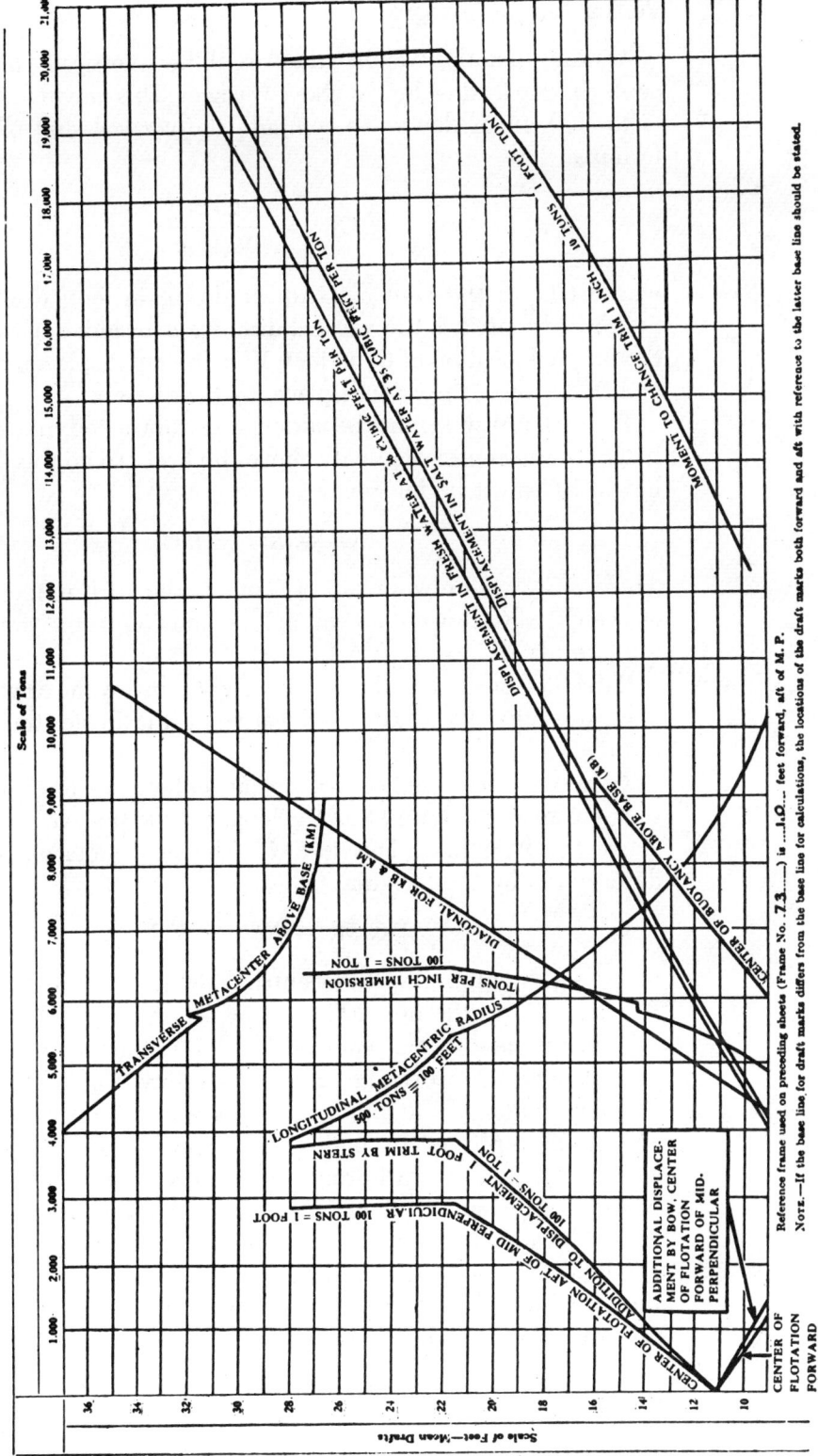

Figure 23–10. Curves of form.

Vertical Weight Changes

Assume that the weight added is shifted vertically on the ship's centerline to its final height above the keel. This movement will cause G to shift up or down. To compute the vertical shift of G use the formula:

$$GG_1 = \frac{(w)(z)}{(W + w)}$$

where GG_1 is the shift of G up or down, in feet; w is the added weight, in tons; z is the vertical distance w is added above or below the original location of G, in feet; W is the old displacement, in tons; and $(W + w)$ is the new displacement, in tons.

This vertical shift must be added to or subtracted from the original height of the center of gravity above the keel. To do this, the original height KG must be known:

$$KG_1 = KG + GG_1$$

where KG_1 is the new height of G above the keel, in feet; KG is the old height of G above the keel, in feet; and GG_1 is the shift of G from the formula $GG_1 = (w)(z)/(W + w)$.

If the final position of the added weight is below the original position of G, then the sign of GG_1 is negative; if it is above, then its sign is positive.

To find the new metacentric height, enter the curves of form with the new mean draft, and find the height of the transverse metacenter above the base line. This is KM_1. The new metacentric height is determined by the formula:

$$G_1M_1 = KM_1 - KG_1$$

where G_1M_1 is the new metacentric height, in feet; KM_1 is the new KM; and KG_1 is the new KG.

With the new displacement $(W + w)$, enter the cross curves and pick out a new, uncorrected curve of stability. Correct this curve for the new height of the ship's center of gravity above the base line. This is accomplished by finding AG_1 (which is KG_1 minus KA) and subtracting $AG_1 \times \sin \theta$ from every vertical on the stability curve, provided G_1 is above A. If G_1 is below A, the values of $AG_1 \times \sin \theta$ must be added to the curve, as previously explained. The resulting curve of righting arms is now correct for the loss of freeboard due to the added weight and for the final height of the ship's center of gravity resulting from weight addition.

Example: Add four gun mounts topside to a ship with the curves of form shown in figure 23–10. Assume an initial KG of 24.5 feet. Assume that the gun mounts weigh 28 tons each, and that their center of gravity is located 48 feet above the keel. What is the effect on stability?

1. New displacement = $W + w$ = 11,500 + (4 × 28) = 11,612 tons.
2. New mean draft = 19.7 feet (figure 23–10).
3. $GG_1 = \dfrac{(w)(z)}{(W+w)}$

    $w = 4 \times 28 = 112$ tons

    $z = 48 - 24.5 = 23.5$ feet

    $GG_1 = \dfrac{112 \times 23.5}{11,612} = 0.23$ feet.
4. $KG_1 = 24.50 + 0.23 = 24.73$ feet.
5. New $KM_1 = 28.4$ feet (figure 23–10).
6. New $G_1M_1 = KM_1 - KG_1 = 28.4 - 24.7 = 3.7$ feet.
7. The values for the angles (0° – 70°) are taken from the cross curves for 11,612 tons displacement (figure 23–1). KA is 20 feet. Corrections are made for $AG_1 \times \sin \theta = (24.73 - 20) \sin \theta = 4.73 \sin \theta$. The corrections are applied to the curve (figure 23–11) as previously explained. Figure 23–11 shows the curve of righting arms corrected for weight addition.

Horizontal Weight Changes

In the previous example of weight addition, suppose the gun mounts are located with their center of gravity 29 feet to starboard of the centerline, and the weight is moved athwartship to its final off-center location. The shift in G may be found by using the proper formula, making the required corrections, and applying the corrections to the curve in figure 23–11. This gives a correct curve of righting arms. To obtain a curve of righting moments, the righting arms are multiplied by the new displacement $(W + w) = 11,612$ tons, and plotted in figure 23–12.

Weight Removal

The results of a weight removal are computed by using the previous procedure, the only difference being that most of the operations and results will be found just the reverse of those that relate to adding a weight.

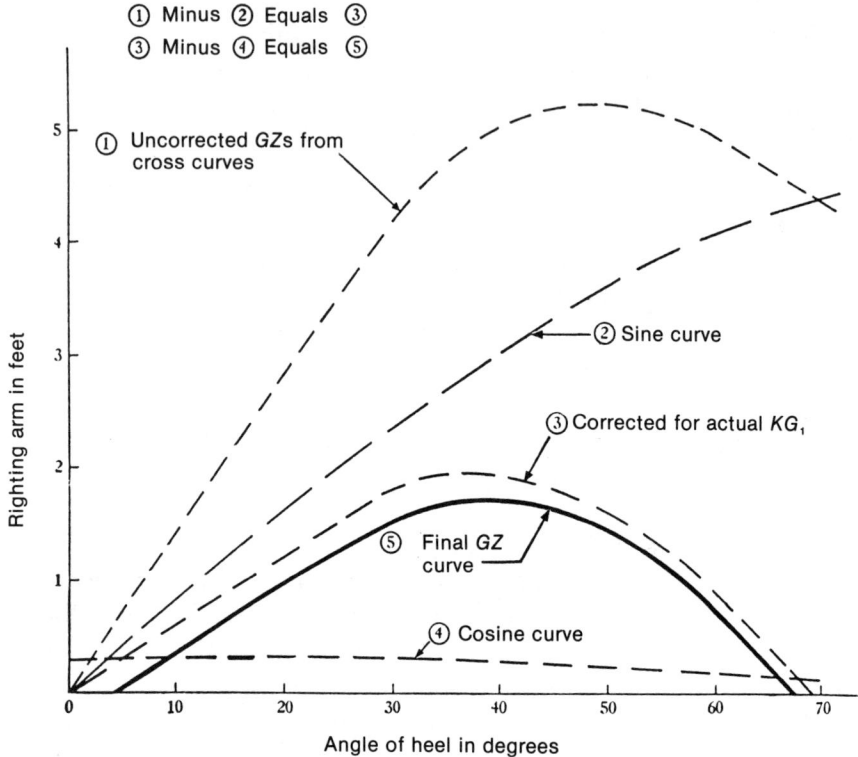

Figure 23–11. Curve of righting arms corrected for a weight addition.

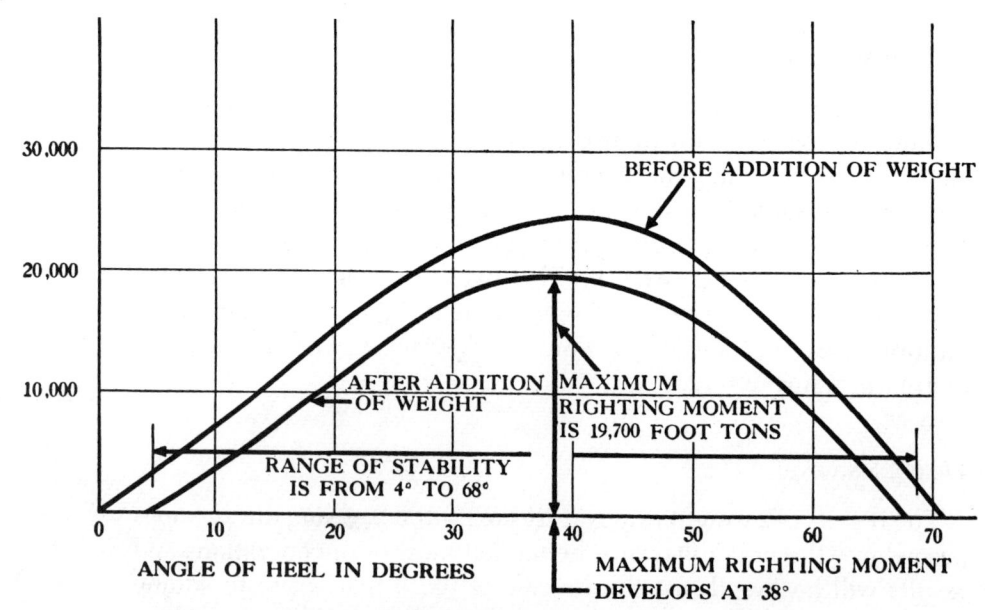

Figure 23–12. Curve of righting moments.

Effects of Loose Water

When a tank or a compartment in a ship is partially full of liquid that is free to move as the ship heels, the surface of the liquid tends to remain level. The surface of the free liquid is referred to as free surface. The tendency of the liquid to remain level as the ship heels is referred to as free surface effect. The term loose water is used to describe water in a compartment that has a free surface; it is not used to describe water or other liquid in a tank intended to hold liquids, regardless of the liquid level within the tank.

Free Surface Effect

Free surface in a ship always causes a reduction in GM with a consequent reduction of stability, superimposed on any additional weight that would be caused by flooding. The flow of the liquid is an athwartship shift of weight which varies with the angle of inclination. Wherever free surface exists, a free surface correction must be applied to any stability calculation. This effect may be considered to cause a reduction in a ship's stability curve in the amount of

$$\frac{i}{V} \times \sin \theta$$

due to a virtual rise in G, where i is the moment of inertia of the surface of water in the tank about a longitudinal axis through the center of area of that surface (or other liquid in ratio of its specific gravity to that of the liquid in which the ship is floating); and V is the existing volume of displacement of the ship in cubic feet. (Note: It is usual to assume all liquids are salt water, and thus neglect density, unless very accurate determinations are required.) For a rectangular compartment, i may be found from

$$i = \frac{b^3 l}{12}$$

where b is the athwartship breadth of the free surface (with the ship upright), in feet; and l is the fore-and-aft length of the free surface, in feet.

To understand what is meant by a virtual rise in G, refer to figure 23–13. This figure shows a compartment in a ship partially filled with water, which has a free surface, fs, with the ship upright. When the ship heels to any small angle, such as θ, the free surface shifts to $f_1 s_1$, remaining parallel to the waterline. The result of the inclination is the movement of a wedge of water from $f_0 s_1$ to $s_0 s_1$. Calling g_1 the

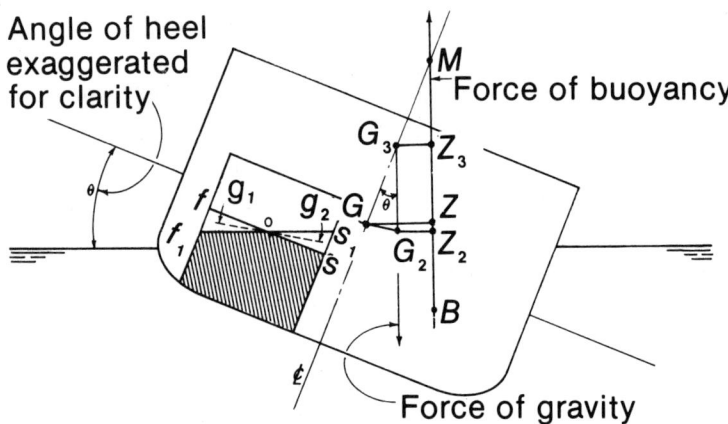

Figure 23–13. Diagram showing virtual rise in G.

center of gravity of this wedge when the ship was upright, and g_2 its center of gravity with the ship inclined, it is evident that a small weight has been moved from g_1 to g_2.

Point G is the center of gravity of the ship when upright, and G would remain at this position if the compartments contained solids rather than a liquid. As the ship heels, however, the shift of a wedge of water along the path g_1g_2 causes the center of gravity of the ship to shift from G to G_2. This reduces the righting arm, at this angle, from GZ to G_2Z_2.

To compute GG_2 and the loss of GZ for each angle of heel is a laborious and complicated task. However, an equivalent righting arm G_3Z_3 (which equals G_2Z_2), can be obtained by extending the line of action of the force of gravity up to intersect the ship's centerline at point G_3. Raising the ship's center of gravity from G to G_3 would have the same effect on stability at this angle as shifting it from G to G_2.

The distance G_3Z_3 is the righting arm the ship would have if the center of gravity had risen from G to G_3, and this *virtual rise* of G may be computed from the formula:

$$GG_3 = \frac{i}{V}$$

Referring to the formula, loss in $GZ = \dfrac{i}{V} \times \sin \theta$. This formula is accurate for small angles of heel only, because of the pocketing effect as the angle increases. In case several compartments or tanks have free surface, their surface moments of inertia are calculated individually, and their sum used in the correction for free surface. The effect of a given area of loose liquid at a given angle of heel is entirely independent of the depth of the liquid in the compartment, as is apparent in the formula:

$$i = \frac{b^3 l}{12}$$

where the only factors are the dimensions of the surface and the displacement of the ship. The free surface effect is also independent of the free surface location in the ship, whether it is high or low, forward or aft, on the centerline or off, as long as the boundaries remain intact.

The loss of metacentric height can obviously be reduced by reducing the breadth of the free surface, as by the installation of longitudinal bulkheads. However, off-center flooding after damage then becomes possible, causing the ship to take on a permanent list and usually bringing about a greater loss in stability than if the bulkhead were not present.

The loss of GZ due to free surface is always lessened to some extent by *pocketing*—the contact of the liquid with the top of the compartment or the exposure of the bottom surface of the compartment, either of which takes place at some definite angle and reduces the breadth of the free surface area. To understand how pocketing of the free surface reduces the free surface effect, study figure 23–14. Part A shows a compartment in which the free surface effect is not influenced by the depth of the loose water. The compartment shown in part B, however, contains only a small amount of water; when the ship heels sufficiently to reduce the waterline in the compartment from wl to $w_1 l_1$, the breadth of the free surface is reduced, and the free surface effect is thereby reduced. A similar reduction in free surface effect occurs in the almost full compartment shown in part C, again because of the reduction in the breadth of the free surface.

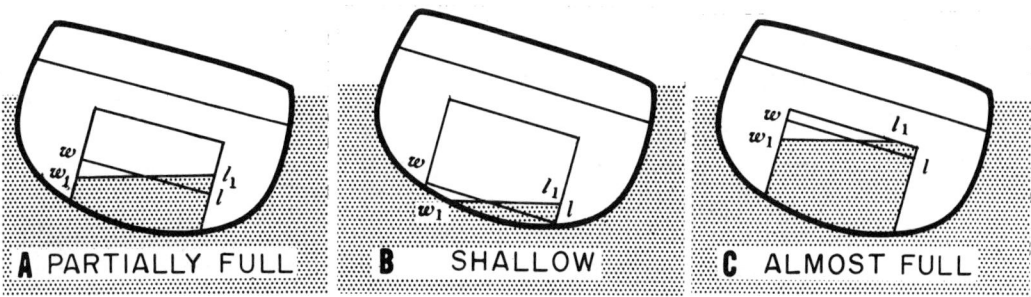

Figure 23–14. *Pocketing of free surface.*

As figure 23–14 shows, the beneficial effect of pocketing is greater at large angles of heel.

The effect of pocketing in reducing the overall free surface effect is extremely variable and not easily determined. In practice, therefore, it is usually ignored and tends to provide a margin for safety when computing stability.

Most compartments of a ship contain some solid objects, such as machinery and stores, that would project through and above the surface of any loose water. If these objects are secured so that they do not float or move about, and if they are not permeable, then the free surface area and the free surface effect are reduced by their presence. The actual value of the reduction (surface permeability effect) is difficult to calculate and, like the value of pocketing, if ignored when calculating stability will provide a further margin of safety.

Swash bulkheads (nontight bulkheads pierced by drain holes) are fitted in deep tanks and double bottoms to hinder the flow of liquid in its attempt to remain continuously parallel to the waterline as the ship rolls. They diminish the free surface effect if the roll is quick, but they have no effect when the roll is slow. A ship taking on a permanent list will incline just as far as if the swash plate were not there. When a fore-and-aft bulkhead separating two adjacent compartments is holed (ruptured) so that any flooding water present in one is free to flow athwartship from one compartment to the other, a casualty duplicating the effect of a swash bulkhead has occurred. In this case, it is incorrect to add the free surface effects of the two compartments together; an entirely new figure for the flooding effect must be computed, regarding the two as one large compartment.

In summary, the addition of a liquid weight with a free surface has two effects on the metacentric height of a ship. First, there is the effect on GM and GZ of the weight addition (considered as a solid), which influences the vertical position of the ship's center of gravity and the location of the transverse metacenter, M. Second, there is a reduction in GM and GZ that is due to the free surface effect.

Free Communication Effect

If one or more of the boundaries of an off-center compartment are ruptured so that the sea may flow freely in and out with a minimum of restriction as the ship rolls, a condition of partial flooding with free communication with the sea exists. The added weight of the flooding water and the virtual rise in G due to the free surface effect cause what is known as the free communication effect. With an off-center space flooded, a ship will assume a list which will be further aggravated by the free surface effect. As the ship lists, more water will flow into the compartment from the sea and will tend to level off at the height of the external waterline. The additional weight causes the ship to sink farther, allowing more water to enter and causing more list until some final list is reached. The reduction of GM due to the free communication effect is approximately equal in magnitude to

$$\frac{ay^2}{V}$$

where a is the area of the free surface, in square feet; y is the perpendicular distance from the geometric center of the free surface area to the fore-and-aft centerline of the intact waterline plane, in feet; and V is the new volume of the ship's displacement after flooding, in cubic feet. Thus reduction in GM is additional to and separate from the free surface effect.

The approximate reduction in GZ may be computed from

$$GZ = \frac{ay^2}{V} \sin \theta$$

This may be considered as a virtual rise in G, superimposed upon the virtual rise in G due to the free surface effect.

Summary of Effects of Loose Water

The addition of loose water to a ship alters the stability characteristics by means of three effects that must be considered separately: (1) the effect of added weight; (2) the effect of free surface; and (3) the effect of free communication.

Figure 23–15 shows the development of a stability curve with corrections for added weight, free surface, and free communication. Curve A is the ship's original stability curve before flooding. Curve B represents the situation after flooding; this curve shows the effect of added weight (increased stability), but it does not show the effects of free surface or free communication. Curve C is curve B corrected for free surface effect only. Curve D is curve B corrected for both free surface effect and free communication effect. Curve D, therefore, is the final stability curve; it incorporates corrections for all three effects of loose water.

Longitudinal Stability and Effects of Trim

The important phases of longitudinal inclination are changes in trim and longitudinal stability. A ship *pitches* longitudinally in con-

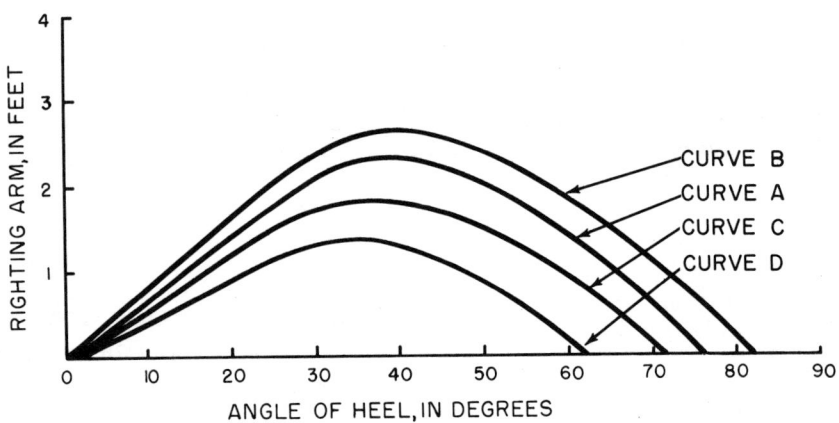

Figure 23–15. Stability curve corrected for effects of added weight, free surface, and free communication.

trast to *rolling* transversely; and it *trims* fore and aft, whereas it *lists* transversely. The difference in forward and after draft is defined as trim.

Center of Flotation

When a ship trims, it inclines about an axis through the geometric center of the waterline plane. This point is known as the center of flotation. The position for the center of flotation aft of the midperpendicular for various drafts may be found from a curve on the curves of form (figure 23–10). When a center of flotation curve is not available, or when precise calculations are not required, the midperpendicular may be used in lieu of the center of flotation.

Change of Trim

Change of trim may be defined as the change in the difference between the drafts forward and aft. If in changing the trim, the draft forward becomes greater, then the change is said to be by the bow. Conversely, if the draft aft becomes greater, the change of trim is by the stern.

Changes of trim are produced by shifting weights forward or aft or by adding or subtracting weights forward of or abaft of the center of flotation.

24

Causes of Impaired Stability

The stability of a ship may be impaired by several causes, resulting from mistakes or from enemy action. A summary of these causes and their effects follows.

Addition of Topside Weight

The addition of appreciable amounts of topside weight may be occasioned by unauthorized alterations; icing conditions; provisions, ammunition, or stores not struck down; deck cargo; and other conditions of load. Whenever a weight of considerable magnitude is added above the ship's existing center of gravity the effects are:

1. Reduction of reserve buoyancy.
2. Reduction of GM and righting arms due to raising G.
3. Reduction of GM and righting arms due to loss of freeboard (change of waterplane).
4. Reduction of righting arms if G is pulled away from the centerline.
5. Increase in righting moment due to increased displacement.

The net effect of added high weight is always a reduction in stability. The reserve buoyancy loss is added weight in tons. The new metacentric height can be obtained from:

$$G_1M_1 = KM_1 - KG_1$$

Stability is determined by selecting a new stability curve from the cross curves and correcting it for $AG_1 \sin \theta$ and $G_1G_2 \cos \theta$.

Loss of Reserve Buoyancy

Reserve buoyancy may be lost because of errors, such as poor maintenance, failure to close fittings properly, improper classification of fittings, and overloading the ship; or it may be lost as a result of enemy action such as fragment or missile holes in boundaries, a blast that carries away boundaries or blows open or warps fittings, and flooding that overloads the ship. When the above-water body is holed, some reserve buoyancy is lost. The immersion of buoyant volume is necessary to the development of a righting arm as the ship rolls; if the hull is riddled, it can no longer be immersed on the damaged side, toward which it will roll. In effect, the riddling of the above-water hull is analogous to losing part of the freeboard, thus reducing stability. When this happens, if the ship takes on water on the roll, the combined effects of high added weight and free surface operate to cut down the righting moment. Therefore, the underwater hull and body should be plugged and patched, and every effort should be made to restore the watertightness of external and internal boundaries in the above-water body.

Flooding

Flooding may take place because of underwater damage, shell or bomb burst below decks, collision, topside hit near the waterline, firefighting water, ruptured piping, sprinkling of magazines, counterflooding, or leakage. Regardless of how it takes place, it can be classified in three general categories, each of which can be further broken down, as follows:

1. With respect to boundaries
   a. Solid flooding
   b. Partial flooding
   c. Partial flooding in free communication with the sea
2. With respect to height in the ship
   a. Center of gravity of the flooding water above G
   b. Center of gravity of the flooding water below G
3. With respect to the ship's centerline
   a. Symmetrical flooding
   b. Off-center flooding

Solid Flooding

The term solid flooding designates the situation in which a compartment is completely filled from deck to overhead. In order for this to occur, the compartment must be vented, as by an air escape, by an open scuttle or vent fitting, or through fragment holes in the overhead. Solid-flooding water behaves exactly like an added weight and has the effect of so many tons placed exactly at the center of gravity of the flooding water. It is more likely to occur below the waterline, where it has the effect of any added low weight. Inasmuch as G is usually a little above the waterline in warships, the net effect of solid flooding below the waterline is most frequently a gain in stability, unless a sizable list or a serious loss of freeboard results in a net reduction of stability. The reserve buoyancy consumed is the weight of flooding water in tons, and the new GM and stability characteristics are found as previously explained.

Partial Flooding with Boundaries Intact

The term partial flooding refers to a condition in which the surface of the flooding water lies somewhere between the deck and the overhead of a compartment. The boundaries of the compartment remain watertight, and the compartment remains partially but not completely filled. Partial flooding can be brought about by leakage from other damaged compartments or through defective fittings, seepage, shipping water on the roll, downward drainage of water, loose water from firefighting, sprinklers, ruptured piping, and other damage.

Partial flooding of a compartment that has intact flooding boundaries affects the stability of the ship because of (1) the effect of added weight and (2) the effect of free surface. The effect of the added weight will depend upon whether the weight is high in the ship or low, and whether it is symmetrical about the centerline or is off center. The effect of free surface will depend primarily upon the athwartship breadth of the free surface. Unless the free surface is relatively narrow and the weight is added low in the ship, the net effect of partial flooding in a compartment with intact boundaries is likely to be a very definite loss in overall stability.

Partial Flooding in Free Communication with the Sea

Free communication can exist only in partially flooded compartments in which it is possible for the sea to flow in and out as the ship rolls. Partial flooding with free communication is most likely to occur when there is a large hole that extends above and below the waterline. It may also occur in a waterline compartment when there is a large hole in the shell below the waterline, if the compartment is vented as the ship rolls. Where free communication exists, the water level in the compartment tends to remain at sea level as the ship rolls.

When a compartment is partially flooded and in free communication with the sea, the ship's stability is affected by (1) added weight, (2) the free surface effect, and (3) the free communication effect. In general, the net effect of partial flooding with free communication is a decided loss in stability.

Fuel Consumption

As fuel is consumed on board ship, a reduction in stability will occur owing to the removal of weight (the fuel) from fuel tanks which are below the center of gravity. As more and more fuel tanks are emptied, the effect on stability is greatly increased.

To restore the ship to its previous level of stability, the empty fuel tanks are filled with salt water from the surrounding seawater. The system that fills the fuel tanks with salt water is called the *ballast system*. Salt water from the ballast system is pumped into the empty fuel oil *storage* tanks only. Salt water is never purposely pumped into the fuel oil service tanks. This ballast water must be removed from the storage tanks prior to refueling.

Some of the newer combatants have an automatic ballast system called an *automatic compensation system*. In this type system salt water is simultaneously pumped into the fuel oil storage system as fuel is pumped out. A ship with this system does not experience the change in stability associated with the more conventional ballast system found on board most naval vessels. An automatic compensation system requires greater scrutiny to ensure that no salt water is accidentally pumped into the fuel oil service tanks when fuel is being transferred.

Problems Involved in Running Aground

Not only do ships occasionally run aground unintentionally, but in cases where loss of stability or structural failure threatens the ship in deep water, the commanding officer may deliberately beach his ship pending salvage operations. In selecting a spot on a friendly shore at which to beach the ship, preference should be given to a locality where the water shoals gradually and the bottom is flat, soft, and of even texture, avoiding rock ledges or pinnacles. It is also desirable to avoid cross-currents and heavy surf. Running aground, whether deliberate or unintentional, involves three major problems:

1. Ability to get off again
2. Structural strength
3. Stability

Ability to Get off Again. The following is general information relating to factors involved in the ability to come clear after running aground:

If the ship runs aground unintentionally, there is a strong temptation to use the engines in an effort to get off again. However, ship's screws become less effective in shallow water, and attempting to back down may do more harm than good, as propeller wash may drive sand in around the hull. If there is any possibility of wind or tide carrying the ship farther aground, her chances of getting off will be jeopardized by attempting to maneuver with screws and rudder. If the propellers are reversed and there is no tendency of the ship to back away from the beach, no further attempts to move the ship by means of the screws should be made. Instead, the ship should be weighed down hard by flooding all tanks, and, if necessary, one or more hold compartments. This will fix her firmly in position until she is ready for hauling off again. Weighing the ship down is especially important if the tide is rising, and becomes absolutely imperative if there is a possibility of a heavy surf driving her harder aground, pounding the ship on the beach, or causing her to broach. In such situations every action must be taken to get the ship down firmly. In such a case it would be preferable to use demolition charges to blow small holes in the side and flood the ship down firmly rather than to permit her to rise and fall against the bottom. Heavy pounding in the surf may result in breaking up, broaching, driving higher aground, and even capsizing.

Once the ship has been weighed down successfully, a careful investigation should be made, sounding all voids, checking fuel tanks for leakage, and examining the entire interior of the hull for signs of structural damage. Soundings should be made all around to determine the slope and nature of the bottom. The boats should be used to continue these soundings in the direction toward which she is to be hauled off, in order to locate rock formations, coral ledges, or other underwater obstructions. Currents that might take charge as the ship comes off should be noted.

Meanwhile ground tackle should be rigged, and kedge anchors laid to seaward as quickly as possible. Ship's boats can be used to carry out a small anchor or two. One merchant captain even used cargo booms to lay both bower anchors abaft the stern. Then, leading cables forward and inboard over sheaves at the bow, he slowly worked himself off the beach. Tugs may also be useful in pulling ships off and in taking charge once they come free. However, one good kedge is worth three or four tugs.

When the beach gear is ready, and the tide is right, the ship can be safely lightened again while a strain is being taken on all cables. Engines in the stranded ship should be warmed up if no tugs are at hand, but should not be turned over, to avoid filling the main condensers with mud and sand, and to avoid damaging the propellers on the bottom or fouling the ground tackle. Shifting fuel oil or ammunition fore-and-aft might produce a change of trim that would aid in coming clear. If suction seems to hold her fast in mud or sand, it will be helpful to have the crew sally ship to start her rolling. Jets from air or steam hoses are also useful for this purpose.

Hull Strength. When the ship is aground, the beach exerts an upward force on the hull equal to that portion of the ship's weight that is not supported by the buoyancy of the water. Its effect on stress is the reverse of flooding. If the ship is aground at one end, sagging stresses are increased; hence weight additions should preferably be made at the ends of the ship, while weight removals should be from the midship region. If the ship is aground on a ledge or pinnacle amidships, then hogging stresses are increased, and weight additions are better made amidship, while weight removals are preferable from the ends. Irregular rock or coral formations or sharp changes of section (as at ledges) produce concentrated pressures that often crush hull plating and result in flooding. Such local damage is intensified if the ship works or shifts her position.

Stability. The following discussions relate to running aground and subsequent loss of stability resulting from upward grounding force:

The magnitude of the ground pressure, or number of tons aground, is the reduction in displacement caused by beaching. It may be found as the difference between the displacement when floating free and the displacement corresponding to the mean draft after grounding. (To be very accurate these displacements must be corrected for trim, list, and draft.) The ground pressure is applied at the keel, or other point of contact, and has all the effects on draft, list, trim, and stability of removing that many tons of weight from the location of the point of contact. If the point of contact is at the keel, it constitutes a low weight removal. The virtual center of gravity goes up by an amount:

$$GG_1 = \frac{P(KG)}{(W-P)}$$

where GG_1 is rise of G, in feet; P is tons aground; KG is height of G above the keel, in feet, if floating free; and $W-P$ is displacement in tons, corresponding to mean draft after grounding.

The shift in the metacenter (MM_1) may be found in the curves of form as the difference in KM for the drafts floating free and after grounding. The metacentric height when aground is then:

$$G_1M_1 = GM \pm MM_1 - GG_1$$

If the point of contact is at one end, the ship will trim by an amount equivalent to removing the tons aground from that end. If the point of contact is off center, the ship will list to an angle corresponding to the off-center weight removal (with metacentric height as found above). If the *GM* calculated is negative, the effect of this condition on list must be considered. The result may be capsizing if the ship is aground only at the bow where she is narrow athwartship. However, if grounded throughout their length on a level surface, ships do not usually capsize even if high and dry.

To check this, take a cross section of the hull at the point of contact and draw a tangent to the skin of the ship at the point of contact; then draw successive tangents to the shell of the ship corresponding to greater angles of heel. If a perpendicular to each of these tangents passes above the center of gravity (ship floating free), there is no danger of capsizing. If the normal intersection of the perpendicular and these tangents lies below the center of gravity, then the possibility of capsizing will depend on how much stability is lost through the application of the upward grounding force.

Figure 24–1 illustrates the forces acting on a ship grounded at the bow. It is noted that the grounding force *P* acts perpendicular to the waterline at the angle of heel shown. If the moment of the force *P* about *G* (ship floating freely) equals the moment of the net buoyant force *B*, the ship will remain at this angle of heel. If the moment of force *P* is greater than the righting moment, the ship will continue to heel. Unless a net positive righting moment develops at some larger angle of heel, the ship will capsize.

The moment $(W-P)GZ$ must exceed $P(GR)$ at some angle of heel to assure the development of a net positive righting moment. The size of righting arms corresponding to the displacement at the lowest mean draft to be expected may be obtained from the cross curves of stability. These righting arms must be corrected for the actual position of *G* (ship floating freely) to determine *GZ*. The distance *GR* is measured on the section drawn.

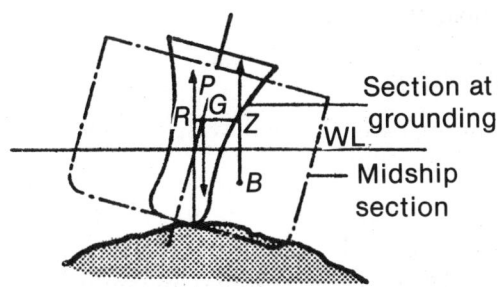

Figure 24–1. Stability forces acting on a ship grounded by the bow.

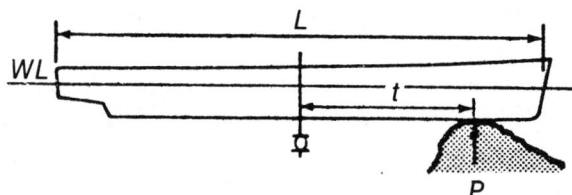

Figure 24–2. Buoyancy and grounding forces, effects of tide action.

The grounding force, P, will be the difference between the displacement of the ship floating freely and that when aground. The greatest groundings force will occur when the waterline of a ship grounded at high tide drops to that of a low tide (figure 24–2). As the tide drops, the buoyant force of water lessens while the grounding force gets proportionately larger.

The grounding force, P, may be determined for any expected fall of tide from the following relationship:

$$\frac{P + Pt^2}{TPI \times (MTI)L} = \text{inches fall of tide}$$

or:

$$P = \frac{\text{inches fall of tide}}{\dfrac{1 + t^2}{TPI \times (MTI)L}}$$

where t is the distance in feet from the grounding point to midships; L is the length of the ship, in feet; TPI is tons per inch immersion; and MTI is the approximate moment to change the trim 1 inch. (Use the mean draft of the ship freely floating for TPI and MTI.)

To improve stability, where it appears that capsizing is imminent, the corrective measures indicated may be undertaken. Use of these measures should take into account their effect on hull strength and the ability to float the ship again.

Summary of Grounding Problem

In most stranding cases, the following considerations will ordinarily constitute good procedure:

1. No attempt should be made to refloat the ship under her own power if wind and sea conditions indicate the possibility of the ship's working harder aground, pounding, or broaching to sea.

2. Anchors to seaward should be quickly laid if possible, to prevent the ship from working farther ashore.

3. The ship should be weighed down, not lightened, in an effort to help keep her from working harder on the beach and, second, to prevent damage caused by working and pounding of the ship on the bottom.

25

Damage Control and the Damage Control Organization

On board ship, the overall damage and casualty control function is composed of two separate but related phases: the engineering casualty control phase and the damage control phase. The engineering officer is responsible for both phases.

The engineering casualty control phase is concerned with the prevention, minimization, and correction of the effects of operational and battle casualties to the machinery, electrical systems, and piping installations, to the end that all engineering services may be maintained in a state of maximum reliability under all conditions of operation. Engineering casualty control is handled almost entirely by personnel of the engineering department.

The damage control phase, on the other hand, involves *every person in the ship*. The damage control phase is concerned with such things as the preservation of stability and watertight integrity, the control of fires, the control of flooding, the repair of structural damage, and the control of nuclear, biological, and chemical warfare agents. Although under the control of the engineer officer, damage control is an all-hands responsibility.

Objectives of Damage Control

The two basic objectives of shipboard damage control are: (1) to take all practical preliminary measures, before damage occurs, to prevent it, by ensuring the maintenance of watertight integrity and fumetight integrity, maintaining reserve buoyancy and stability, removing fire hazards, and maintaining and distributing emergency equipment; (2) to cut down and localize damage, when it does occur, by such measures as controlling flooding, preserving stability and buoyancy, replacing essential structures, and manning essential equipment.

The ship's ability to inflict punishment upon or destroy the enemy, or to perform any other assigned mission, may well depend upon the effectiveness of damage control. Damage control then must be considered as an offensive, as well as a defensive, function.

Damage control is concerned not only with battle damage but also with nonbattle damage, such as fire, collision, grounding, or explosion. It may be necessary in port as well as at sea, and may involve the use of personnel and facilities of an undamaged ship.

Damage control requires a detailed knowledge of ship construction characteristics, compartmentation, stability, and that equipment placed on board a ship to prevent or control damage should a ship be endangered. Basically, the control of damage depends upon the ability of personnel to take prompt corrective action, using material that is available.

Damage Control Organization

In order to ensure damage control training and to provide prompt control of damage, a damage control organization must be set up and kept alive. This organization is in turn a part of the ship's battle organization which readies the ship for battle conditions or major emergencies. Since relatively minor periods of a ship's total time are spent under battle conditions, this organization is more or less temporary. The word temporary should not be misconstrued, for the battle organization is the most important one on board ship and must be kept up to the highest possible standards of readiness.

The damage control organization consists of damage control central and the repair parties. The engineer officer is responsible (as is the damage control assistant who is under the engineer officer) for establishing and maintaining an effective damage control organization. The DCA reports to the engineering officer for administration matters. During General Quarters, in his capacity as the officer in charge of damage control central, he reports to the commanding officer. The following personnel report to the DCA: (1) repair party officers during General Quarters, and (2) division damage control petty officers for proper maintenance and operation of damage control equipment.

Damage Control Central

In any ship's damage control organization, arrangements are made for a designated repair station to take charge of damage control activities. Should this central station be destroyed or rendered unable to retain control, other repair stations, in designated order, take over these same functions. Provisions are also made for passing the control of each repair party and its operation down through the officers, petty officers, and nonrated men, so that at no time will any group be without a leader.

The primary purpose of damage control central is to collect and compare reports from the various repair parties in order to determine the condition of the ship and the action that should be taken. The commanding officer is kept posted on the condition of the ship and on important corrective measures taken. Repair party reports are carefully checked so that immediate action can be taken to isolate damaged systems and to make emergency repairs in the most logical manner. Graphic records of the damage are made on various damage control diagrams or status boards, as the reports are received. For example, reports concerning flooding are marked as they come in on a status board that indicates liquid distribution before damage. With this information, the stability and buoyancy of the ship can be estimated, and the necessary corrective measures determined. Orders can then be sent for the required action.

A ship is a very complex structure, and it would be practically impossible for an individual to learn all the details of the built-in systems such as piping and ventilation. Therefore, there must be some means used to facilitate the location of any part of these systems. This is accomplished by means of diagrams that are kept at damage control central and at various repair party stations.

The damage control library contains publications, diagrams, damage control books, and pamphlets which form the basis for the effective operation of the damage control organization. It is important that all damage control personnel be familiar with these publications and drawings.

Diagrams and plans for piping systems and wiring circuits are maintained at repair stations and many unit lockers (large ships) throughout the ship. Decisions to isolate or cross-connect damage control systems are made on the basis of information contained in these diagrams. They also serve as aids in tracing lines and circuits throughout an area.

Repair Parties

The organization of repair stations is basically the same in all types of ships; however, more men are available for manning repair stations in large than in small ships. The number and the ratings of men assigned to a repair station, as specified in the battle bill, are determined by the location of the station, the portion of the ship assigned to that station, and the total number of men available.

On board a destroyer or a frigate there are generally three repair parties. Repair party II is responsible for the forward portion of the ship; repair party III is responsible for the after portion of the ship;

and repair party V is responsible for the main engineering spaces. Each repair party has an officer in charge, who may in some cases be a chief petty officer. The second in charge is usually a chief petty officer who is qualified in damage control and who is capable of taking over the supervision of the repair party.

Many repair stations have unit patrol stations at key locations in their assigned areas to supplement the repair station. Operating instructions should be posted at each repair station. In general, instructions should include the purpose of the repair station; the specific assignments of space for which that station is responsible; instructions for assigning and stationing personnel; methods and procedures for damage control communications; instructions for handling machinery and equipment located in the area; procedures for nuclear, biological, and chemical (NBC) warfare defense; sequence and procedure for passing control from one station to another; a list of current damage control bills; and a list of all damage control equipment and gear provided for the repair station.

Fire Parties

Repair parties provide the only personnel immediately available to fight fires during combat action. It is essential that a plan of action that includes a systematic procedure for fighting fires be established. Loss of valuable time will be the inevitable result if the decision as to the method to be used in fighting a fire is allowed to remain open until a fire is actually under way. Large repair parties may be divided into several firefighting groups or parties. In small ships, an entire repair party could be required to make up a firefighting party. Where possible, at least two fire parties are to be organized from any one repair party. Such groups should be trained so that each member can quickly undertake any of the detailed duties if necessary.

The following personnel are generally required to be in a fire party:

1. Hose men (number determined by size of hose)
2. Plug man
3. Access men
4. Foam generator operator
5. Foam supply men
6. Portable CO_2 men
7. Oxygen breathing apparatus (OBA) wearers
8. Ventilation detail
9. Electrician

In port when many of the repair party personnel are ashore, it is necessary to establish a fire party from the duty section that remains on board. The inport fire party must be capable of performing all the duties required of the fire parties during combat action.

Preparations to Resist Damage

Naval ships are designed to resist accidental and battle damage. Damage-resistant features include structural strength, watertight compartmentation, and buoyancy. Maintaining these damage-resistant features and maintaining a high state of material and personnel readiness before damage is far more important for survival than are any damage control measures that can be taken after the ship has been damaged. It has been said that 90 percent of the damage control needed to save a ship takes place before the ship is damaged, and only 10 percent can be done after the damage has occurred. In spite of all precautions and all preparatory measures, however, the survival of a ship sometimes depends upon prompt and effective damage control measures taken after damage has occurred. It is essential, therefore, that all shipboard personnel be trained in damage control procedures.

A major factor in the success of damage control is the proper utilization of the watertight-integrity features of a ship. Part of watertight integrity is the compartmentation of the ship. The ship is divided into compartments to control flooding; to restrict chemical, biological, and radiological agents; to segregate activities of personnel; to provide underwater protection by means of tanks and voids; and to strengthen the structure of the ship. In some cases, where an increase in plating would reduce speed or have an adverse effect on the operation of the ship (as in the case of aircraft carriers), compartmentation has been increased to compensate for the reduction of plating.

Every naval ship is subdivided, by decks and bulkheads both above and below the waterline, into as many watertight compartments as are compatible with the ship's mission. In general, the more minute this subdivision, the greater the ship's resistance to sinking from damage. A modern combatant ship has over 600 watertight compartments. The condition of this subdivision (watertight integrity), which is of greatest importance, must be maintained in the highest degree of perfection. Originally, watertight integrity is established by the skill and thoroughness of the builders who strive to make the boundaries of these subdivisions strong and watertight in accordance with specified requirements.

Material Conditions of Readiness

In order to use compartmentation to its fullest advantage, and to provide further for maximum preparedness, all the doors, hatches, scuttles, accesses, valves, and fittings of damage control value are classified and marked. Naval ships maintain different material conditions of readiness according to whether contact with the enemy is improbable, probable, or imminent. Each condition represents a different degree of tightness and guarantees the maximum protection for the proximity of the enemy, with due regard for the health and comfort of personnel. Maximum closure is not maintained at all times because it interferes with the normal operation of the ship.

NWIP 50–1 (battle control) requires that all surface ships be classified as three-condition ships, that is, X-RAY, YOKE, and ZEBRA. Condition X-RAY provides the least protection. It is set when the ship is no longer in danger from attack, such as at anchor in a well-protected harbor or secured at a home base during regular working hours. Condition YOKE is set whenever a ship is in an unprotected harbor or when the ship is under way for normal peacetime steaming. Condition ZEBRA is set prior to going to sea and entering port, during wartime. It is set immediately and without further orders, when manning General Quarters stations. It is set to localize and control fire or flooding when not at General Quarters stations.

The setting of material conditions is normally a departmental responsibility and is carried out by utilizing the compartment check-off lists provided by the DCA. When General Quarters is sounded, the setting and maintaining of condition ZEBRA is the responsibility of all hands. Tables of material conditions and classifications affected by each are posted throughout the ship. The chart illustrates the material conditions of closure.

Condition	Damage is	Close Fittings Marked		
X-RAY	Improbable (well-protected harbor)	X		
YOKE	Probable (cruising conditions, unprotected harbors)	X	Y	
ZEBRA	Imminent (battle conditions, maximum protection)	X	Y	Z

The responsibility for the setting and maintaining of conditions X-RAY and YOKE belongs to the divisions concerned. The setting of condition ZEBRA is the responsibility of the repair parties. Proper setting of condition ZEBRA, however, cannot be accomplished without assistance of the individual divisions.

Circle X and Y fittings may be opened without special permission when proceeding to battle stations or during action, if necessary. They must be kept closed when not in use. Red Circle Z fittings may be opened during prolonged periods of General Quarters; however, these fittings are guarded for immediate closure if necessary.

Closure Classifications of Fittings

Closures involved in setting material conditions are labeled as follows:

X-RAY (marked X)—closed during conditions X-RAY, YOKE, and ZEBRA.
YOKE (marked Y)—closed during conditions YOKE and ZEBRA.
ZEBRA (marked Z)—closed during condition ZEBRA or General Quarters.
Circle WILLIAM (marked Ⓦ)—closed during NBC warfare only.

Once the material condition is set, no fitting marked as indicated previously may be opened without permission of the commanding officer through the damage control assistant or the officer of the deck (OOD). Other fitting markings, which are modifications to the three basic conditions, are as follows:

Circle X-RAY (marked Ⓧ) and Circle YOKE (marked Ⓨ)—closed during condition X-RAY and YOKE, respectively: This applies to access fittings which must later be returned to their material condition.
DOG ZEBRA: These fittings are closed during darken ship as well as General Quarters. Fittings of this type include portholes, doors, and hatches leading to the main deck (marked ⊡).

Maintenance of Watertight Integrity and Damage Control Fittings

The proper setting of material conditions of readiness is important for maintaining a ship's watertight integrity. However, integrity can be lost by improper maintenance of watertight fittings and compartment boundaries. To ensure the rigid maintenance of watertight integrity, a thorough system of tests and inspections is prescribed. The condition of watertight boundaries, compartments, and fittings is determined by visual observation and by various tests, including chalk tests and air tests. All defects discovered by any test or inspection must be remedied immediately.

Piping System Identification

To aid in damage control all piping systems in a surface ship are marked according to function by color, number, letters, and symbol for ready identification. These markings occur at intervals considered necessary. Color identification is applied to the following systems:

1. Fireplugs—red
2. Gasoline—yellow
3. JP–5 fuel—purple
4. Oxygen—light green
5. Foam discharge—striped red/green

All markings on piping systems are black letters on a white background except for oxygen piping, which is marked with white letters on a black background.

Reports on Damage Control Equipment

A great many items of damage control equipment must be inspected and tested at frequent intervals, and written reports on their condition should be submitted to the DCA. NavSea requires that the following items be used or operated at least once a month: electric submersible pumps, portable pumps (fire and drainage), shallow-water diving

COMPARTMENT CHECKOFF LIST

NAVSHIPS 184

COMP'T NO. 1-125-0-L   NAME Passage

ITEM	FITTING	NUMBER	LOCATION AND PURPOSE	CLASSIFICATION	DIVISION RESPONSIBLE
	ACCESS				
1	F.T. Door	1-126-2	To 1-125-2-L	Z	R
2	W.T. Hatch	01-138-1	To Weather Deck	Y	2
3	Escape Scuttle	01-27-1	To 01-125-0M	Y	2
	FIREMAIN SYS.				
4	Valve	1-128-1	Foam Proportioner	X	2
5	Valve	1-140-1	Fireplug 01-139-1 Cutout	W	2

Figure 25–1. *Compartment checkoff list.*

equipment, air (hose-line) masks, oxygen breathing apparatus, and various kinds of power tools. A locally prepared weekly report of damage control equipment may be required on board some ships by the DCA. Specific requirements for maintenance and testing of damage control equipment are included in the PMS (planned maintenance system). Much of this routine work is the responsibility of the individual damage control petty officer. All damage control equipment in repair lockers should be inventoried at least once a month. A complete list of all allotted equipment should be posted at each locker, and the inventory should be checked against this list. Replacements for any missing equipment should be ordered as soon as possible.

Compartment Checkoff Lists

The compartment checkoff list (figure 25–1) is an itemized list of all classified fittings and all equipment or facilities useful for damage control within a compartment or an area. The list shows the name, location, purpose, and classification of each fitting within the compartment or area, and states who is responsible for its proper closure. A compartment checkoff list is permanently posted in each compartment. A master copy of each list must be kept in the damage control office.

26
Principles of Practical Damage Control

The DCA must be given all available information concerning the nature and extent of damage so that he will be able to analyze the damage and decide upon appropriate measures of control. The repair parties that are investigating the damage at the scene are normally in the best position to give dependable information on the nature and extent of the damage. All repair party personnel should be trained to make prompt, accurate, and complete reports to damage control central. Items that should normally be reported to damage control central include:

1. Description of important things seen, heard, or felt by personnel.
2. Location and nature of fires, smoke, and toxic gases.
3. Location and nature of progressive flooding.
4. Overall extent and nature of flooding.
5. Structural damage to longitudinal strength members.
6. Location and nature of damage to vital piping and electrical systems.
7. Local progress made in contolling fire; halting flooding; isolating damage systems; and rigging jury piping, casualty power, and emergency communications.
8. Compartment-by-compartment information on flooding, including depth of liquid in each flooded compartment.
9. Condition of boundaries (decks, bulkheads, and closures) surrounding each flooded compartment.
10. Local progress made in reclaiming compartments by plugging, patching, shoring, and removing loose water.
11. Areas in which damage is suspected but cannot be reached or verified. These areas must receive repetitive inspections until issues are resolved.

The DCA must ascertain just what information the commanding officer desires concerning the extent of the damage incurred and the repairs needed.

Immediate Local Measures

Immediate local measures are those actions taken by repair parties at the scene of the damage. In general, these measures include all on-scene efforts to investigate the damage, to report to damage control central, and to accomplish the following:

1. Establish flooding boundaries by selecting a line of intact bulkheads and decks to which the flooding may be held and by rapidly plugging, patching, and shoring to make these boundaries watertight and dependable.
2. Control and extinguish fires.
3. Establish secondary flooding boundaries by selecting a second line of bulkheads and decks to which the flooding may be held if the first flooding boundaries fail.
4. Advance flooding boundaries by moving in toward the scene of the damage, plugging, patching, shoring, and removing loose water.
5. Isolate damage to machinery, piping, and electrical systems.
6. Restore piping systems to service by the use of patches, jumpers, clamps, couplings, and so on.
7. Rig casualty power.
8. Rig emergency communications and lighting.
9. Rescue personnel and care for the wounded.
10. Remove wreckage and debris.
11. Cover or barricade dangerous areas.
12. Ventilate compartments that are filled with smoke or toxic gases.
13. Take measures to counteract the effects of nuclear, biological, and chemical contamination or weapons.

Immediate local measures for the control of damage are of vital importance. It is not necessary for damage control central to decide on these measures; rather, they should be carried out automatically and rapidly by repair parties. However, damage control central should be continuously and accurately advised of the progress made by each party so that the efforts of all repair parties may be coordinated to the best advantage.

Practical Damage Control

Both the immediate local measures and the overall ship survival measures have, of course, the common aim of saving the ship and restoring her to service. The following subsections deal with the practical methods used to achieve this aim: controlling fires, controlling flooding, repairing structural damage, and restoring vital services.

It should be noted that controlling the effects of nuclear, biological, and chemical warfare weapons or agents may in some situations take precedence over other damage control measures. Because of the complex nature of NBC defense, this subject is treated separately in a later section of this text.

Control of Fires

Fire is a constant potential hazard on board ship. All possible measures must be taken to prevent its occurrence or to bring about its rapid control and extinguishment. In many cases, fire occurs in conjunction with other damage, as a result of enemy action, weather, or accident. Unless fire is rapidly and effectively extinguished, it may cause more damage than the initial casualty, and it may, in fact, cause the loss of a ship even after other damage has been repaired or minimized.

Fires are classified according to the nature of the combustible material. Class Alfa fires are those that involve ordinary combustible material such as wood, paper, mattresses, canvas, and so on. Class Bravo fires are those that involve the burning of oils, greases, gasoline, and similar materials. Class Charlie fires are those that occur in electrical equipment. Class Delta fires are those that involve certain metals such as magnesium, potassium, powdered aluminum zinc, sodium, titanium, zirconium, and others.

Class Alfa fires are extinguished by the use of water. Class Bravo fires are extinguished chiefly by smothering with foam, fog, steam, or purple K powder dry chemical agent (as appropriate for the particular fire). Class Charlie fires are preferably extinguished by the use of carbon dioxide. Because of the danger of electric shock, a solid stream of water must never be used to extinguish a Class Charlie fire. Class Delta fires are presently extinguished by using large amounts of water. Personnel safety is of prime concern when fighting this class of fire; toxic gasses, possible hydrogen explosions, splattering of molten metal, and intense heat are its prime characteristics. Presently, intensive research is being conducted on better methods of attack and more suitable extinguishing agents.

Firefighting Equipment

Because of the large amounts of explosives, fuels, and other flammable materials on board ship, getting a fire under control and then extinguished is of prime importance. As previously mentioned, firefighting (and other aspects of damage control) is not the responsibility of just the hull maintenance technician. At any time you may be called upon to serve on a repair party, or you may be the only person present to combat a fire. If you don't know how to use the equipment available, or what equipment to use, the result could be disastrous. In the pages that follow, we will describe basic firefighting equipment and give some applicable safety precautions to be followed.

Firemain Systems

The firemain system receives water pumped from the sea and distributes it to fireplugs, sprinkling systems, flushing systems, auxiliary machinery cooling water systems, washdown systems, and other systems as required.

There are three basic types of firemain systems used on board naval ships: the single main system, the horizontal loop system, and the vertical loop system. The type of firemain system installed in any particular ship depends upon the characteristics and functions of the ship. Small ships generally have single main firemain systems; large ships usually have one of the loop systems or a composite system which is some combination or variation of the three basic types.

The single main firemain system consists of one main that extends fore and aft. The main is generally installed near the centerline of the ship, extending as far forward and as far aft as necessary. The horizontal loop firemain system consists of two single fore-and-aft cross-connected mains. The two mains are installed in the same horizontal plane but are separated athwartships as far as practicable. In general, the two mains are installed on the damage control deck. The vertical loop firemain system consists of two single fore-and-aft cross-connected mains. The two mains are separated both athwartship and vertically. As a rule, the lower main is located below the highest complete watertight deck.

In the larger ships most fireplug outlets are 2½ inches in diameter. A wye gate provides two 1½-inch outlets, or a single reducing fitting can be used to provide a single 1½-inch outlet. In destroyers and smaller ships, fireplug outlets are 1½ inches throughout the ship.

A quick-cleaning strainer is attached directly to the fireplug, and to it is attached a 2½-inch hose or a wye gate. The strainer (figure 26–1) collects foreign matter, such as marine growth and incrustation particles that otherwise would pass through the hose and possibly clog the nozzle. When the handle is in the open position (parallel to the strainer), the water is forced through the strainer into the hose.

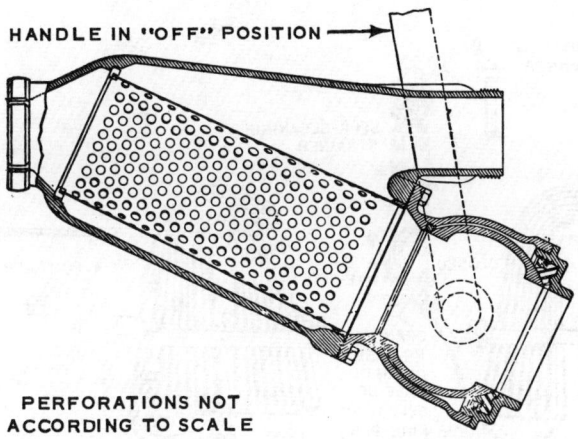

Figure 26–1. Quick-cleaning strainer.

When the handle is in the off position, the strainer is open, and the water flushes accumulated matter out of the strainer.

Sprinkler Systems

Sprinkler systems are installed in magazines, turrets, ammunition handling rooms, spaces where flammable materials are stored, and hangar bays of aircraft carriers. Water for these systems is piped from the firemain. Some sprinkler systems are automatically triggered when the protected compartment reaches a certain temperature, but most are operated by manually controlled valves, either locally or from remote locations.

Hoses

The standard naval firehose has an interior lining of rubber covered with cotton jackets. It comes in 50-foot lengths with a female coupling at one end and a male coupling at the other end. The female coupling is connected to the fireplug; the male coupling is connected to another length of hose or receives a nozzle.

Ships of destroyer size and smaller use 1½-inch hose. Larger ships use 2½-inch hose on the weather deck, and 1½-inch below deck and in the superstructure.

One or more racks are provided at each fireplug for stowage of firehose. The hose must be faked on the rack so that it is free-running and with the ends hanging downward so that the couplings are ready for instant use. On board large ships, each weather deck fire station has 100 feet of 2½-inch hose faked on a rack and connected to the plug. Below deck, 200 feet (two lines) of 1½-inch hose is stowed by

Principles of Practical Damage Control   **385**

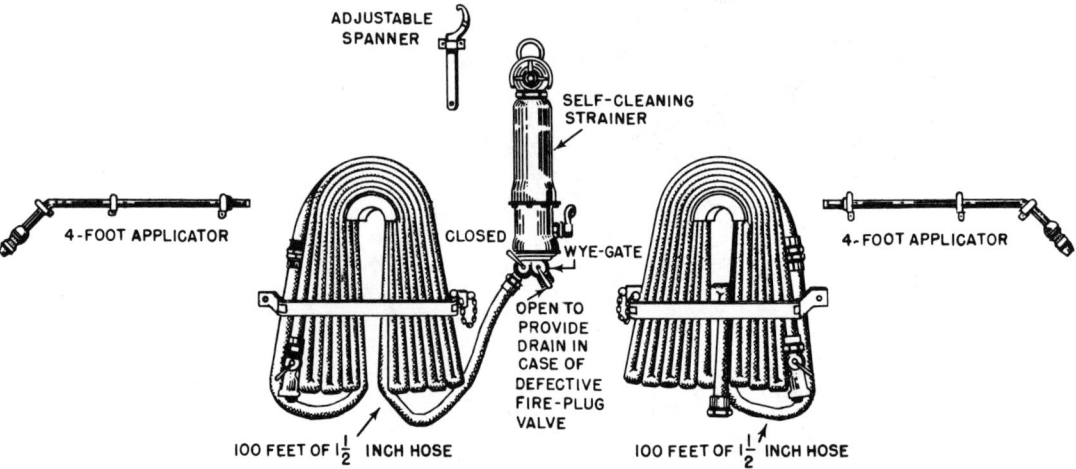

Figure 26–2. Fire station equipment.

each plug, but only one line is connected to the plug. In smaller ships, 100 feet of 1½-inch hose is faked on the racks, with 50 feet connected to the plug. A spanner wrench for tightening connections, and one or two applicators also are stowed at each fire station (figure 26–2).

Spare lengths of hose are rolled and stowed in repair lockers. To roll a hose, lay it out straight, then fold it back so that the male end is on top and about 4 feet from the female end. Starting at the fold, roll the hose, ensuring that the male end stays inside the roll so that its threads are protected. When it is necessary to use the hose, set it on deck (female end down), and simply give it a shove to unroll it. (Note: Always take the male end to the scene of action.)

Before connecting two lengths of hose, make sure that the coupling gaskets are in place. One man can make a connection by stepping on the hose just behind the male fitting, which will cause the end to point upward. Grasp the female end of the other hose and couple it with the male end, take a half-turn to the right to seat the threads, and then turn it in the other direction to tighten it.

Double female couplings are available for connecting two male ends, as when making a jumper across a broken fire main or for connecting a male coupling to a plug. Double male couplings are used to connect two female ends (e.g., to connect a nozzle to a female coupling).

All-purpose Nozzle

The all-purpose nozzle can produce a solid stream of water, high-velocity fog, or low-velocity fog. It is available for both 1½-inch and 2½-inch hose. The nozzle can be adjusted easily and quickly by means of a handle (figure 26–3).

The solid stream comes out of the topmost of two openings. High-velocity fog is produced by a nozzle tip in the lower opening. Low-velocity fog is produced by replacing the nozzle tip with an applicator. Types of applicators and the nozzles they are used with are shown in figure 26–4.

Keep the threads of all couplings and fog heads free of dirt and vertigris by using a wire brush. Never use brightwork polish on applicator heads or nozzle tips, as the polish will harden and could clog the orifices.

Never pick up a charged hose by the handle of the all-purpose nozzle. (A hose is charged when it has water available at the nozzle). The handle could easily move to the fog or open position, and the high water pressure (about 100 psi) could cause the hose to whiplash dangerously, possibly injuring personnel or damaging equipment.

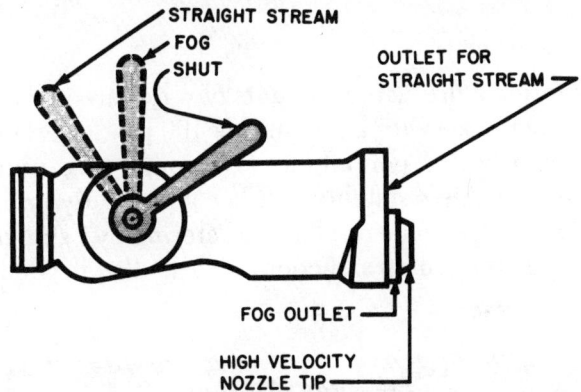

Figure 26–3. Naval all-purpose nozzle.

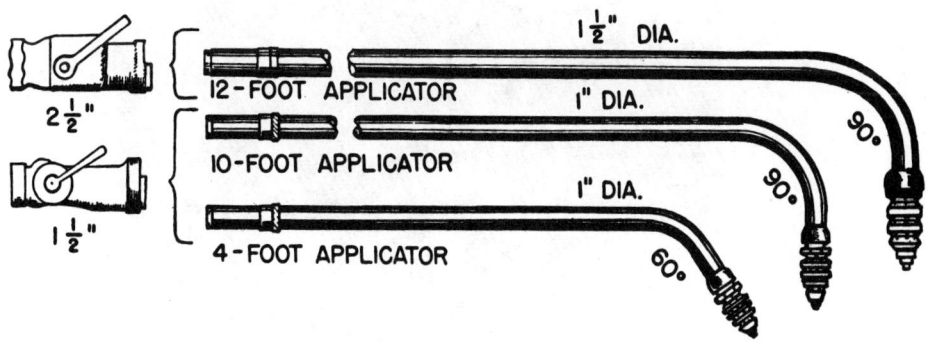

Figure 26–4. Standard applicators.

Principles of Practical Damage Control 387

Pumps

There are three types of portable gasoline-driven pumps, but our discussion is confined to the model P–250. This pump is replacing the P–60 (handy billy) and P–500 pumps, which are considered obsolete.

The P–250 (figure 26–5) is a self-priming centrifugal pump designed for firefighting and dewatering operations. It has a pumping capacity of 250 gallons per minute, with a suction lift of between 16 and 20 feet.

The pump has a 3-inch intake and a 2½-inch outlet, to which may be attached a 2½-inch hose, or a trigate having either three 1½-inch outlets for firefighting, or two 1½-inch and one 2½-inch outlets for dewatering with an eductor.

Like any other gasoline engine, the P–250 produces carbon monoxide. When it is used below deck, its exhaust must be led outside the ship through ports provided for that purpose. Another precaution is never to run the engine in a space containing explosive vapors.

Eductors

When fighting a fire, large amounts of water are introduced in the ship. A 2½-inch hose with a pressure of 100 psi, for example, pumps nearly a ton of water per minute. Obviously this water must be removed, or the ship's stability will become greatly impaired.

The P–250 pump can be used for dewatering by a straight pumping action (placing the pump's suction hose in the flooded space and

Figure 26–5. P–250 portable pump.

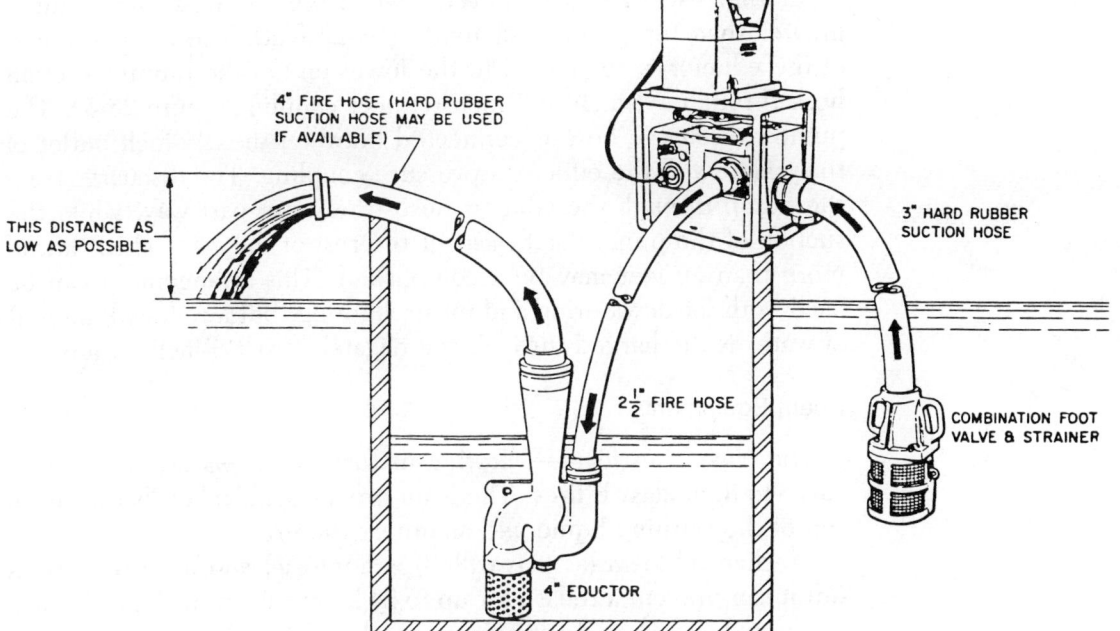

Figure 26–6. P–250 and eductor rigged for dewatering.

directing the discharge overboard), at a rate of about 300 gallons per minute. (The rate is higher than for firefighting because of lower discharge pressure requirements.)

While the foregoing pumping rate may be sufficient in some instances, it usually is desirable to pump at a greater rate. The dewatering rate of a single pump can be nearly doubled by utilizing a type of pump called an eductor.

Figure 26–6 shows how an eductor operates. Water drawn in through the foot valve is discharged through the pump's 2½-inch discharge hose, which is connected to the eductor. Force of the water passing through a jet in the eductor creates a partial vacuum that draws in additional water through the eductor's 4-inch intake. Thus, the total amount of water discharged overboard is the normal capacity of the pump plus the water picked up by the eductor.

Eductors also are used when the liquid to be pumped (such as gasoline or other flammables) cannot be handled directly by the pump itself. In such instances, the pump's suction hose is placed overboard or in any source of uncontaminated water. This practice eliminates chances of damaging the pump or of igniting the flammable liquid. However, the dewatering rate is reduced by an amount equal to the pump's capacity.

Another use for eductors occurs when the required suction lift is greater than the pump's capability (16–20 feet). The discharge end of the eductor is connected to the lower end of the pump's suction hose through a 4-inch to 3-inch reducer coupling (figure 26–7). The pump's discharge hose is connected between the 2½-inch outlet of the trigate and the eductor's pressure coupling. The discharge from the pump through the eductor raises the water part way, while the suction of the pump itself raises it the rest of the way. Thus, lifts of more than 20 feet may be accomplished. This arrangement can be used both for dewatering and for firefighting, but a reduced amount of water is discharged through the trigate's two 1½-inch outlets.

Foam Equipment

The Navy uses foam—a frothy mixture of air, water, and chemicals—to fight class B fires. The foam provides a blanket that floats on top of the burning liquid and smothers the fire.

Mechanical foam (as it is called) is nontoxic, and once the fire is out it can prevent a reflash for up to 24 hours. Foam will not damage surfaces, but it should not be used on class C fires because of obvious cleanup problems.

Mechanical foam is produced by supplying a mixture of foam-forming concentrate and water under pressure to the mixing area of a special nozzle where air is added to the mixture. A proportioner is required to induct the correct proportion of concentrate into the water stream (6 percent concentrate, 94 percent water). Both a straight-type suction proportioner and a water motor proportioner may be used.

Simplest of the proportioners is the straight type shown in figure 26–8. The mechanical foam nozzle is a 21-inch length of 2-inch-di-

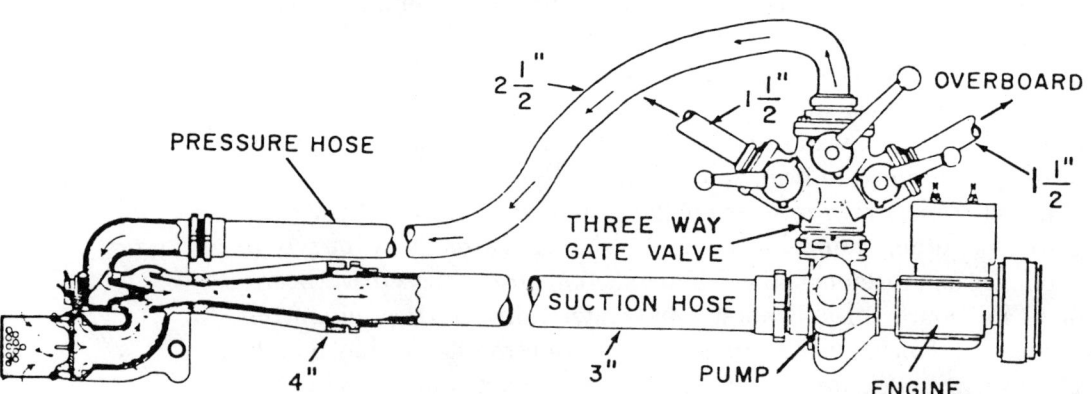

Figure 26–7. *Eductor and pump arrangement for increasing suction lift.*

390 Introduction to Naval Engineering

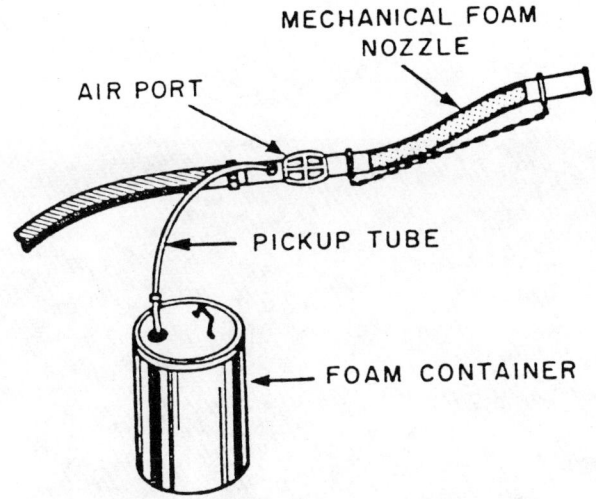

Figure 26–8. Straight-type suction proportioner.

ameter flexible metal or asbestos hose, with a suction chamber (the proportioner) and an air port at the pump end. The pickup tube siphons the correct amount of foam concentrate into the water stream. As the stream crosses the air port, air is mixed with the solution to produce the foam. Foam concentrate comes in 5-gallon cans, which will last about 1½ minutes and produce 660 gallons of foam.

The FP–180 water motor proportioner (figure 26–9) consists of a foam liquid pump driven by a water motor. It has 2½-inch connections at both the inlet and outlet sides, and uses two ½-inch pickup tubes. When the foam valve is in either of two foam positions, water flow through the motor causes the foam pump to inject the proper amount of foam liquid into the water stream. With the valve in the off position, no foam is delivered, and the fireline (hose) is available for conventional firefighting. The valve should always be in the off position except when actually drafting foam.

Foam can be produced with the water motor proportioner by using any one of four discharge combinations:

    1. One 1½-inch discharge line, reduced from the normal 2½-inch outlet. The proportioner may be supplied by either a 1½-inch or 2½-inch hose.

    2. Two 1½-inch discharge lines through use of a wye gate outlet.

    3. Three 1½-inch discharge lines, utilizing a trigate at the proportioner's outlet.

    4. One 2½-inch line equipped with a high-capacity fog-foam nozzle.

Figure 26-9. FP-180 water motor proportioner.

Except where noted otherwise, the mechanical foam nozzle is used (without its pickup tube).

The best-quality foam is produced by the first two of the above methods; the poorest by the third. Discharge lines should not exceed 100 feet in length. When the need for foam is over, always flush the foam pump by running the proportioner for 2 or 3 minutes in the off position.

Portable Extinguishers

Two types of portable fire extinguishers are used by the Navy: carbon dioxide (CO_2) and dry chemical. Each is effective in fighting both class B and class C fires.

Carbon dioxide extinguishers are used mainly for putting out electrical fires, but they are effective on any small fire, including burning oil, gasoline, paint, and trash cans. The carbon dioxide, being heavier than air, forms a smothering blanket over the fire. Maximum range of the extinguisher is 5 feet from the horn.

To use the extinguisher, remove the locking pin from the valve, grasp the insulated handle of the horn with one hand, and squeeze the grip with the other. If in the open, approach the fire from the windward side. Direct the discharge at the base of the flames, sweeping the horn back and forth. The valve may be opened and closed as necessary. For continous operation, a D-ring is provided that slips over the operating handle.

When CO_2 is released from the container, it expands rapidly to 450 times its stored volume. This rapid expansion causes the gas's temperature to drop to $-110°F$, and forms carbon dioxide "snow." Don't permit the snow to come in contact with your skin, as it will cause painful blisters.

Dry chemical extinguishers are provided primarily for use on class B fires. The chemical used is potassium bicarbonate (similar to baking soda), called purple K powder, or simply PKP. It is nontoxic and is four times as powerful as CO_2 for extinguishing fuel fires. PKP is effective on class C fires, but it should not be used on internal fires in gas turbines or jet engines, as it leaves a residue that cannot be completely removed without disassembly of the engine.

The dry chemical extinguisher (figure 26–10) is of the 18-pound size, and uses CO_2 as the expellent gas. The extinguisher shell is not pressurized until it is to be used. Operating procedures are as follows:

1. Pull the locking pin from the seal cutter assembly.
2. Sharply strike the puncture lever to cut the gas cartridge seal. The extinguisher is now charged and ready for use.

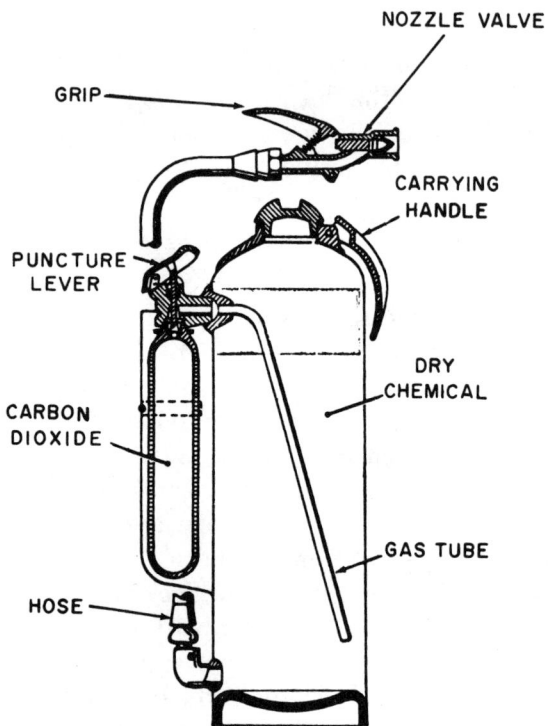

Figure 26–10. Portable dry chemical extinguisher.

3. Discharge the chemical in short bursts by squeezing the grip on the nozzle. Aim the discharge at the base of the flames and sweep it rapidly from side to side. If the fire's heat is intense, a short burst of powder into the air will provide a heat shield.

4. When finished, invert the cylinder, squeeze the discharge lever, and tap the nozzle on the deck. This action releases all pressure and clears the hose and nozzle of powder. If not cleared, the PKP could cake and cause difficulty the next time the extinguisher is used.

Dry chemical is an excellent firefighting agent, but its effects are temporary. It has no cooling effect and provides no protection against reflash of the fire. Therefore, it should always be backed up by foam. In confined spaces, PKP should be used sparingly, consistent with extinguishing the fire, as unnecessarily long discharges reduce visibility, render breathing difficult, and induce coughing.

Fire Extinguishing Systems

Fixed fire extinguisher installations are provided in several loca-

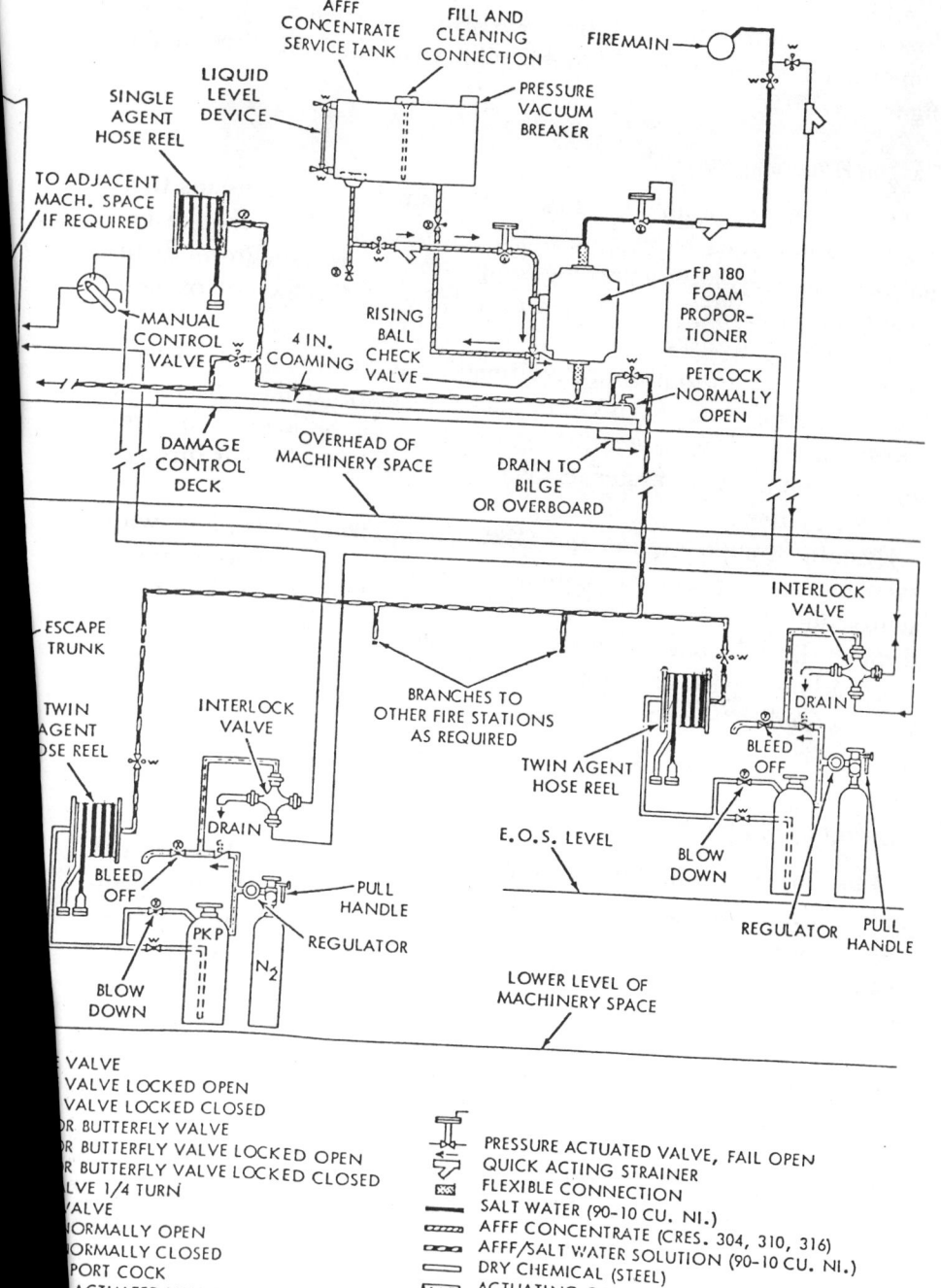

Figure 26–12. *Typical shipboard AFFF/PKP layout.*

tions, such as machinery spaces and hangar decks. Some have already been discussed, such as the sprinkler system. Here we consider carbon dioxide and light water installations.

Carbon Dioxide. Fixed installations of carbon dioxide extinguishers provide a dependable, ready means of flooding spaces that present a greater than ordinary fire hazard. Cylinders of the system have a 50-pound capacity and are mounted either singly or in banks of two or more. Installed CO_2 extinguishers are either the hose-and-reel type for machinery spaces, or the flooding system for spaces not continually occupied by personnel, such as paint lockers.

The hose-and-reel installation consists of two cylinders, a length of CO_2 hose on a reel, and a horn-shaped nozzle equipped with a second control valve. A cable runs from the release control on the cylinders to an operating handle near the reeled hose. There may be local control at the cylinders.

The CO_2 flooding system consists of two or more cylinders connected by leads from their valve outlets to a manifold. Fixed piping extends from the manifold to various parts of the space to be flooded. Cables run from the valve control mechanism to pull boxes located outside of the space containing the cylinders. To release the CO_2, the firefighter breaks the glass in the front of the pull box, reaches in, and pulls the handle. Most of these systems are equipped with visible and audible alarms to warn personnel that the release valve has been opened and the protected compartment is flooding with, or has been flooded with, CO_2.

Caution: The very qualities that make CO_2 a valuable fire-extinguishing agent also make it dangerous to life. If there is insufficient oxygen in a compartment to sustain a fire, there is not enough to sustain breathing. You should not enter a CO_2-flooded compartment without protection unless the ventilation system has been operating for at least 15 minutes.

A CO_2-flooded compartment may be entered if you use an approved naval oxygen breathing apparatus or hose (air line) mask. Otherwise, you should not work in it until a safety lamp, placed in the compartment, burns without interruption. Do not use a canister-type gas mask in place of an oxygen breathing apparatus because it merely filters the air without adding to it the oxygen you need.

Light Water. As mentioned previously, PKP cannot prevent reflash of a fire because it has no vapor suppression capability. Neither can PKP be used with protein (mechanical) foam, as it causes the foam to break down. It is fully compatible, however, with a fluorochemical agent known as aqueous film forming foam (AFFF), commonly called light water. Light water is a 6 percent concentration

Figure 26–11. Twinned agent unit.

that, when mixed with water, produces a foam. As the water drains from the foam, a vapor-tight film is formed on top of the fuel.

A light water/PKP combination will extinguish a fuel fire 1½ to 3 times faster than protein foam. The dry chemical is used to beat down the fire, and light water prevents its reflash.

Aircraft carriers have a portable light water/PKP system mounted on a truck (figure 26–11), known as a twinned agent unit (TAU). The TAU has a sphere containing 200 pounds of PKP, and a cylinder containing 80 gallons of light water concentrate. Nitrogen is used for pressurization. Discharge hoses are 1 inch for light water and ¾ inch for PKP. They are joined together near their nozzles, and may be operated independently or concurrently by one man.

A fixed TAU installation has been developed for shipboard use and is being installed in each machinery space as ships go into a yard for overhaul. The CO_2 hose reel system will no longer be used except in certain ships having a heavy concentration of electrical equipment, such as those with electric propulsion. Protein foam also is being phased out in favor of light water.

There will be a minimum of two twinned agent stations in each machinery space, each station consisting of 50 feet of twinned AFFF and PKP hoses mounted on reels, and one 125-pound PKP cylinder.

396 Introduction to Naval Engineering

As with the aircraft carrier TAU, one man can operate the unit the damage control deck will be the AFFF tank and a fixed FF proportioner for mixing light water with water from the fire (figure 26–12).

Oxygen Breathing Apparatus

Fires almost invariably produce a lot of smoke, which me the firefighter must have some means of protection from s halation when he enters a burning compartment. Air in the i vicinity of a fire may also contain toxic fumes or insufficie for breathing purposes.

The device used to permit the firefighter to breathe is breathing apparatus, or OBA as it is usually called. Unlike which only filters outside air and, therefore, cannot be fighting, the OBA generates its own oxygen and circula a closed system.

Essential components of the OBA are an airtight eyepieces and a speaking diaphragm, exhalation and in an oxygen-producing canister, and a breathing bag. F through the exhalation tube to the bottom of the ca loses carbon dioxide and moisture and gains oxygen) ing bag, and then into the lungs through the inhal 26–13).

Control of Flooding

Flooding may occur from a number of different or waterline damage, ruptured water piping, th tities of water for firefighting or counterfloodir maintenance of boundaries are all possible caus ship. It should be noted that ballasting fuel o after the oil has been removed is not conside ballasting merely consists of replacing one order to maintain the ship in a condition of damage.

If a ship suffers such extensive damage th trimming, and settling in the water, the cl down within a very few minutes. If, on th listing, trimming, and settling shortly af is not likely to sink at all unless progre occur. Thus there is an excellent chance not sink immediately. There is no case

Figure 26–13. OBA.

suddenly after she stopped listing, trimming, and settling, except in cases where progressive flooding occurred.

The control of flooding requires that the amount of water entering the hull be restricted or entirely stopped. The removal of flooding water cannot be accomplished until flooding boundaries have been established. Pump capacity should never be wasted on compartments that cannot be quickly and effectively made tight. If a compartment fills rapidly, it is a sign that pumping capacity will be wasted until the openings have been plugged or patched. The futility of merely circulating seawater should be obvious.

Once flooding boundaries have been established, the removal of the flooding water should be undertaken on a systematic basis. Loose water—that is, water with free surface—and water that is located high in the ship should be removed first. Compartments that are solidly flooded and are low in the ship are generally dewatered last, unless the flooding is sufficiently off center to cause a serious list. Compartments must always be dewatered in a sequence that will contribute to the overall stability of the ship. For example, a ship could be capsized if low, solidly flooded compartments were dewatered while water still remained in high, partially flooded compartments.

In order to know which compartments should be dewatered first, it is necessary to know the effect of flooding on all ship's compartments. This information is given in the flooding effect diagram in the ship's Damage Control Book. The flooding effect diagram consists of a series of plan views of the ship at various levels, showing all watertight, oiltight, airtight, fumetight, and fire-retarding subdivisions. Compartments on the flooding effect diagram are colored in the following way:

1. If flooding the compartment results in a decrease in stability because of high weight, free surface effect, or both, the compartment is colored pink.

2. If flooding the compartment improves stability even though free surface exists, the compartment is colored green.

3. If flooding the compartment improves stability when the compartment is solidly flooded but impairs stability when a free surface exists, the compartment is colored yellow.

4. If flooding the compartment has no very definite effect on stability, the compartment is left uncolored.

The flooding effect diagram also shows the weight of salt water (in tons) required to fill the compartment; this is indicated by a numeral in the left-hand corner. In addition, the transverse moment of the weight (in foot-tons) about the centerline of the ship is indicated for all compartments that are not symmetrical about the centerline.

Facilities for dewatering compartments consist of the fixed drainage systems of the ship and portable equipment such as electric submersible pumps, P–250 pumps, and eductors. In a large combat ship, the fixed drainage systems have a total pumping capacity of about 12,200 gallons per minute—less, it might be noted, than the amount of water admitted by a hole 1 square foot in size in an area located 15 feet below the waterline.

Portable submersible pumps used on board naval ships are centrifugal pumps driven by a water-jacketed constant-speed AC or DC motor. When a submersible pump is being used to dewater a compartment, the pump is lowered into the water, and a discharge hose is led to the nearest point of discharge. Since the delivery of the pump increases as the discharge head decreases, dewatering can be accomplished faster if the water is discharged at the lowest practicable point, and if the discharge hose is short and free from kinks. When it is necessary to dewater against a high discharge head, two submersible pumps can be used in tandem, as shown in figure 26–14. The pump at the lower level lifts water to the suction side of the pump at the higher level.

tions, such as machinery spaces and hangar decks. Some have already been discussed, such as the sprinkler system. Here we consider carbon dioxide and light water installations.

Carbon Dioxide. Fixed installations of carbon dioxide extinguishers provide a dependable, ready means of flooding spaces that present a greater than ordinary fire hazard. Cylinders of the system have a 50-pound capacity and are mounted either singly or in banks of two or more. Installed CO_2 extinguishers are either the hose-and-reel type for machinery spaces, or the flooding system for spaces not continually occupied by personnel, such as paint lockers.

The hose-and-reel installation consists of two cylinders, a length of CO_2 hose on a reel, and a horn-shaped nozzle equipped with a second control valve. A cable runs from the release control on the cylinders to an operating handle near the reeled hose. There may be local control at the cylinders.

The CO_2 flooding system consists of two or more cylinders connected by leads from their valve outlets to a manifold. Fixed piping extends from the manifold to various parts of the space to be flooded. Cables run from the valve control mechanism to pull boxes located outside of the space containing the cylinders. To release the CO_2, the firefighter breaks the glass in the front of the pull box, reaches in, and pulls the handle. Most of these systems are equipped with visible and audible alarms to warn personnel that the release valve has been opened and the protected compartment is flooding with, or has been flooded with, CO_2.

Caution: The very qualities that make CO_2 a valuable fire-extinguishing agent also make it dangerous to life. If there is insufficient oxygen in a compartment to sustain a fire, there is not enough to sustain breathing. You should not enter a CO_2-flooded compartment without protection unless the ventilation system has been operating for at least 15 minutes.

A CO_2-flooded compartment may be entered if you use an approved naval oxygen breathing apparatus or hose (air line) mask. Otherwise, you should not work in it until a safety lamp, placed in the compartment, burns without interruption. Do not use a canister-type gas mask in place of an oxygen breathing apparatus because it merely filters the air without adding to it the oxygen you need.

Light Water. As mentioned previously, PKP cannot prevent reflash of a fire because it has no vapor suppression capability. Neither can PKP be used with protein (mechanical) foam, as it causes the foam to break down. It is fully compatible, however, with a fluorochemical agent known as aqueous film forming foam (AFFF), commonly called light water. Light water is a 6 percent concentration

Figure 26–11. Twinned agent unit.

that, when mixed with water, produces a foam. As the water drains from the foam, a vapor-tight film is formed on top of the fuel.

A light water/PKP combination will extinguish a fuel fire 1½ to 3 times faster than protein foam. The dry chemical is used to beat down the fire, and light water prevents its reflash.

Aircraft carriers have a portable light water/PKP system mounted on a truck (figure 26–11), known as a twinned agent unit (TAU). The TAU has a sphere containing 200 pounds of PKP, and a cylinder containing 80 gallons of light water concentrate. Nitrogen is used for pressurization. Discharge hoses are 1 inch for light water and ¾ inch for PKP. They are joined together near their nozzles, and may be operated independently or concurrently by one man.

A fixed TAU installation has been developed for shipboard use and is being installed in each machinery space as ships go into a yard for overhaul. The CO_2 hose reel system will no longer be used except in certain ships having a heavy concentration of electrical equipment, such as those with electric propulsion. Protein foam also is being phased out in favor of light water.

There will be a minimum of two twinned agent stations in each machinery space, each station consisting of 50 feet of twinned AFFF and PKP hoses mounted on reels, and one 125-pound PKP cylinder.

As with the aircraft carrier TAU, one man can operate the unit. On the damage control deck will be the AFFF tank and a fixed FP–180 proportioner for mixing light water with water from the fire main (figure 26–12).

Oxygen Breathing Apparatus

Fires almost invariably produce a lot of smoke, which means that the firefighter must have some means of protection from smoke inhalation when he enters a burning compartment. Air in the immediate vicinity of a fire may also contain toxic fumes or insufficient oxygen for breathing purposes.

The device used to permit the firefighter to breathe is the oxygen breathing apparatus, or OBA as it is usually called. Unlike a gas mask, which only filters outside air and, therefore, cannot be used for firefighting, the OBA generates its own oxygen and circulates it through a closed system.

Essential components of the OBA are an airtight faceplate with eyepieces and a speaking diaphragm, exhalation and inhalation tubes, an oxygen-producing canister, and a breathing bag. Exhaled air flows through the exhalation tube to the bottom of the canister (where it loses carbon dioxide and moisture and gains oxygen), into the breathing bag, and then into the lungs through the inhalation tube (figure 26–13).

Control of Flooding

Flooding may occur from a number of different causes. Underwater or waterline damage, ruptured water piping, the use of large quantities of water for firefighting or counterflooding, and the improper maintenance of boundaries are all possible causes of flooding on board ship. It should be noted that ballasting fuel oil tanks with seawater after the oil has been removed is not considered a form of flooding; ballasting merely consists of replacing one liquid with another in order to maintain the ship in a condition of maximum resistance to damage.

If a ship suffers such extensive damage that she never stops listing, trimming, and settling in the water, the chances are that she will go down within a very few minutes. If, on the other hand, a ship stops listing, trimming, and settling shortly after the damage occurs, she is not likely to sink at all unless progressive flooding is allowed to occur. Thus there is an excellent chance of saving any ship that does not sink immediately. There is no case on record of a ship sinking

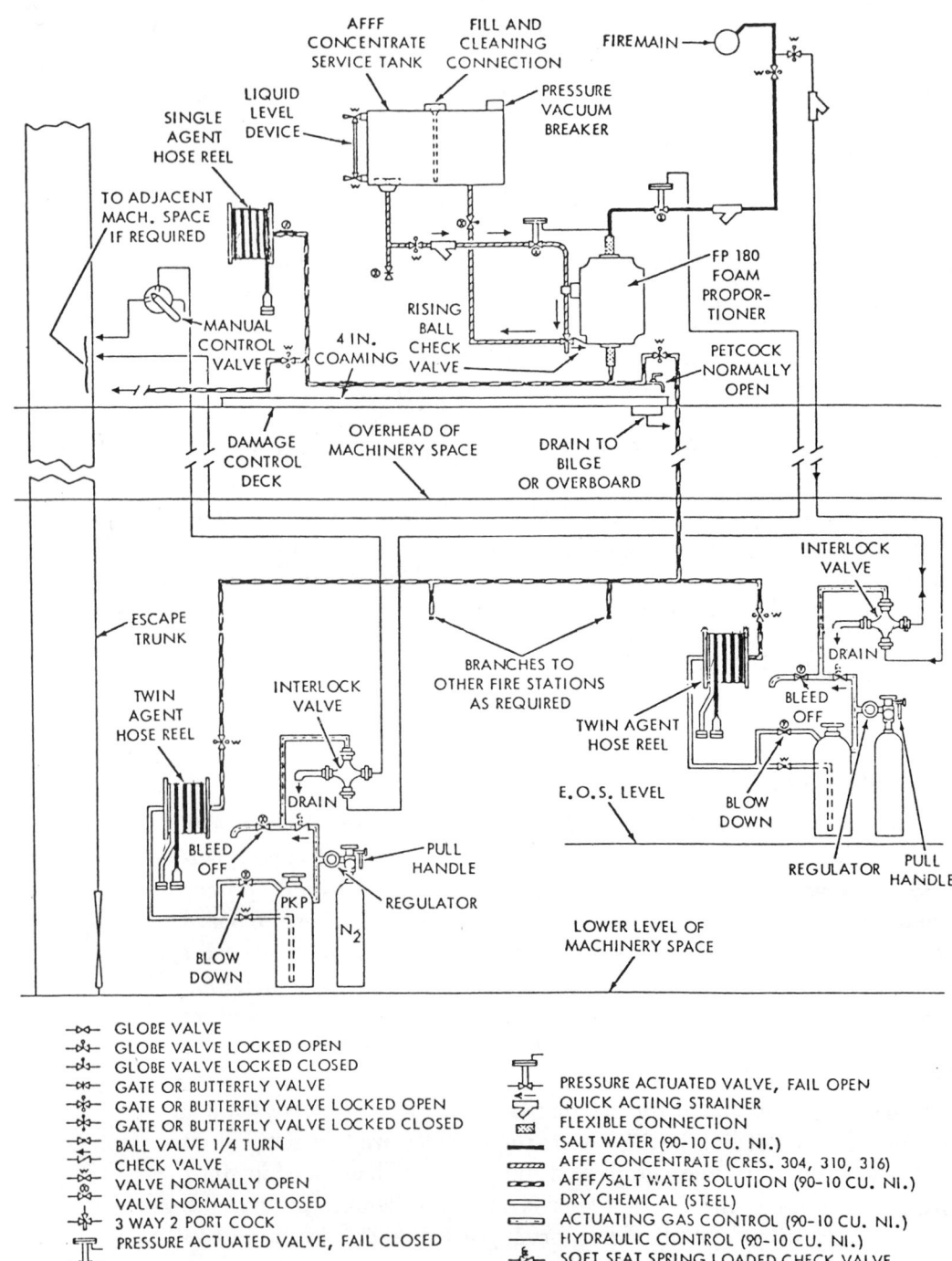

Figure 26–12. Typical shipboard AFFF/PKP layout.

Figure 26–13. OBA.

suddenly after she stopped listing, trimming, and settling, except in cases where progressive flooding occurred.

The control of flooding requires that the amount of water entering the hull be restricted or entirely stopped. The removal of flooding water cannot be accomplished until flooding boundaries have been established. Pump capacity should never be wasted on compartments that cannot be quickly and effectively made tight. If a compartment fills rapidly, it is a sign that pumping capacity will be wasted until the openings have been plugged or patched. The futility of merely circulating seawater should be obvious.

Once flooding boundaries have been established, the removal of the flooding water should be undertaken on a systematic basis. Loose water—that is, water with free surface—and water that is located high in the ship should be removed first. Compartments that are solidly flooded and are low in the ship are generally dewatered last, unless the flooding is sufficiently off center to cause a serious list. Compartments must always be dewatered in a sequence that will contribute to the overall stability of the ship. For example, a ship could be capsized if low, solidly flooded compartments were dewatered while water still remained in high, partially flooded compartments.

In order to know which compartments should be dewatered first, it is necessary to know the effect of flooding on all ship's compartments. This information is given in the flooding effect diagram in the ship's Damage Control Book. The flooding effect diagram consists of a series of plan views of the ship at various levels, showing all watertight, oiltight, airtight, fumetight, and fire-retarding subdivisions. Compartments on the flooding effect diagram are colored in the following way:

1. If flooding the compartment results in a decrease in stability because of high weight, free surface effect, or both, the compartment is colored pink.

2. If flooding the compartment improves stability even though free surface exists, the compartment is colored green.

3. If flooding the compartment improves stability when the compartment is solidly flooded but impairs stability when a free surface exists, the compartment is colored yellow.

4. If flooding the compartment has no very definite effect on stability, the compartment is left uncolored.

The flooding effect diagram also shows the weight of salt water (in tons) required to fill the compartment; this is indicated by a numeral in the left-hand corner. In addition, the transverse moment of the weight (in foot-tons) about the centerline of the ship is indicated for all compartments that are not symmetrical about the centerline.

Facilities for dewatering compartments consist of the fixed drainage systems of the ship and portable equipment such as electric submersible pumps, P–250 pumps, and eductors. In a large combat ship, the fixed drainage systems have a total pumping capacity of about 12,200 gallons per minute—less, it might be noted, than the amount of water admitted by a hole 1 square foot in size in an area located 15 feet below the waterline.

Portable submersible pumps used on board naval ships are centrifugal pumps driven by a water-jacketed constant-speed AC or DC motor. When a submersible pump is being used to dewater a compartment, the pump is lowered into the water, and a discharge hose is led to the nearest point of discharge. Since the delivery of the pump increases as the discharge head decreases, dewatering can be accomplished faster if the water is discharged at the lowest practicable point, and if the discharge hose is short and free from kinks. When it is necessary to dewater against a high discharge head, two submersible pumps can be used in tandem, as shown in figure 26–14. The pump at the lower level lifts water to the suction side of the pump at the higher level.

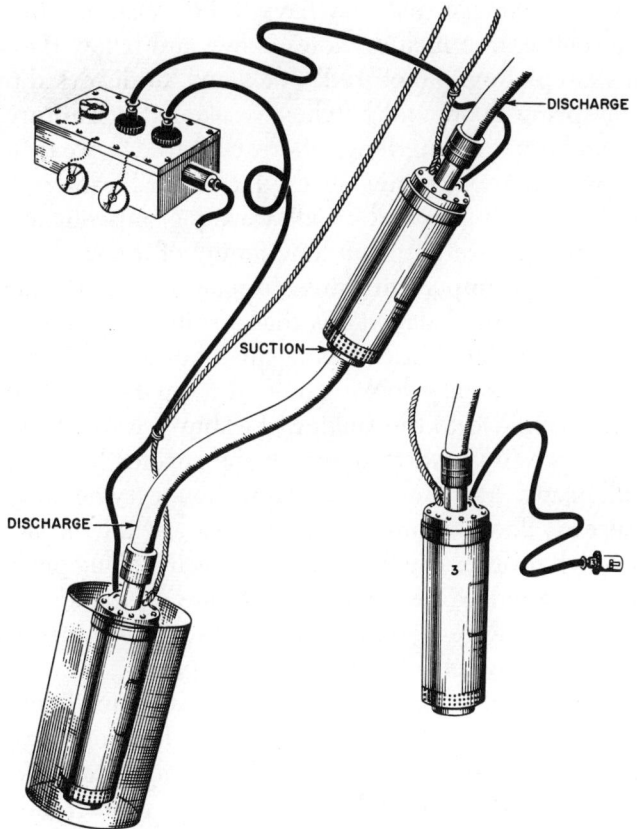

Figure 26–14. Tandem connections for submersible pumps.

The P–250 pump is of the centrifugal type; it is driven by a water-cooled gasoline engine of special design. The pump delivers 250 gallons per minute at 100 pounds per square inch pressure, with a suction lift of 16 feet. The capacity may be increased by decreasing the discharge pressure.

In order to estimate the number of pumps required to handle a flooding situation, it is necessary to consider the amount and location of the water to be removed, the capacity and availability of the installed drainage systems, and the capacity of the available portable pumps. It is also necessary to know whether the leaks are completely plugged, partially plugged, or not plugged at all—in short, it is necessary to know how much water is coming in while water is being pumped out.

Repair of Structural Damage

The kinds of damage that may have to be repaired while a ship is still in the battle area include holes above and below the waterline; cracks in steel plating; punctured, weakened, or distorted bulkheads; warped or spring doors and hatches; weakened or ruptured beams, supports, and other strength members; ruptured or weakened decks; ruptured or cracked piping; severed electrical cables; broken or pierced machinery units; and a wide variety or miscellaneous wreckage that may interfere with the functioning of the ship.

One of the most important things to remember in connection with the repair of structural damage is that a ship can sink just as easily from a series of insignificant-looking small holes as she can from one larger and more dramatic-looking hole. A natural-enough tendency— and one that can lead to the sinking of a ship— is to attack the large, obvious damage first and to overlook the smaller holes through interior bulkheads. Men sometime waste hours trying to patch large holes in already flooded compartments, disregarding the smaller holes through which progressive flooding is gradually taking place. In many situations, it would be better to concentrate on the smaller interior holes; as a rule, the really large holes in the underwater hull cannot be repaired anyway until the ship is drydocked.

Holes in the hull at or just above the waterline should be given immediate attention. Although holes in this location may appear to be relatively harmless, they are actually extremely hazardous. As the ship rolls or loses buoyancy, the holes become submerged and admit water at a level that is dangerously high above the ship's center of gravity.

The methods and materials used to repair holes above the waterline are also used, for the most part, for the repair of underwater holes. The greatest difficulty encountered in repairing underwater damage is usually the inaccessibility of the damage. If an inboard compartment is flooded, opening doors or hatches to get to the damage would result in further flooding of other compartments. In such a case, it is usually necessary to send a man wearing a shallow-water diving apparatus down into the compartment. His repair work is likely to be hampered by tangled wreckage in the water, by the absence of light to work by, and by the difficulties of trying to keep buoyant repair materials submerged.

Shoring is often used on board ship to support ruptured decks, to strengthen weakened bulkheads and decks, to build up temporary decks and bulkheads against the sea, to support hatches and doors, and to provide support for equipment that has broken loose.

The basic materials required for shoring are shores, wedges, sholes, and strongbacks. A *shore* is a portable beam. A *wedge* is a block, triangular on the sides and rectangular on the butt end. A *shole* is a flat block that may be placed under the end of a shore for the purpose of distributing the pressure. A *strongback* is a piece used to distribute pressure or to serve as an anchor for a patch.

The problem of when to shore cannot be solved by the application of any one set of rules. Sometimes the need for shoring is obvious, as in the case of damaged hatches; but sometimes dangerously weakened supports under guns or machinery may not be so readily noticed. Although shoring is sometimes done when it is not really necessary, the best general rule to follow is this: in case of doubt, it is always better to shore than to gamble on the strength of an important deck, bulkhead, hatch, or other member.

Some examples of shoring are illustrated in figure 26-15.

Restoration of Vital Services

Thus far we have considered practical damage control operations from the point of view of combatting fires, getting rid of flooding water, repairing structural damage, and in general restoring the ship to a stable and seaworthy condition. To function as a fighting unit, however, a ship must be more than stable and seaworthy—she must also be able to move. The restoration of vital services is therefore an integral part of damage control, even though it must often be accomplished after fires and flooding have been controlled.

The restoration of vital services includes making repairs to machinery and piping systems and reestablishing a source of electrical power. The casualty power system, developed as a result of war experience, has proved to be one of the most important damage control devices. The casualty power system is a simple electrical distribution system used to maintain a source of electrical supply for the most vital machinery. It is used to supply power *only* in emergencies.

Damage Control Precautions

The urgent nature of damage control operations can lead to a dangerous neglect of necessary safety precautions. Driven by the need to act rapidly, men sometimes take chances they would not even consider taking in less hazardous situations. This is unfortunate, since there are few areas in which safety precautions are as important as

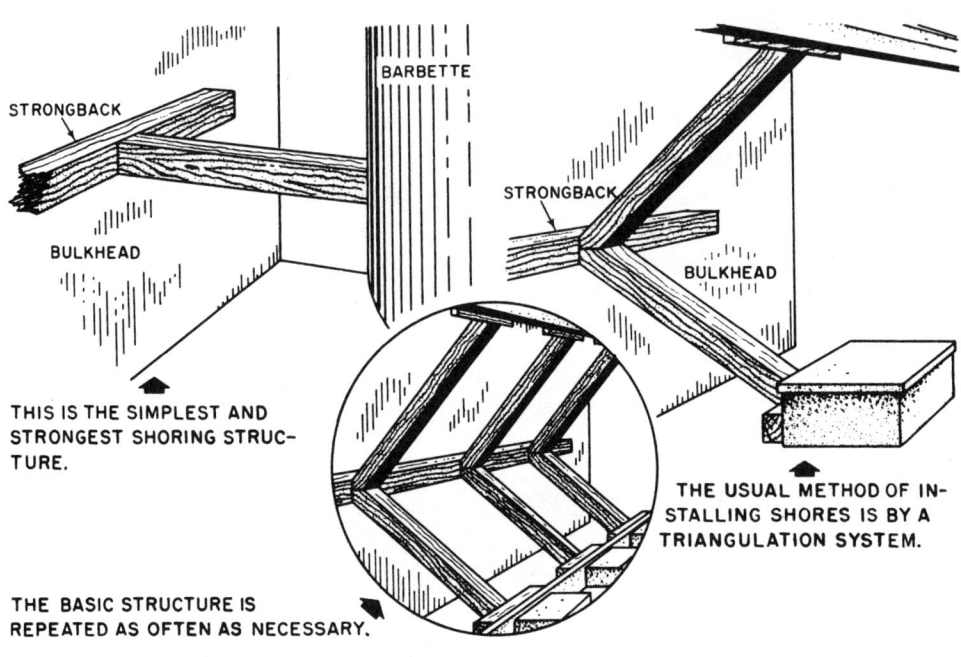

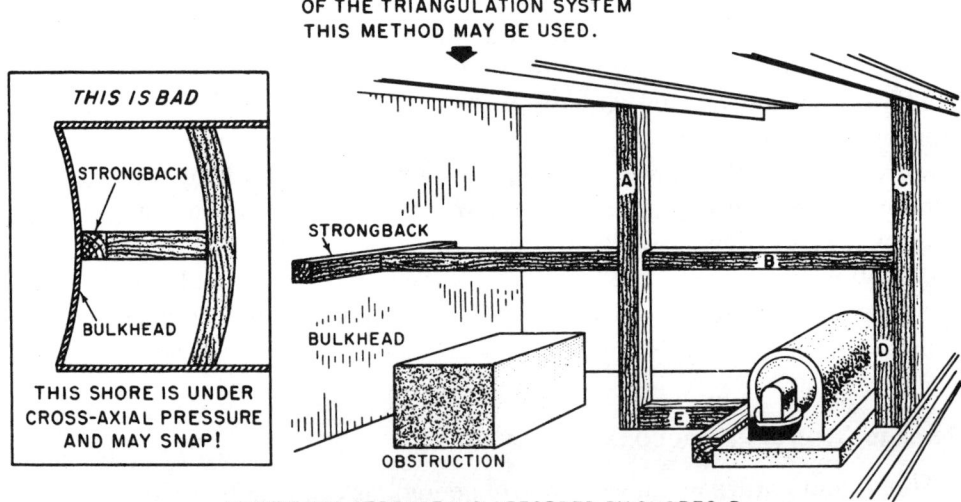

Figure 26–15. Examples of shoring.

they are in damage control. Failure to observe safety precautions can lead—and, in fact, has led—to the loss of ships.

Because damage control includes so many operations and involves the use of so many items of equipment, it is not feasible to list all the detailed precautions that must be observed. Some of the basic precautions that apply to practically all damage control work are noted briefly in the following paragraphs.

No one should be allowed to take any action to control fires, flooding, or other damage until the situation has been investigated and analyzed. Although speed is essential for effective damage control, correct action is even more important.

The extent of damage must not be underestimated. It is always necessary to remember that hidden damage may be even more severe than visible damage. Very real dangers may exist from damage that is not giving immediate trouble. For example, small holes at or just above the waterline may appear to be relatively minor; but they have been known to sink a ship.

It is extremely dangerous to assume that damage has been permanently controlled merely because fires have been put out, leaks plugged, and compartments dewatered. Fires may flare up again, plugs may work out of holes, and compartments may spring new leaks. Constant checking is required for quite some time after the damage appears to be controlled.

Doors, hatches, and other accesses should be kept open only as long as necessary while repairs are being made. Wartime records of naval ships show many cases of progressive flooding that were the direct result of failure to close doors or hatches.

No person should attempt to be a one-man damage control organization. All damage must be reported to damage control center or to a repair party before any individual action is taken. The damage control *organization* is the key to successful damage control. Separate, uncoordinated actions by individuals may actually do more harm than good.

Many actions taken to control damage can have a definite effect on ship's characteristics such as watertight integrity, stability, and weight and moment. The dangers involved in pumping large quantities of water into the ship to combat fires should be obvious. Less obvious, perhaps, is the fact that the repair of structural damage may also affect the ship's characteristics. For example, the addition of high or off-center weight produces the same general effect as high or off-center solid flooding.

While most repairs made in action would not amount to much in terms of weight shifts or additions, it is possible that a number of relatively small changes could add up sufficiently to endanger an already damaged and unstable ship. The only way to control this kind of hazard is by making sure that all damage control personnel report fully and accurately to damage control central. Ship stability problems are worked out in damage control central, but the information must come from repair personnel.

In all aspects of damage control, it is important to make full use of all available devices for the detection of hazards. Several types of instruments are available in most ships for detecting dangerous concentrations of explosive, flammable, toxic, or asphyxiating gases. Personnel should be trained to use these devices before entering potentially hazardous compartments or spaces.

27

Defense Against NBC Attack

The basic guidelines for defensive and protective actions to be taken in the event of nuclear, biological, or chemical (NBC) attack are set forth in the Repair Party Manual. On board ship, the engineering officer is responsible for maintaining this bill and ensuring that it is current and ready for immediate execution.

NBC defense measures may be divided into two phases: (1) preparatory measures taken in anticipation of attack, and (2) active measures taken immediately following an attack.

Preparatory measures to be taken before an attack include:

1. Thorough indoctrination and training of ship's personnel.
2. Removal of material that may constitute contamination hazards.
3. Masking of personnel who may be exposed (and of other personnel, as ordered).
4. Establishment of ship closure, including closing of CIRCLE WILLIAM fittings.
5. Donning of protective clothing by exposed personnel, as ordered.
6. Evasive action by the ship.
7. Activation of water washdown systems.

Active measures to be taken immediately following an attack include the following:

1. Evasive and self-protective action by personnel.
2. Evacuation and remanning of exposed stations, as ordered, by the commanding officer.
3. Decontamination of personnel.
4. Detection and prediction of contaminated areas.
5. Ventilation of contaminated spaces, once the ship is in a clean atmosphere.

From the above list, it becomes apparent that NBC defense is an enormously complex and wide-ranging subject, and one in which policies and procedures are subject to constant change. The present discussion is limited to a few of the aspects of NBC defense that are of primary practical importance on board ship. More detailed information on all aspects of NBC defense may be obtained from chapters 070 and 079 of the *Naval Ships Technical Manual.*

Protective Clothing

There are three types of clothing that are useful in NBC defense: permeable, impregnated, protective clothing; foul weather clothing; and ordinary work clothing.

Permeable protective clothing is supplied to ships in quantities sufficient to outfit 25 percent or more of the ship's complement. Permeable clothing is olive green in color. A complete outfit includes impregnated socks, gloves, trousers with attached suspenders, and jumper (parka) with attached hood. Permeable clothing is treated with a chemical agent that neutralizes possible chemical warfare agents; a chlorinated paraffin is used as a binder. The presence of these chemicals gives the permeable clothing a slight odor of chlorine and a slightly greasy or clammy feel. It is believed that the impregnation treatment should remain effective from 5 to 10 years (or possibly longer) if the clothing is stowed in unopened containers in a dry place with cool to warm temperatures and no exposure to sunlight.

Permeable protective clothing should not be worn longer than necessary, especially in hot weather; prolonged wearing may cause a rash to develop where the skin comes in contact with impregnated material.

Foul weather clothing of stock issue serves to protect ordinary clothing and the skin against penetration by liquid chemical agents and radioactive alpha particles. It also reduces the amount of vapor that penetrates to the skin. Foul weather clothing, which includes a parka, trousers, rubber boots, and gloves, is easily decontaminated.

Ordinary work clothing (including long underwear, field socks, coverall, field boots, and watch cap) is partially effective in preventing droplets of liquid chemical agents and vapors from reaching the skin. However, ordinary work clothing is not as effective as the other two types of clothing in preventing contamination. Under some conditions, personnel may wear two layers of ordinary work clothing to achieve greater protection than can be obtained with one layer.

Protective Masks

The protective mask is a very important item of protective equipment, since it protects such vulnerable areas as the eyes, the face, and the respiratory tract. The protective mask provides protection against NBC contamination by filtering the air before it is inhaled.

In general, all protective masks operate on the same principles. As the wearer inhales, air is drawn into a filtering system. This system consists of a mechanical filter that clears the air of solid or liquid particles and a chemical filling (usually activated charcoal) that absorbs or neutralizes toxic and irritating vapors. The purified air then passes to the region of the mask, where it can be inhaled. Exhaled air is expelled from the mask through an outlet valve which is so constructed that it opens only to permit exhaled air to escape.

Protective masks do not afford protection against ammonia or carbon monoxide, nor are they effective in confined spaces where the oxygen content of the atmosphere is too low (less than about 16 percent) to sustain life. When it is necessary to enter spaces where there is a deficiency of oxygen, the naval oxygen breathing apparatus (OBA) can usually be used.

Detection of NBC Contamination

The very nature of NBC contamination makes detection and identification difficult. Nuclear radiation cannot be seen, heard, felt, or otherwise perceived through the senses. Biological agents are small in size and have no characteristic color or odor to help in their identification. Although some chemical agents do have a characteristic color and odor, recently developed nerve agents are usually colorless and odorless.

It is obvious, then, that for contamination that cannot be seen, smelled, felt, tasted, or heard, specialized methods of detection are required. Mechanical, chemical, and electronic devices are available or are under development for the detection of NBC contamination.

Nuclear Radiation

The instruments used for detecting radiological contamination are known as *radiacs*, the name being an abbreviation of radiation, detection, indication, and computation. Various types of radiacs are used on board ship and at shore stations, since no single type of radiac can make all the radiological measurements that may be required.

The radiacs used on board ship include: (1) intensity meters for measuring gamma radiation; (2) intensity meters for measuring beta and gamma radiation; (3) survey meters for measuring alpha radiation; and (4) dosimeters for measuring accumulated doses of radiation received by individuals. These basic types of radiacs are described briefly here. Specific information on operating principles and detailed instructions for operating the instruments may be obtained from the manufacturer's technical manual furnished with each instrument.

Gamma Meters. Intensity meters for measuring gamma radiation include both portable instruments and fixed systems installed on board ship. The intensity of gamma radiation is measured in roentgens per hour (r/hr) or milliroentgens per hour (mr/hr). The roentgen is a unit of measurement for expressing the amount of gamma radiation or X-ray radiation present. A milliroentgen is 1/1000 of a roentgen. Radiacs used for measuring large amounts of gamma radiation are called high-range intensity meters; these instruments are usually calibrated in roentgens per hour. Radiacs designed for measuring smaller amounts of gamma radiation are called low-range intensity meters; they are usually calibrated in milliroentgens per hour. Both high-range and low-range instruments are likely to have several scales; a range selector switch allows selection of the appropriate scale for each monitoring survey.

Beta and Gamma Meters. Intensity meters that measure gamma radiation and also detect or measure beta radiation are usually of the Geiger-Mueller type. These instruments can measure gamma radiation alone, or they can measure combined gamma radiation and beta radiation; an indirect measure of beta radiation can be obtained by subtracting the gamma radiation from the gamma–beta radiation.

Alpha Survey Meters. Meters for measuring alpha radiation are usually calibrated to give a meter reading in counts per minute (c/m, or cpm). However, some alpha survey meters give a reading in a unit called disintegrations per minute (d/m). The two units are not the same numerically.

Dosimeters. There are two basic types of dosimeters. Self-reading dosimeters can be read by the person wearing the instrument. Non-self-reading dosimeters cannot be read directly by the wearer but must be read with the aid of special instruments. Some dosimeters are calibrated in roentgens, others in milliroentgens. Both self-reading and non-self-reading dosimeters measure exposure to radiation over a period of time—in other words, they measure accumulated radiation exposure.

Self-reading dosimeters are provided in various ranges for use by personnel on board ship. Some of these self-reading dosimeters indicate accumulated gamma radiation from 0 to 200 mr; others indicate doses from 0 to 100 r; others from 0 to 200 r; and still others from 0 to 600 r. The dosimeter selected for any particular use will depend on the radiological situation existing at the time. Self-reading dosimeters must be charged before they are used. A special charging unit is furnished for shipboard use.

High-range non-self-reading dosimeters of the DT–60/PD type are furnished for use on board ship. A dosimeter of this type consists of a special phosphor glass between lead filters, encased in a bakelite housing. The dosimeter, which is small, lightweight, and rugged, is worn on a chain around the neck. This dosimeter will measure accumulated doses of gamma radiation from 25 to 600 r. A special instrument, the CP–95/PD computer-indicator, is required to read the DT–60/PD dosimeter.

Film badge dosimeters are non-self-reading devices for measuring both gamma radiation and beta radiation in low or moderate ranges. A film badge uses a special photographic film surrounded with moisture-proof and light-proof paper and shielded with lead, cadmium, plastic, or other shielding material. By the use of different shielding materials, the badge can be made to differentiate between gamma radiation and beta radiation. Laboratory techniques are required for the development and reading of the film.

Biological Agents

Basically, there are two possible approaches to the problem of detecting biological agents. Physical detection is based on the measurement of particles within a specified size range (and possibly the simultaneous measurement of other physical properties of the particles). Research is currently being done with a view to developing effective methods of physical detection. Biological detection involves growing the organisms, examining them under a microscope, and subjecting them to a variety of biochemical and biological tests. Although positive identification can frequently be made by biological detection methods, the procedure is difficult, exacting, and relatively slow. By the time a biological agent has been detected and identified in this fashion, personnel may well be showing symptoms of illness.

Biological detection may be divided into two phases: the sampling phase and the laboratory phase. The sampling phase may be a joint responsibility of damage control personnel and the medical department. The laboratory phase is obviously a medical department responsibility.

Chemical Agents

Various detection devices have been developed for the detection and identification of chemical agents. Most of these devices indicate the presence of chemical agents by color changes that are chemically produced. To date, no single detector has been developed that is effective under all conditions for all chemical agents. A number of devices, including air sampling kits, papers, crayons, silica gel tubes, and indicator solutions, are in naval use. Some of these devices are also useful in establishing the completeness of decontamination and in estimating the hazards of operating in contaminated areas.

Monitoring and Surveying

The monitoring and surveying of any area contaminated with NBC contamination is a vital part of NBC defense. In general, monitoring and surveying are done for the purposes of locating the hazards, isolating the contaminated areas, recording the results of the survey, and reporting the findings through the appropriate chain of command.

Specifically, the purpose of a radiological monitoring survey is to determine the location, type, and intensity of radiological contamination. The type of monitoring survey made at any given time depends on the radiological situation and on the tactical situation. Gross or rapid surveys are made as soon as possible after a nuclear weapon has been exploded, to get a general idea of the extent of contamination. Detailed surveys are made later, to obtain a more complete picture of the radiological situation.

On board ship, two main types of radiological surveys would be required after a nuclear attack: ship surveys (first gross, then detailed), including surveys of all weather decks, interior spaces, machinery, circulating systems, equipment, and so forth; and personnel safety surveys (usually detailed), concerned with protecting personnel from skin contamination and internal contamination. Personnel safety surveys include the monitoring of skin, clothing, food, and water, and the measurement of concentrations of radioactive material in the air (aerosols). Both ship surveys and personnel safety surveys are made on board ship by members of the damage control organization. The medical department makes clinical tests, maintains dosage records, and makes specific recommendations concerning the monitoring of food, water, air, and so on; but the actual surveys are made by damage control personnel of the engineering department.

Detailed instructions for making monitoring surveys cannot be specified for all situations, since a great many factors (type ship, distance from blast, extent of damage, tactical situation, etc.) must necessarily be considered before monitoring procedures can be decided upon. However, certain basic guidelines that apply to monitoring situations may be stated as follows:

    1. Monitors must be thoroughly trained before the need for monitoring arises. Learning to operate radiacs takes time. Sim-

ulated practice—as, for example, walking through a drill using a block of wood to represent a radiac—may teach a man something about the general movements made by a monitoring team, but it will not prepare him for actually using the instruments. All personnel who may be required to perform monitoring operations must be given adequate instruction and training in the use of radiac equipment.

    2. Standard measuring techniques must be used. A measurement of radiation is meaningless unless the distance between the source of radiation and the point of measurement is known. For example, a radiac held 2 feet away from a source of radiation will indicate only one-fourth as much radiation as the same instrument would indicate if it were held 1 foot away from the same source. A radiac held 3 feet from the source will indicate only one-ninth as much radiation as it would when held 1 foot from the source. As may be seen, therefore, the distance between the source of radiation and the radiac must be known before the radiac reading can have any significance.

    3. All necessary information must be recorded and reported. The information obtained by monitoring parties is forwarded to damage control central, where the measurements are plotted according to location and time. In order to develop an accurate overall picture of the radiological condition of the ship, damage control central must have precise and complete information from all monitoring parties. Each monitoring party must record and report the object or area monitored, the location of the object or area in relation to some fixed point, the intensity and type of radiation, the distance between the radiac and the source of radiation, the time and date of the measurements, the name of the man in charge of the monitoring party (or other identification of the party), and the type and serial number of the instrument used.

Contamination Markers

A standard system for marking areas contaminated by nuclear, biological, or chemical contamination has been adopted by nations included in the North Atlantic Treaty Organization. These standard survey markers are illustrated in figure 27–1.

NBC Decontamination

The basic purpose of decontamination is to minimize NBC contamination through removal or neutralization so that the mission of

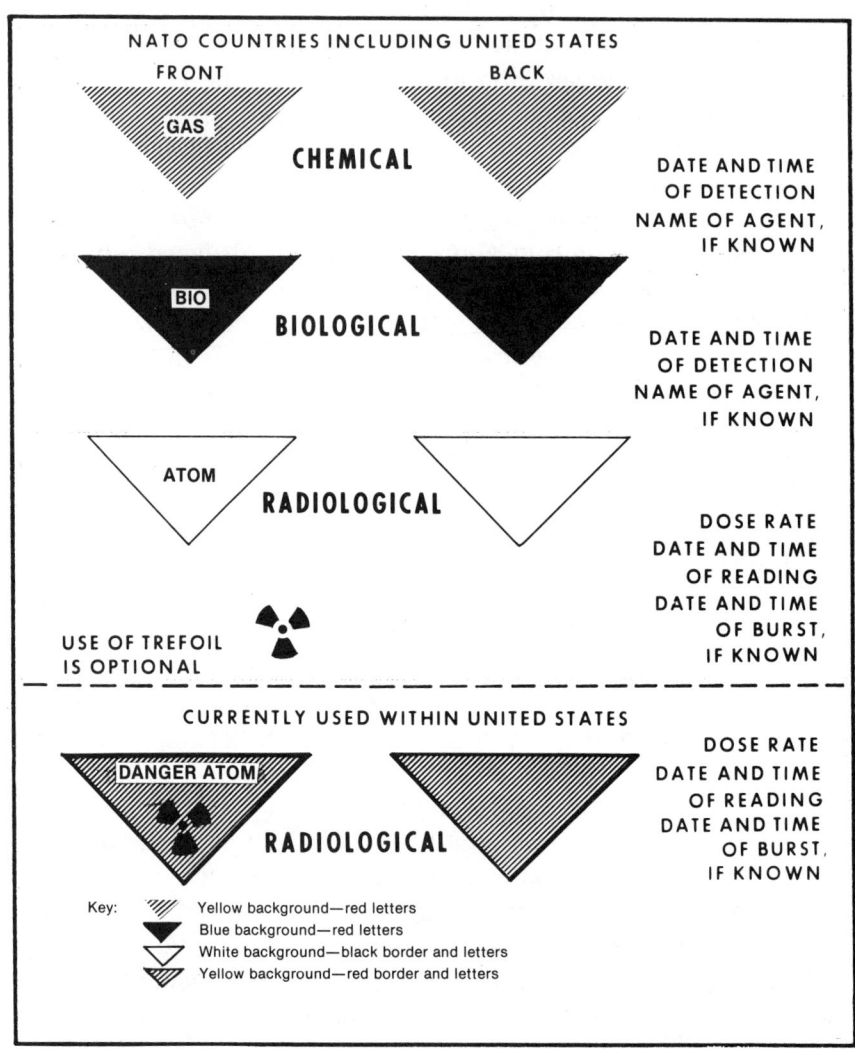

Figure 27-1. NBC contamination markers.

the ship or activity can be carried out without endangering the life or health of assigned personnel. The purpose of radiological decontamination is to remove contamination and shield personnel who are required to work in contaminated areas. The purpose of chemical decontamination is to remove or neutralize the chemical agents so that they will no longer be a hazard to personnel.

Decontamination operations are both difficult and dangerous, and personnel engaged in these operations must be thoroughly trained in the proper techniques. Certain operations, such as the decontamination of food and water, should be done only by experts qualified in such work. However, all members of a ship's company should receive adequate training in the elementary principles of decontamination so that they can perform emergency decontamination operations.

After an attack, data from NBC surveys will be used to determine the extent and degree of contamination. Contaminated personnel must be decontaminated as soon as possible. Before decontamination of installations, machinery, and gear is undertaken, appraisals of urgency must be made in light of the tactical situation.

Radiological Decontamination

Radiological decontamination neither neutralizes nor destroys the contamination; instead, it merely removes the contamination from one particular area and transfer it to an area in which it presents less of a hazard. At sea, radioactive waste is disposed of directly over the side. At shore installations, the problem is more difficult.

Several methods of radiological decontamination have been developed; they differ in effectiveness in removing contamination, in applicability to given surfaces, and in the speed with which they may be applied. Some methods are particularly suited for rapid gross decontamination; others are better suited for detailed decontamination.

Gross Decontamination. The purpose of gross decontamination is to reduce the radiation intensity as quickly as possible to a safe level—or at least to a level that will be safe for a limited period of time. In gross decontamination, speed is the major consideration.

Flushing with water, preferably under high pressure, is the most practicable way of accomplishing gross decontamination. On board ship, a water washdown system is used to wash down all the ship's surfaces, from high to low and from bow to stern. The washdown system consists of piping and a series of nozzles that are specially designed to throw a large spray pattern on weather decks and other surfaces. The washdown system is particularly effective if it is activated before the ship is exposed to contamination; a film of water covering the ship's surfaces keeps the contaminating material from sticking to the surfaces. Figure 27–2 shows a water washdown system in operation.

Manual methods may be used to accomplish gross decontamination, but they are slower and less effective than the ship's washdown system. Manual methods that may be used by ship's force include: (1) firehosing the surfaces with salt water, and (2) scrubbing the surfaces

Figure 27–2. Water washdown system in operation.

Figure 27–3. Decontamination by manual scrubbing.

with detergent, firehosing the surfaces, and flushing the contaminating material over the side. Figure 27–3 shows men performing gross decontamination operations by manual scrubbing.

Steam is also a useful agent for gross decontamination, particularly where it is necessary to remove greasy or oily films. Steam decontamination is usually followed by hosing with hot water and detergents.

Detailed Decontamination. As time and facilities permit, detailed decontamination is carried out. The main purpose of detailed decontamination is to reduce the contamination to such an extent that only a minimum of radiological hazard to personnel would persist.

Three basic methods of detailed decontamination may be used—surface decontamination, aging and sealing, and disposal. Each of these methods has a specific purpose, and one method can often be used to supplement another. Surface decontamination reduces the contamination without destroying the utility of the object. In aging and sealing, radioactivity is allowed to decrease by natural decay, and any remaining contamination is then sealed onto the surface. The disposal method merely consists of removing contaminated objects and materials to a place where they can do little or no harm.

Biological Decontamination

The methods available for biological decontamination include scrubbing, flushing, heating, and the use of disinfectant sprays, disinfectant vapors, and sterilizing gases. The method to be used in any particular case depends upon the nature of the area or equipment to be decontaminated and upon the nature of the agent (if this is known).

Chemical Decontamination

The major problem in chemical decontamination is to decontaminate successfully after an attack by any of the blister or nerve agents. The general methods used in chemical decontamination include natural weathering, chemical action, the use of heat, the use of sealing, and physical removal.

Natural weathering relies on the effects of sun, rain, and wind to dissipate, evaporate, or decompose chemical agents. Weathering is by far the simplest and most widely applicable method of chemical decontamination; in some cases, it offers the only practicable means of neutralizing the effects of chemical agents, particularly where large areas are contaminated.

Decontamination by chemical action involves a chemical reaction between the chemical agent and the chemical decontaminant. The reaction usually results in the formation of a harmless new compound or a compound that can be removed more easily than the original agent. Neutralization of chemical agents can result from chemical reactions of oxidation, chlorination, reduction, or hydrolysis.

Expendable objects or objects of little value may be burned if they become contaminated. This procedure should not be used except as an emergency measure or as a means of disposing of material that has been highly contaminated. If this method is used, a very hot fire must be used. Intense heat is necessary for destruction of chemical agents; moderate or low heat may serve only to volatilize the agent and spread it by means of secondary aerosols. When a large amount of highly contaminated material is being burned, downwind areas may contain a dangerous concentration of toxic vapors; personnel should be kept away from such areas.

Hot air may be blown over a contaminated surface to decontaminate it. Steam, especially high-pressure steam, is also a useful decontaminating agent; the steam hydrolyzes and evaporates chemical agents and flushes them from the surfaces. Chemical decontamination may also be accomplished by sealing off porous surfaces to prevent the absorption of chemical agents or to prevent volatilization of agents already on the surface.

Decontamination can also be effected by physically removing the toxic agents from the contaminated surfaces. This can be done by washing or flushing the surfaces with water, steam, or various solvents. Figure 27–4 shows a decontamination party hosing down a gun mount in order physically to remove toxic agents.

Figure 27–4. Decontamination team hosing down a gun mount.

Reference Books

Principles of Naval Engineering, Bureau of Naval Personnel, NavPers 10788–B, first edition, 1958; revised, 1966, 1970.

Basic Shipboard Engineering, Naval Officer Candidate School Naval Base, Newport, R.I., 1970.

Introduction to Naval Engineering Systems, Division of Engineering and Weapons, U. S. Naval Academy.

Propulsion Plant Manual for Spruance *Class*, by direction of Commander, NavSeaSysCom, NavSea 0941–LP–054–1010, 9 Sept. 1976.

Propulsion Plant Manual for Ocean Escorts, NavSeaSysCom (PMS301) DE 1078–DE 1097, NavSea 0941–LP–051–6010, 26 Dec. 1974.

Various individualized learning systems, Propulsion Engineering School, U. S. Navy, Great Lakes, Ill. for BT "A" School, MM "A" School, and EN "A" School.

Boiler Technician 3 & 2, Naval Education and Training Command, NavEdTra 10535–G, revised, 1977.

Machinist's Mate 3 & 2, Naval Education and Training Command, NavEdTra 10524–E, revised, 1978.

Fireman, Bureau of Naval Personnel, NavPers 10520–D, first edition, 1949; revised 1954, 1960, 1966, 1970.

Basic Military Requirements, Naval Training Command, NavTra 10054–D, revised, 1973.

ComNavSurfLant/Pac Inst. 3541.1A/3541.4, Repair Party Manual.

Introduction to Naval Architecture by T. C. Gillmer and B. Johnson. Annapolis: Naval Institute Press, 1982.

Elements of Applied Thermodynamics by R. M. Johnston, W. A. Brockett, A. E. Bock, & E. L. Keating. Annapolis: Naval Institute Press, 1978.

Shipboard Damage Control by A. M. Bissell, E. J. Oertel, & D. J. Livingston. Annapolis: Naval Institute Press, 1976.

Engineering for the OOD by CDR D. Felger USN. Annapolis: Naval Institute Press, 1979.

Naval Engineer's Guide by CDR J.V. Folliff and CDR H.E. Robertson, USN. Naval Institute Press, 1972.

Index

Abrasion, in engine, 208
Absolute pressure, defined, 23
Absolute temperature, defined, 18
Accelerated natural circulation boiler, 38
Accumulators, of compressed air plant, 261
Air ejector, 31, 104, 107–10, 151
   condenser, 31
Air exhaust system, engine, 203–4
Air intake system, engine, 203–4
Air lock system, of ABC system, 130, 132
Air pilots, excess feed system, 122
Air system, engine, 203
All-purpose firehose nozzle, 386–87
Alpha radiation, 170, 409
Alpha survey meters, 410
Alternating current (AC), 292, 302, 306
   generators, 295
   synchronous transmission operation, 289
A/M selector station, 130
Annular combustor, gas turbine engine, 222
Anchor windlass, 281, 285
Archimedes' law, 331
Armature, D.C. generator, 293
Astern turbine, 83
Atmospheric air, 19
Atmospheric pressure, 10, 21–23
Atomic mass number, defined, 168
Atomic rotation, 169
Atomic number, defined, 168
Atomization, of fuel, 206
Atomizer assembly, 56–59, 69
Atomizing valve, DFT, 114
   assembly, 114
Automatic boiler control system, 69, 123, 129, 130, 132–33
Automatic combustion control system, 123–25, 130–32
Auxiliary condenser, 14
Auxiliary/emergency power, 302, 306–7
Auxiliary exhaust steam, 63, 114, 116
   inlet, 114

Auxiliary exhaust system, 63, 90, 92, 136, 147, 151, 153
Auxiliary gland
   exhaust condenser, 116
   leak-off condenser, 120
Auxiliary pumps, 36
Auxiliary steam systems, 136, 147, 150–51, 153
   150 psi, 31
   1200 psi, 61, 65
Auxiliary turbines, 86, 89–92
Axial flow compressors, 219, 220, 222
Axial steam flow, 81, 83, 90

Baffles, heat exchanger, 15
Ballasting systems, 163
Barometric pressure, defined, 23
Barrel (bore), of engine, 198, 199
Batch process, lube oil purification, 88–89
Battery ignition system, 208
Bearings
   lubrication, 98
   use of, 96–98
Bedplate (base), of engine, 197
Beta meters, 410
Beta particle, 170–71
Beta radiation, 409
Binding energy, atomic, 173
Biological agents, 409, 411
Biological decontamination, 417
Blister agents, 417
Blower drive shaft, 61
Boiler, 12, 19, 27, 33, 34, 36, 43, 46, 48, 50, 52
   600 psi, 40, 42
   750 psi, 40
   1200 psi, 40, 42
   control console, 130
   D-type, 54
   efficiency, 36
   feedwater, 19, 45
   foundation, 55–56
   pressure, 49
   steam drum, 44, 142

Bottom blow system, 45
  connection, 50
  nozzles, 53
Brake horsepower (BHP), 195
Breaker assembly, 207
Brickwork, boiler, 56–57
British Thermal Unit (BTU), defined, 10
Bullgear, of reduction gear, 94
Buoyancy, ship, 331–42
Burners, 56–59, 68
  Ce-wallsend, 57
Bus-tie, 303
  breaker, 160, 162

Can combustor, gas turbine engine, 222
  can annular combustor, 222
Carnot cycle, 18–19
Casings, boiler, 41, 212
Casualty power, 302, 307
Cavitation, propellers, 100
Center of gravity
  change in, 343–52
  horizontal shift, 348–52
  vertical shift, 345–48, 358
Centrifugal compressors, 219, 220, 222, 258
Centrifugal pump, 31, 32
Chemical agents, 409, 411
Chemical decontamination, 417
Chemical energy, 2, 3
Chilled-water circulating system, 276
Circle fittings
  WILLIAM, 407
  X-RAY, 378
  YOKE, 378
  ZEBRA, 378
Circuit breaker, 299, 304, 305
Circulating pump, main, 102, 104–5, 107
Circulating tubes, boiler, 41
Closed cooling system, 208
Closed cycle, 19
  unheated engine cycle, 20
CO_2 extinguisher
  types, 395
  use, 393
Color code
  DC diagrams, 400
  piping, 379
Combustion, 42, 200–9
Combustion chamber, 193, 199, 210, 212–13, 218, 220, 222, 224–225
Combustor, of gas turbine engine, 210
Commutator, generator, 293

Compartment checkoff list, 379
Compartmentation, ship, 376
Compressing element, of air compressor, 259–60
Compression, stress, 315–16
Compression ignition, 190
Compression ratio, 220
Compression rings, engine, 201
Compression stroke, engine, 191, 193
Compressor, gas turbine engine, 210, 212, 218–20, 225
  refrigeration plant, 274
Compressor classifications, 258
Compressor stages, 258–61
Condensate, 30, 31, 93, 117, 119
Condensate depression, 105, 107
Condensate inlet
  connection, 114, 122
  chamber, 114, 116
Condensate pump, main, 102, 104, 107
  system, 102–12
Condensation phase, steam cycle, 27, 30
Condenser, 36, 134–35, 139–40, 150, 153, 163
  R–12, 274
Conditions of closure, 377, 378
Conditions of readiness, 377
Conductor, electrical, 291, 293–96, 298
Connecting rod, engine, 191, 196, 199, 202
Conservation of energy principle, 15–16
Constant-pitch propellers, 99
Constant-pressure pump governors, 91
Constant-speed governors, 91
Constant-speed turbine, 91
Contamination markers, NBC, 413–14
Control rods, reactor, 177
Controllable reverse pitch propeller (CRP system), 100, 101, 153, 159
Controlled circulation boiler, 37, 38
Controlled superheat, 40
Convection, 6–9, 27
Converter, reactor, 180
Crankshaft, engine, 191–93, 196, 199, 202
Cross curves of stability, 343
Cross flow, heat exchanger, 14, 102
Cross head, 196
  assembly, 200
Crossover pipe, 52
Cruising range, 315
Current, 291
  direct, 292

single-phase AC, 301
three-phase AC, 301–2
Curtis stage, turbine, 79, 83
Cyclonic separator, 256
Cylinder, 191–208, 259–61
   assembly, 197, 198
   assembly gasket, 198, 199
   assembly head, 198–200, 207
   assembly studs, 198–99
   block, 197–99
   liners, 197–98
   wall, 201

Damage control
   central station, 374
   equipment, 379
   library, 374
   objectives, 372–73
   organization, 373
   precautions, 405
Deaerating feed tank (DFT), 31, 32, 107, 110–11, 113–14, 116–17, 121–22, 134, 153
Density, defined, 8
Desuperheated steam, 34, 36, 41
Desuperheater, 41, 43, 51
Detailed decontamination, NBC, 416
Detailed surveys, NBC, 412
Diaphragms (baffles), 50, 271
Diesel engine, 193–209
Direct contact vent condenser, 114
Direct current (DC), 292
   generators, 293, 303
Discharge regulating valve, automatic, 120
Displacement, ship, 331, 333, 336, 341, 342, 352
Distillation, 251–52, 253
   plants, 153, 163, 251–57   vertical basket, 255–57
Dosimeters, NBC, 409–11
Double-acting compressor, 259
Double-acting gas-action engine, 195
Double furnace, 40
Downcomers, 38, 43, 51–52
Draft, ship 332–33, 341
Drum type boilers, 39, 41
Dry chemical extinguishers, 393
D–type boilers. *See* Drum type boilers
Dump valve, solenoid-operated, 257
Duplex strainers, 68

Economizer, 32, 33, 41, 43, 45, 46, 117
   outlet header, 48

surface, 37
tubes, 43, 45
Eductor, 389–90
Electrical distribution system, 302
Electrical grounding, 308–9
Electrical safety rules, 307–13
   circuits, 309–10
   portable tools, 310–12
   personal equipment, 312–13
Electrohydraulic speed gear, 281–83
Electrohydraulic steering gear
   development, 286
   operation, 286–87
   parts, 286–87
Electromagnetic induction, 293
Electromagnetic wave phenomenon, 7
Electromotive force (emf), 292
Elliptical casing, compressor, 263
Emergency steering, 287
Energy. *See also* Mechanical energy
   conversions, 73, 75
   defined, 1, 2, 4, 9, 18
   redistribution, 19
   system, 2, 5, 13
   transformation, 19, 20, 24
   in transition, 2–4
   types, 2
Engineering casualty control, 372
Enginerooms, 139, 150, 153, 160, 162–65
Equilibrium, 25, 332
Erosion, 29, 32, 36
Evaporation, 253, 273
Evaporator (R–12), 273
Excess feedwater. *See* makeup/excess feed system
Exhaust gases, 208
Exhaust stroke, engine, 193
Exhaust valve, 191–93, 195–96
Expansion joint, 63, 67
Expansion phase, steam cycle, 27, 30

Factors for lube oil efficiency, 86
Feed phase, steam cycle, 27, 31
Feedwater, 9, 31–33, 43, 45–46, 48, 102–3, 114–18, 121–22, 126–28, 133, 141, 163
   check valve, automatic, 128
   cooler, 31, 116
   regulator control, 123, 126–28, 132–33
Fire classifications
   alfa, 383
   bravo, 383
   charlie, 383
   delta, 383

Fireman system types
  single, 384
  horizontal loop, 384
  vertical loop, 384
Fireroom, 136, 139, 150
First law of thermodynamics, 15, 17
First-stage air ejector, 109–11
Fixed blades, turbines, 77, 79, 81
Flash tank, distilling plants, 256
Flash-type distilling units, 254–55
Flooding, ship, 364
Fluid film, 13
Fluid flow, 14
Fluid inlets, 15
Fluid lubrication, 84
Fluid torque converter, 225
Flywheel, 202–3
Foot pounds, defined, 4
Forced convection, 9
Forced draft blower, 19, 60–67, 134, 135, 150, 151
  automatic shutters, 63–64
  turbine, 125
Four-stroke cycle, 191, 193, 195, 203–4
FP–180 foam proportioner, 391
Frame, of engine, 197
Frame number, ship, 146
Framing, ship construction, 320–21
Freeboard, 334–35, 341
Free communication effect, 360
Free surface effect, 357
Free water drains, 119, 120
Fresh water drain collecting system, 119
  discharge, 120
  tank, 109–10, 112, 119–20, 122
  tank pump, 120
Friction, 84–85, 208
Fuel oil, 56–59, 67–69
  accumulators, 67
  burners, 41
  consumption, 366
  control valve, 69, 125, 132
  heaters, 12, 68
  injection system, 205–6
  micrometer valves, 69
  nozzle, 212
  pump, port, 67
  pump, service, 67, 69
  quick-closing valve, 68
  service, 147, 160, 162–63
  supply manifold, 69
  supply valve, 125
  tanks, 67, 69
  unloading valve, 67
Funnel drains, 119

Furnace, 27, 33, 35, 43, 51
Furnace refractories. *See* Refractories
Fuse, 299

Gamma meters, 410
Gamma radiation, 170–71, 409
Gage pressure, 22
Gas generator section, gas turbine, 225
Gases of combustion, 9, 37
Gas turbine, 153, 159, 160, 163–64, 210–42
  fuel system, 214
  lubricating system, 215
  starting system, 215
Geiger-Mueller meters, 410
General energy equation, 15
General principles of generation, 25
Generating surface, 37
Generating tubes, 27, 41, 43
Generation phase, 27
Generator, electrical, 160, 162, 202
Gland exhauster, 84, 112
  condenser, 109–12
Gland seal, 84
  system, 151, 153
Gravity, forces of, 331–32. *See also* Center of gravity
Grounding problems, electrical, summary of, 370

Hagevap, 253
Handhole plate, 50, 52–53
Header, 34, 150, 164
Header-type boiler, 39
Heat exchanger, 12–15
  counter flow, 14
Heat flow, 6, 9, 12
Heat receivers, 18, 20
Heat transfer, 2, 6, 8, 9, 12–14, 41–42, 251–54
Heated engine cycle, 19–20
Heavy water, defined, 32
Heel, of ship, 338, 339, 341
Helical steam flow, 81
Hertz (Hz), 293
High pressure comnpressors, 258
High pressure drains, 114
Hogging, stress, 317
Hull members, 318
Hydraulic accumulators, 283–84
  types, 283–85
  uses, 283
Hydraulic principle, 279–80
Hydraulic system problems, 281
Hydrogen isotopes, 169

Ignition system, engines, 206–7
Impulse turbine, 75, 78, 79, 83, 90
  pressure compounded, 79
  velocity compounded, 79
  velocity-pressure compounded, 80, 83
Inclination angle, defined, 339
Inclining experiment, 331
Inclining moments, 336–37
Induced current, 293
Induced emf, 293
Inlet headers, 45, 49, 50–51
Inner/outer casings, 52
Intake screen, 63
Intake stroke, engine, 191–93
Intake valve, engine, 191–93, 195
Integral superheater, 40
Inter-condenser, 109–10, 112
Intercoolers, 261
Internal access door, 114
Internal combustion engine, 19, 20, 189–93, 195, 196–97, 199, 209–11
Internal energy, 5, 6, 16, 24
  kinetic, 4, 5, 6, 9, 10
  potential, 5, 9, 10
Internal feed pipe, 48
Internal steam supply pipe, 114
Isolation valve, 121
Isotope, defined, 168

Jacking gear, 95

KB, distance, 333–34
Keel, ship 319, 333, 338

Latent heat, 9, 10, 12, 29–31
  of condensation, 11, 109
  of vaporization, 10, 11, 267
Light steam, 32
Light water, 395
Liquids, 9, 10, 25
  saturated, 25
  subcooled, 25
Local manual control, 129–30, 132
Longitudinal, 314
Loop seal, 109–10
Loose water
  definition, 357
  effects, 361
Low pressure boiler, 41
Low pressure compressor, 258
Low pressure steam, 149, 150, 256
Low water casualty, 32
Lubricating oil system, 84–90, 160, 163–64, 208–9, 260

centrifugal purifier, 86, 88–89
contamination of, 87–88
coolers, 12
filters, 86
pumps, 86, 90
settling tank, 86
strainers, 86
sump, 64

Machinery room, 139, 153, 163
Magneto ignition system, 207, 208
Main condensate system, 31, 120. *See also* Main condenser pump, 31
Main condenser, 12 14, 30, 31, 92, 102–4, 117, 120
Main engines, 29–30, 132
Main feed system
  booster pump, 32, 115–18
  check valve, 130
  piping system, 128
  pump, 8–9, 29, 32, 116–18, 128–29, 133
  pump differential control, 123, 128–29, 133
Main reduction gears, 88, 90, 93–97, 134, 135, 140, 153, 159, 160, 163, 165
Main shaft bearing, 64
Main steam system, 30, 136, 141, 149, 150
Main thrust bearing, 97
Makeup/emergency feed tanks, 121–22
Makeup/excess feed system, 120–21
Manhole plate, 46, 51–53
Manifold, engine, 39, 203–4
Masker air, 164
Mass energy equation, 172
Mechanical energy, 2–4, 15–16. *See also* Energy
  kinetic, 3, 4
  potential, 3, 4
Mechanical foam, 390
Medium pressure compressor, 258
Metacenter, 337–38
  height, 345
  radius, 338
Moderator, nuclear, 177–78
Moment, 336
  arm, 336
  couple, 336
Monitoring and surveying, 412–13
Monitoring parties, NBC, 413
M-type boiler, 39
Multistage axial flow compressor, 219
Multistage centrifugal compressor, 219

Natural circulation boiler, 37, 39
NBC
  attack, 407
  contamination detection, 409
  defense measures, 407
Negative temperature coefficient, 183–85
Nerve agents, 417
Newton's third law, 217
Nonreturn check valve, 114, 116
Nozzles, 73, 74, 81, 206, 212–13, 217, 229–30
  bottom blow, 50, 53
  drum level transmitter, 48
  flanged, 40
  gas turbine, 212–13, 217, 229–30
  outlet, 48, 50–51
  steam outlet, 49
  water gage glass, 48
Nuclear plants, 29, 30, 186–87

Oil rings, engine, 201
Open cooling system, 208
Open cycle, 19, 218
Operating pressure, 24, 36
Opposed-piston gas-action engine, 195–96, 199, 201
Outlet headers, 45, 48–51
Overspeed trips, 92, 232
Oxygen breathing apparatus (OBA), 397, 409

P–250 pump, 388
Parallel flow heat exchanger, 14
Partial flooding, 365
Perpetual motion, 16
Personnel safety surveys, 412
Physical properties of liquids, 279–80
Pinions, 94, 95, 140, 159
Piston, 191–93, 195–96, 199–200, 202–4, 206, 207, 259–60
  cylinder arrangement, 21
  pin, 200–201
  rings, 200–201
  rod, 196, 199, 200
Pitch angle, 99
Plenum chamber, boiler, 66–67
Power stroke, engine, 192, 193
Power transmission systems, 216
Prairie air, 164
Preheating, feedwater, 32
Pressure closed feed system, 117
Pressure gages, 21, 22
Pressurized furnace boiler, 42
Pressurized water reactor (PWR), 179–83

Primary winding, transformer, 298
Propellers, ship, 98–101, 202
  shafts, 202
Proportioners, foam, 390
Propulsion boilers, 134–36, 140–41, 150, 163
Propulsion cycle, 36
Propulsion plant, defined, 1
Propulsion power coupling, 224
Propulsion shafts, 96, 98, 135, 139, 147, 153, 159, 162, 164
Propulsion turbines, 84, 134, 135, 139, 140, 150. *See also* Turbines
Protective clothing, NBC, 408
Protective masks, NBC, 408
Pulsating emf, 295
Purification process, 88–89
Purifiers, 209

R–12 refrigerant, 268–78
Radiacs, 409
Radial steam flow, 81
Radiant energy, 7
Radiation, 6, 7, 8, 27
Radioactive decomposition, 170–71
  waste, 415
Radioactivity, 169–72
Radiological decontamination, 414–15
Rateau stage, turbine, 79–80
Reaction turbine, 75, 79, 80, 81, 83. *See also* Turbines
  pressure compounded, 80, 83
Reactor coolant, 178
Rear wall header, 43
Rear wall tubes, 52
Reciprocating compressors, 258, 259–62
Reciprocating internal combustion engine, 196
Reciprocating motion, engine, 199, 201, 202
Reduction gear, 202
Reentry turbine, 82, 83
Reflector, reactor, 178
Refractories, 52, 56–57
Refrigerant circulating system, 275
Refrigerating effect, 267
Refrigeration ton, defined, 267
Regenerator, gas turbine engine, 217, 218
Relief valve, spring loaded, 115
Repair parties, 374–76
Reserve buoyancy, 334–35
Resistance, electrical, 4, 292
Retractable soot blower, 70, 72
Revolving-field AC generator, 296

Righting arm, 336, 339, 341, 342
  moment, 336, 337, 339, 342
Riser pipes, 52
Roentgen, defined, 410
Rotary centrifugal compressors, 258, 262–64
Rotor
  gas turbine engine, 213–14
  generator, 296
Rotor chamber, 263
Rutherford-Bohr atom, 167–68

Safety valves, boiler, 44, 49
Saturated steam, 28, 29, 34, 35, 43, 49, 90, 150
Scale, boiler, 252, 253
Scavenging, 193, 195–96, 203–4
  air, 204
  ports, 204
Scoop injection system, 104–5, 107
Screw. See Propellers, ship
Secondary refrigerant, 277
Secondary winding, transformer, 298
Second-effect distillate, 256
Second law of thermodynamics, 17
Second-stage air ejector, 109–11
Sensible heat, 9, 10, 12, 29, 36, 267
Shaft rotation, 202–3
Shaft turning gear, 95
Shear, stress, 315
Shielding, reactor, 178–79
Ship's service distribution system, 302–3
Ship's service turbogenerators, 302–3
Sidewall header, 43, 45, 46, 51, 52
Single-acting gas-action engine, 195, 196, 201
Single-acting compressor, 259
Single-entry turbine, 82, 83
Single flow, turbine, 82
Single furnace, boiler, 40–42
Single-stage centrifugal compressor, 219
Sliding foot, of boiler, 56
Smoke indicator, 56, 59–60
Smoke pipe, 33
Snail, vapor separator, 256
Soot blowers, 56, 69, 150
  stationary type, 70
Spark ignition system, 190
Specific fuel consumption, 195
Specific heat, 12, 267
Speed control devices, auxiliary turbines, 91
Split shaft, gas turbines, 214, 216
Spray nozzles, 114

Spring bearings, 97
Sprinkler systems, 385
Stability, ship, 314, 331–42
Stability curve, 331, 340–41, 343
Stator, gas turbine, 213, 296
Steam atomizing valve, 114
Steam cycle, 120–22
Steam drum, 27, 29, 32, 34, 41, 43, 45–48, 50–52, 123–28, 130, 133
  head, 46
  wrapper sheet, 49
Steam flow, 73
Steam generator, 34
Steam lance, superheater, 72
Steam propulsion plant, 120
Steam separators, 47
  primary, 47
  secondary, 47
Steam supply valve, 125
Steering gear, 286
Stern tube bearing, 98
Strainers, lube oil, 209
Strut bearing, 98
Stuffing box, 196
Submersible pumps, damage control, 400
Sump, oil, 86, 89, 197, 209
Supercharging, diesel engine, 203, 204
Superheated steam, 11, 25, 29, 34, 90, 149, 150
Superheater, 25, 29, 34, 35, 41
  headers, 43, 49, 50
  inlet, 41
  surface, 37
Surface blow, 45
  piping, 47, 48
Surface heat exchanger, 12, 14
Swash bulkheads, 360
Swash plates, 115
Swing check valves, 119
Switchboard, ship's service, 60, 298, 303

Tank suction valves, 121
TAU system, 396
Tension, stress, 315–16
Test tanks, 255
Thermal efficiency, Carnot cycle, 18, 19
Thermal energy, 1–10, 12, 15, 16, 20, 27, 29, 32
Thermal equilibrium, defined, 7
Thermal radiation, defined, 7
Thermodynamics, 1, 6, 15
  cycles, 19, 20, 34
  processes, 18, 19, 73

Thermostatic expansion valve, 271
Thermostatic temperature regulating valve, 65
Torsional stress, 316
Torque, defined, 216
Toxic agents, 418
Transformation of energy, 1
Transformers, electric 297
Trim, of ship, 362
Trips, 305–6. *See also* Overspeed trips
  overcurrent, 305
  reverse power, 305
  underfrequency, 306
  undervoltage, 305
Tube arrangement, boiler, 15
Turbines, 29, 30–32, 42, 74, 78–80, 83, 94, 210–11, 213, 244. *See also* Gas turbine
  casing, 84
  exhaust, 30–31
  high pressure, 30, 83, 94
  low pressure, 30, 83, 91, 94
  rotor, 29
Turbogenerator, 29, 150
Two stroke cycle, engine, 191, 193, 195, 196, 200, 203, 204

Unloading system, air compressor, 262

Vacuum, 23, 30–31, 102, 104, 252, 253
Vacuum breaker, 115, 116
Vacuum drag line, 120
Vacuum gage scale, 23
Variable pitch propellers, 99
Variable speed turbine, 91
Venturi assembly, oil burner, 57
Viscosity, of liquid, 86
Voltage transformer, 207

Water baffle, conical, 114
Water drum, boiler, 41, 43, 45, 50–51
Water-regulating valve, 275
Water screen header, 43, 48, 51
Water tube boiler, 37, 41
Watertight markings. *See also* Circle fittings
  X-RAY, 377
  YOKE, 377
  ZEBRA, 377
Weight shift
  horizontal, 348
  vertical, 345
Wet cell storage battery, 299
Wet steam, 28, 29
Work, defined, 2, 17
Working fluid, defined, 211
Working substance, 20